PROCEEDINGS OF THE 14th AND 15th

# ASIAN LOGIC CONFERENCES

PROCEEDINGS OF THE 14th AND 15th

# ASIAN LOGIC CONFERENCES

Mumbai, India 5 – 8 January 2015
Daejeon, South Korea 10 – 14 July 2017

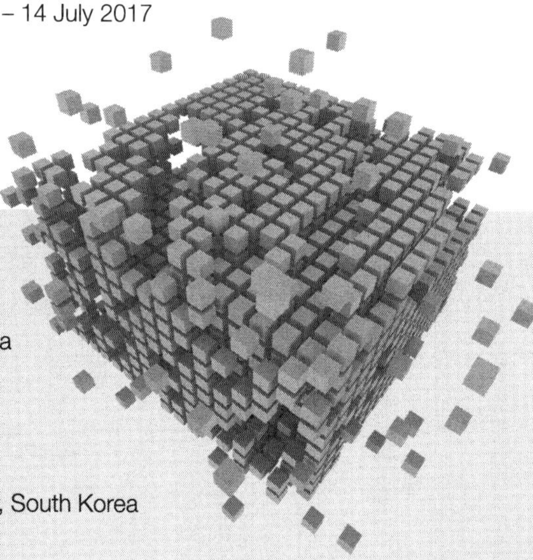


edited by

**Byunghan Kim**
Yonsei University, South Korea

**Jörg Brendle**
Kobe University, Japan

**Gyesik Lee**
Hankyong National University, South Korea

**Fenrong Liu**
Tsinghua University, China

**R. Ramanujam**
IMSc, India

**Shashi M. Srivastava**
ISI-Kolkata, India

**Akito Tsuboi**
University of Tsukuba, Japan

**Liang Yu**
Nanjing University, China



**World Scientific**

NEW JERSEY • LONDON • SINGAPORE • BEIJING • SHANGHAI • HONG KONG • TAIPEI • CHENNAI • TOKYO

*Published by*

World Scientific Publishing Co. Pte. Ltd.
5 Toh Tuck Link, Singapore 596224
*USA office:* 27 Warren Street, Suite 401-402, Hackensack, NJ 07601
*UK office:* 57 Shelton Street, Covent Garden, London WC2H 9HE

**British Library Cataloguing-in-Publication Data**
A catalogue record for this book is available from the British Library.

**PROCEEDINGS OF THE 14TH AND 15TH ASIAN LOGIC CONFERENCES**



ISBN 978-981-3237-54-4

For any available supplementary material, please visit
https://www.worldscientific.com/worldscibooks/10.1142/10919#t=suppl

Printed in Singapore

## Preface

The *Fourteenth Asian Logic Conference* (ALC 2015) was held at the Indian Institute of Technology Bombay, Mumbai, India during January 5–8, 2015. And the *Fifteenth Asian Logic Conference* (ALC 2017) was held at NIMS, Daejeon, South Korea during July 10–14, 2017.

The Asian Logic Conference (ALC) is a major international event in mathematical logic. It features the latest scientific developments in the fields in mathematical logic and its applications, logic in computer science, and philosophical logic. The ALC series also aims to promote mathematical logic in the Asia-Pacific region and to bring logicians together both from within Asia and elsewhere to exchange information and ideas.

From 1981 to 2008, the Asian Logic Conference has been held triennially and rotated among countries in the Asia-Pacific region. The previous meetings took place in Singapore (1981), Bangkok (1984), Beijing (1987), Tokyo (1990), Singapore (1993), Beijing (1996), Hsi-Tou (1999), Chongqing (2002), Novosibirsk (2005), and Kobe (2008). In 2008, the East Asian and Australasian Committees of the Association for Symbolic Logic decided to shorten the three-year cycle to two. The new two-year cycle began with the meeting in Singapore (2009), and subsequent meetings have been held in Wellington (2011), Guangzhou (2013),

ALC 2015 was held jointly with ICLA 2015, the Sixth Indian Conference on Logic and Its Applications during January 8–10, 2015. The conference was supported by the Indian Institute of Technology Bombay, Association for Symbolic Logic, and Indo-European Research Training Network in Logic (IERTNiL). Plenary key note lectures are given by Sergei Artemov (City University of New York), Anuj Dawar (University of Cambridge Computer Laboratory), Assia Mahboubi (INRIA), Ng Keng Meng (Nanyang Technological University), Rohit Parikh (City University of New York), Anand Pillay (University of Notre Dame), and Dilip Raghavan (National University of Singapore). Twenty one invited lectures were given at special sessions on Logic in Computer Science, Model Theory and Algebraic Logic, Philosophical Logic, Proof Theory, Recursion Theory, and Set Theory.

ALC 2017 was the first Asian Logic Conference since its status changed from an ASL (the Association for Symbolic Logic)-sponsored meeting to an official ASL meeting by ASL Council action in May 2016. The conference was supported by NIMS (National Institute for Mathematical Sciences, South Korea), Samsung Science and Technology Foundation, and Association for Symbolic Logic. There were ninety eight registered participants from five continents. Plenary lectures are given by Johann Makowsky (Technion IIT, Israel), Yinhe Peng (University of Toronto, Canada), Katrin Tent (University of Münster, Germany), Byeong-uk Yi (University of Toronto, Canada), Kwangkeun Yi (Seoul National University, Korea), Yimu Yin (Sun Yat-sen University, China), Liang Yu (Nanjing University, China), and Yizheng Zhu (University of Münster, Germany). Eighteen invited lectures were given at special sessions on Computability, Model Theory, Philosophical Logic, and Set Theory. In the afternoon of July 13, Korean Association for Logic Annual Summer Meeting was held in parallel.

We keep our tradition of choosing World Scientific as the publisher of ALC proceedings, and in this volume 14 refereed papers of speakers in the two conferences are presented. We express our thanks to all who contributed to this volume, specially to authors and reviewers.

**Editors**
Byunghan Kim (Chair), Jörg Brendle, Gyesik Lee, Fenrong Liu,
R. Ramanujam, Shashi M. Srivastava, Akito Tsuboi, Liang Yu

## The 14th Asian Logic Conference
## IIT Bombay, Mumbai, India, January 5–8, 2015

### Organizing Committees

#### Program Chairs

| | |
|---|---|
| Byunghan Kim | – Yonsei University, South Korea |
| R. Ramanujam | – IMSc, India |

#### Program Committee

| | |
|---|---|
| Toshiyasu Arai | – Kobe University, Japan |
| Mohua Banerjee | – IIT-Kanpur, India |
| Mihir K. Chakraborty | – ISI Kolkata and Jadavpur University, India |
| Supratik Chakraborty | – IIT-Bombay, India |
| Rod Downey | – Victoria Univ of Wellington, New Zealand |
| Shier Ju | – Zhongshan University, China |
| Fenrong Liu | – Tsinghua Univ, China |
| Kamal Lodaya | – IMSc, India |
| S. M. Srivastava | – ISI-Kolkata, India |
| Liang Yu | – Nanjing Univ, China |
| Yang Yue | – NUS, Singapore |

# The 15$^{th}$ Asian Logic Conference
# NIMS, Daejeon, Korea, July 10–14, 2017

## Organizing Committees

### Program Committee

| | |
|---|---|
| Byunghan Kim | – (Chair) Yonsei University, Korea |
| Jörg Brendle | – Kobe University, Japan |
| Rod Downey | – Victoria University of Wellington, New Zealand |
| Qi Feng | – Chinese Academy of Sciences, China |
| Fenrong Liu | – Tsinghua University, China |
| R. Ramanujam | – IMSc, India |
| Akito Tsuboi | – University of Tsukuba, Japan |
| Yue Yang | – National University of Singapore, Singapore |

### Local Committee

| | |
|---|---|
| Inkyo Chung | – Korea University, Korea |
| Joongol Kim | – Sungkyunkwan University, Korea |
| Gyesik Lee | – Hankyong National University, Korea |
| Martin Ziegler | – KAIST, Korea |

# Contents

# PART A

# The $14^{th}$ Asian Logic Conference

# The Entropy Function of an Invariant Measure

Nathanael Ackerman

*Department of Mathematics, Harvard University,*
*Cambridge, MA 02138, USA*
*E-mail: nate@math.harvard.edu*

Cameron Freer

*Borelian Corporation,*
*Cambridge, MA 02139, USA*
*E-mail: freer@borelian.com*

Rehana Patel

*Department of Philosophy, Harvard University,*
*Cambridge, MA 02138, USA*
*E-mail: rpatel@fas.harvard.edu*

Given a countable relational language $L$, we consider probability measures on the space of $L$-structures with underlying set $\mathbb{N}$ that are invariant under the logic action. We study the growth rate of the entropy function of such a measure, defined to be the function sending $n \in \mathbb{N}$ to the entropy of the measure induced by restrictions to $L$-structures on $\{0, \ldots, n-1\}$. When $L$ has finitely many relation symbols, all of arity $k \geq 1$, and the measure has a property called non-redundance, we show that the entropy function is of the form $Cn^k + o(n^k)$, generalizing a result of Aldous and Janson. When $k \geq 2$, we show that there are invariant measures whose entropy functions grow arbitrarily fast in $o(n^k)$, extending a result of Hatami–Norine. For possibly infinite languages $L$, we give an explicit upper bound on the entropy functions of non-redundant invariant measures in terms of the number of relation symbols in $L$ of each arity; this implies that finite-valued entropy functions can grow arbitrarily fast.

## 1. Introduction

The information-theoretic notion of entropy captures the idea of the typical "uncertainty" in a sample of a given probability measure. An important class of probability measures in model theory, combinatorics, and probability theory are the Sym($\mathbb{N}$)-invariant measures on the space of structures, in a countable language, with underlying set $\mathbb{N}$. We call such measures *invariant measures*, and consider the entropy of such measures.

When the support of a probability measure is uncountable, as is usually the case for invariant measures, the entropy of the measure is infinite. But any such measure $\mu$ can be approximated by its projections to the spaces of structures on initial segments of $\mathbb{N}$. We define the *entropy function* of $\mu$ to be the function from $\mathbb{N}$ to $\mathbb{R} \cup \{\infty\}$ taking $n$ to the entropy of the measure induced by $\mu$ on structures with underlying set $\{0, \ldots, n-1\}$. In this paper, we study the growth of entropy functions.

The growth of the entropy function of an invariant measure has been studied by Aldous [Ald85] Chapter 15 in the setting of exchangeable arrays and by Janson [Jan13] §10 and §D.2 and Hatami–Norine [HN13] Theorem 1.1 in the case where the measure is concentrated on the space of graphs. The leading coefficient of the entropy function has been used to study large deviations for Erdős–Rényi random graphs and exponential random graphs in [CV11] and [CD13], and phase transitions in exponential random graphs in [RS13]. For additional results on the entropy functions of invariant measures concentrated on the space of graphs, see [HJS18].

Further, ergodic invariant measures are the focus of a project that treats them model-theoretically as a notion of "symmetric probabilistic structures"; see [AFP16, AFNP16, AFKwP17, AFP17], and [AFKrP17]. The entropy function of such a measure is one gauge of its complexity.

In this paper we primarily consider invariant measures, for a countable relational language, that are *non-redundant*, namely those that concentrate on structures in which a relation can hold of a tuple only when the tuple consists of distinct elements. In §1.2 we develop machinery for quantifier-free interdefinitions that allows us to show, in Proposition 1.21, that every entropy function of an invariant measure for a countable language is the entropy function of some non-redundant invariant measure for a countable relational language.

In Section 2, we study the case of invariant measures for countable relational languages $L$ with all relations having the same arity $k \geq 1$. The Aldous–Hoover–Kallenberg theorem provides a representation, which we call an *extended L-hypergraphon*, of a non-redundant invariant measure for $L$. Our main technical result in this section, Theorem 2.7, shows that under a certain condition (which is satisfied, e.g., when $L$ is finite), the entropy function of the invariant measure corresponding to an extended $L$-hypergraphon $W$ grows like $Cn^k + o(n^k)$, where the constant $C$ can be calculated from $W$. When the invariant measure is concentrated on the space of graphs, Theorem 2.7 reduces to a result first observed by Aldous in [Ald85] Chapter 15, and also by Janson in [Jan13] Theorem D.5. The proof of Theorem 2.7 follows closely that of Janson.

In Section 3 we consider invariant measures that arise via sampling from a Borel $k$-uniform hypergraph for $k \geq 2$. The entropy functions of such measures grow like $o(n^k)$, as we show in Lemma 3.2. We prove, in Theorem 3.5, that for every function $\gamma(n) \in o(n^k)$ there is an invariant measure sampled from a Borel $k$-uniform hypergraph whose entropy function grows faster than $\gamma(n)$. Our theorem generalizes a result of Hatami–Norine[HN13] Theorem 1.1, who prove it in the case where $k = 2$ (hence where the Borel hypergraph is simply a Borel graph). Hatami–Norine's proof analyzes the random graph obtained by subsampling from an appropriate "blow up" of the Rado graph, i.e., the unique countable homogeneous-universal graph. Our proof takes an analogous path, using the unique countable homogeneous-universal $k$-uniform hypergraph.

Finally, in Section 4 we consider invariant measures for countable languages that may be of unbounded arity. We provide an upper bound on the entropy functions of non-redundant invariant measures for a given countable relational language, in terms of the number of relations of each arity. We do so by calculating the entropy function of a particular non-redundant invariant measure whose entropy function is maximal among those for that language. This calculation also demonstrates that whereas the growth of the entropy function of an invariant measure for a finite language is polynomial, there are $\mathbb{R}$-valued entropy functions that grow arbitrarily fast.

### 1.1. *Preliminaries*

*In this subsection, let L be a countable language and let* $\mathfrak{n} \in \mathbb{N} \cup \{\mathbb{N}\}$.

For $n \in \mathbb{N}$, write $[n] := \{0, \ldots, n-1\}$, and $[\mathbb{N}] := \mathbb{N}$. Let $\mathrm{Sym}(\mathfrak{n})$ denote the collection of permutations of $[\mathfrak{n}]$.

For $k \in \mathbb{N}$, define $\mathfrak{P}_{<k}(\mathfrak{n}) := \{Y \subseteq [\mathfrak{n}] : |Y| < k\}$ and $\mathfrak{P}_k(\mathfrak{n}) := \{Y \subseteq [\mathfrak{n}] : |Y| = k\}$, and let $\mathfrak{P}_{\leq k}(\mathfrak{n}) := \mathfrak{P}_{<k}(\mathfrak{n}) \cup \mathfrak{P}_k(\mathfrak{n})$. We order each of these sets using *shortlex* order, i.e., ordered by size, with sets of the same size ordered lexicographically.

We write $\mathcal{L}_{\omega,\omega}(L)$ to denote the collection of first-order $L$-formulas. An *L-theory* is a collection of first-order $L$-sentences. A theory is $\Pi_1$ when every sentence is quantifier-free or of the form $(\forall \overline{x})\varphi(\overline{x})$ where $\overline{x}$ is a tuple of variables and $\varphi$ is quantifier-free. The *maximum arity* of $L$ is the maximum arity, when it exists, of a relation symbol or function symbol in $L$, where we consider constant symbols to be function symbols of arity 0.

Fix a probability space $(\Omega, \mathcal{F}, \mathbb{P})$. Suppose $(D, \mathcal{D})$ is a measurable space. A *D-valued random variable Z*, also called a *random element in D*, is an $(\mathcal{F}, \mathcal{D})$-measurable function $Z \colon \Omega \to D$. The *distribution* of $Z$ is the probability measure $\mathbb{P} \circ Z^{-1}$. Given an event $B \in \mathcal{D}$, we say that $B$ holds *almost surely* when $\mathbb{P}(B) = 1$, and abbreviate this *a.s.* Typically $B$ will be specified indirectly via some property of random variables; for example, we say that two random variables are equal a.s. when the subset of $\Omega$ on which they are equal has full measure. For a topological space $S$, let $\mathcal{P}(S)$ denote the space of Borel probability measures on $S$, with $\sigma$-algebra given by the weak topology. We use the symbol $\bigwedge$ for conjunctions of probabilistic events, as well as for conjunctions of logical formulas. See[Kal02] for further background and notation from probability theory.

We write $\lambda$ to denote the uniform (Lebesgue) measure on $[0, 1]$ and on finite powers of $[0, 1]$. Suppose a variable $x$ takes values in some finite power of $[0, 1]$. We say that an expression involving $x$ holds *almost everywhere*, abbreviated *a.e.*, when it holds of $x$ on all but a $\lambda$-null subset.

Let $Str_L(\mathfrak{n})$ denote the collection of $L$-structures that have underlying set $[\mathfrak{n}]$. We will consider $Str_L(\mathfrak{n})$ as a measure space with the $\sigma$-algebra generated by the topology given by basic clopen sets of the form

$$\{\mathcal{M} \in Str_L(\mathfrak{n}) : \mathcal{M} \models \varphi(x_0, \ldots, x_{\ell-1})\}$$

when $\varphi \in \mathcal{L}_{\omega,\omega}(L)$ is a quantifier-free formula with $\ell$ free variables and $x_0, \ldots, x_{\ell-1} \in [\mathfrak{n}]$.

**Definition 1.1.** Suppose $\varphi \in \mathcal{L}_{\omega,\omega}(L)$ has $\ell$ free variables and $r_0, \ldots, r_{\ell-1} \in [\mathfrak{n}]$. Define the **extent** (on $[\mathfrak{n}]$) of $\varphi(r_0, \ldots, r_{\ell-1})$ to be

$$[\![\varphi(r_0, \ldots, r_{\ell-1})]\!]_{[\mathfrak{n}]} := \{\mathcal{M} \in Str_L(\mathfrak{n}) : \mathcal{M} \models \varphi(r_0, \ldots, r_{\ell-1})\}.$$

When $\mathfrak{n} = \mathbb{N}$ we will sometimes omit the subscript $[\mathfrak{n}]$. For an $L$-theory $T$, define $[\![T]\!]_{[\mathfrak{n}]} := \bigcap_{\rho \in T} [\![\rho]\!]_{[\mathfrak{n}]}$.

There is a natural action of $\mathrm{Sym}(\mathfrak{n})$ on $Str_L(\mathfrak{n})$ called **the logic action**, defined as follows: for $\sigma \in \mathrm{Sym}(\mathfrak{n})$ and $\mathcal{M} \in Str_L(\mathfrak{n})$, let $\sigma \cdot \mathcal{M}$ be the structure $\mathcal{N} \in Str_L(\mathfrak{n})$ for which

$$R^{\mathcal{N}}(r_0, \ldots, r_{k-1}) \quad \text{if and only if} \quad R^{\mathcal{M}}(\sigma^{-1}(r_0), \ldots, \sigma^{-1}(r_{k-1}))$$

for all relation symbols $R \in L$ and $r_0, \ldots, r_{k-1} \in \mathfrak{n}$, where $k$ is the arity of $R$, and similarly with constant and function symbols. Note that the orbit under the logic action of any structure in $Str_L(\mathfrak{n})$ is its isomorphism class. By Scott's isomorphism theorem, every such orbit is Borel. For more details on the logic action, see[Kec95] §16.C.

We say a probability measure $\mu$ on $Str_L(\mathfrak{n})$ is **invariant** if it is invariant under the logic action of $\mathrm{Sym}(\mathfrak{n})$, i.e., if $\mu(B) = \mu(\sigma \cdot B)$ for every Borel $B \subseteq Str_L(\mathfrak{n})$ and every $\sigma \in \mathrm{Sym}(\mathfrak{n})$. We call such a probability measure an *invariant measure for $L$*. An invariant measure $\mu$ is **ergodic** if $\mu(X) = 0$ or $\mu(X) = 1$ whenever $\mu(X \triangle \sigma(X)) = 0$ for all $\sigma \in \mathrm{Sym}(\mathbb{N})$. Every ergodic invariant measure on $Str_L(\mathfrak{n})$ is an *extreme point* in the simplex of invariant measures on $Str_L(\mathfrak{n})$, and any invariant measure on $Str_L(\mathfrak{n})$ can be decomposed as a mixture of ergodic ones (see [Kal05] Lemma A1.2 and Theorem A1.3).

For $n \in \mathbb{N}$, any probability measure $\mu$ on $Str_L(\mathbb{N})$ induces a probability measure $\mu_n$ on $Str_L(n)$ such that for any Borel set $B \subseteq Str_L(n)$, we have

$$\mu_n(B) := \mu(\{\mathcal{M} \in Str_L(\mathbb{N}) : \mathcal{M}|_{[n]} \in B\}),$$

where $\mathcal{M}|_{[n]} \in Str_L(n)$ denotes the restriction of $\mathcal{M}$ to $[n]$. Further, by the Kolmogorov consistency theorem, $\mu$ is uniquely determined by the collection $\langle \mu_n \rangle_{n \in \mathbb{N}}$.

Towards defining the entropy function of an invariant measure, we give the standard definition of the entropy of a probability measure. For this definition, we use the convention that $-0 \log_2(0) = 0$.

**Definition 1.2.** Let $\nu$ be a probability measure on a standard Borel space $S$, and let $A := \{s \in S : \nu(\{s\}) > 0\}$ be its (countable) set of atoms. If $\nu$ is purely atomic, i.e., $\nu(A) = 1$, then the **entropy** of $\nu$ is given by

$$h(\nu) := -\sum_{x \in A} \nu(\{x\}) \log_2(\nu(\{x\})).$$

Otherwise, let

$$h(\nu) := \infty.$$

For any random variable $X$ with distribution $\nu$, define $h(X) := h(\nu)$.

**Definition 1.3.** The **joint entropy** of a pair of random variables $X$ and $Y$, written $h(X, Y)$, is defined to be the entropy of the joint distribution of $(X, Y)$. Similarly, $h(\langle X_i \rangle_{i \in I})$ is defined to be the entropy of the joint distribution of the sequence $\langle X_i \rangle_{i \in I}$.

**Definition 1.4.** Given random variables $X$ and $Y$, the **conditional entropy** of $X$ given $Y$, written $h(X \in \cdot \mid Y)$, is defined to be the function

$$y \mapsto h(\mathbb{P}(X \in \cdot \mid Y = y)).$$

We have defined the entropy of a random variable to be the entropy of its distribution. When the random variable takes values in a space of measures, instead of considering the entropy of the random variable directly, we sometimes need the random variable, defined below, that is obtained by taking the entropies of these measures themselves.

**Definition 1.5.** Suppose $\chi$ is a measure-valued random variable. The **random entropy** of $\chi$, written $H(\chi)$, is a random variable defined by

$$H(\chi)(\varpi) = h(\chi(\varpi)).$$

for $\varpi \in \Omega$.

This notion allows us to define random conditional entropy.

**Definition 1.6.** Given random variables $X$ and $Y$, the **random conditional entropy** of $X$ given $Y$, written $H(X \mid Y)$, is defined by

$$H(X \mid Y) = H(\mathbb{E}(X \mid Y)),$$

i.e., the random entropy of the random measure $\mathbb{E}(X \mid Y)$, the conditional expectation of $X$ given $Y$.

The following three lemmas are standard facts about entropy, which we will need later.

**Lemma 1.7** (see[CT06] Theorem 2.2.1)**.** *For random variables $X$ and $Y$, we have*

$$h(X, Y) = h(X) + \mathbb{E}(H(Y \mid X)).$$

**Lemma 1.8** (see[CT06] Theorem 2.6.5)**.** *For random variables $X$ and $Y$, we have*

$$h(X) \geq \mathbb{E}(H(X \mid Y)),$$

*with equality if and only if $X$ and $Y$ are independent.*

**Lemma 1.9** (see[CT06] Theorem 2.6.6)**.** *Let $\langle X_i \rangle_{i \in I}$ be a sequence of random variables. Then*

$$h(\langle X_i \rangle_{i \in I}) \leq \sum_{i \in I} h(X_i),$$

*with equality if and only if the $X_i$ are independent.*

We now define the entropy function of an invariant measure on $Str_L(\mathbb{N})$.

**Definition 1.10.** Let $\mu$ be an invariant measure on $Str_L(\mathbb{N})$. The **entropy function** of $\mu$ is defined to be the function $\mathrm{Ent}(\mu)$ from $\mathbb{N}$ to $\mathbb{R} \cup \{\infty\}$ given by

$$\mathrm{Ent}(\mu)(n) := h(\mu_n).$$

The following basic property of the entropy function is immediate.

**Lemma 1.11.** *Let $\mu$ be an invariant measure on $Str_L(\mathbb{N})$. Then* $\mathrm{Ent}(\mu)$ *is a non-decreasing function.*

In this paper we will mainly be interested in invariant measures whose entropy functions take values in $\mathbb{R}$. The following lemma is immediate because in a finite language, there are only finitely many structures of each finite size.

**Lemma 1.12.** *Suppose $L$ is finite, and let $\mu$ be an invariant measure on $Str_L(\mathbb{N})$. Then* $\mathrm{Ent}(\mu)$ *is* $\mathbb{R}$*-valued.*

In Section 4 we will strengthen this lemma by showing that there is a polynomial upper bound of $O(n^k)$ on $\mathrm{Ent}(\mu)(n)$, where $k$ is the maximum arity of $L$.

The notion of *non-redundance* for an invariant measure, which we introduce next, will be key throughout this paper.

**Definition 1.13.** Suppose $L$ is relational. For each $R \in L$ define $\vartheta_R$ to be the formula

$$(\forall x_0, \ldots, x_{k-1})\Big(R(x_0, \ldots, x_{k-1}) \rightarrow \bigwedge_{i<j<k} (x_i \neq x_j)\Big),$$

where $k$ is the arity of $R$. Define the $\Pi_1$ $L$-theory

$$\mathrm{Th}_{\mathrm{nr}}(L) := \{\vartheta_R : R \in L\}.$$

An $L$-structure $\mathcal{M}$ is **non-redundant** when $\mathcal{M} \models \mathrm{Th}_{\mathrm{nr}}(L)$. An invariant measure $\mu$ on $Str_L(\mathbb{N})$ is **non-redundant** when

$$\mu([\![\mathrm{Th}_{\mathrm{nr}}(L)]\!]) = 1.$$

For example, any $k$-uniform hypergraph is non-redundant, as is any directed graph without self-loops.

The following straightforward lemma provides conditions under which the entropy function takes values in $\mathbb{R}$.

**Lemma 1.14.** *Suppose $L$ has finitely many relations of any given arity, and let $\mu$ be a non-redundant invariant measure on $Str_L(\mathbb{N})$. Then for every $n \in \mathbb{N}$ there is a finite set $A_n \subseteq Str_L(n)$ such that $\mu_n$ concentrates on $A_n$. In particular,* $\mathrm{Ent}(\mu)(n) \in \mathbb{R}$, *and so* $\mathrm{Ent}(\mu)$ *is* $\mathbb{R}$*-valued.*

In Section 4 we will consider the case of non-redundant invariant measures for a relational language, and will strengthen this lemma by providing an explicit upper bound on $\mathrm{Ent}(\mu)(n)$.

## 1.2. *Quantifier-free interdefinitions*

The notion of *quantifier-free interdefinability*, which we define below, is a variant of the standard notion of interdefinability from the setting of $\aleph_0$-categorical theories (see, e.g., [AZ86] §1). It provides a method for translating invariant measures concentrated on the extent of a given $\Pi_1$ theory to invariant measures concentrated on the extent of a target $\Pi_1$ theory, in a way that preserves the entropy function. We use this machinery to show, in Proposition 1.21, that every entropy function of an invariant measure already occurs as the entropy function of a non-redundant invariant measure for a relational language.

*Throughout this subsection, $L_0$ and $L_1$ will be countable languages, sometimes with further restrictions.*

**Definition 1.15.** Suppose $T_0$ is an $L_0$-theory and $T_1$ is an $L_1$-theory. A **quantifier-free interdefinition** between $T_0$ and $T_1$ is a pair $\Psi = (\Psi_0, \Psi_1)$ of maps

$$\begin{aligned} \Psi_0 &: \mathcal{L}_{\omega,\omega}(L_0) \to \mathcal{L}_{\omega,\omega}(L_1) \qquad \text{and} \\ \Psi_1 &: \mathcal{L}_{\omega,\omega}(L_1) \to \mathcal{L}_{\omega,\omega}(L_0) \end{aligned}$$

such that for $j \in \{0, 1\}$, the formula $\Psi_j(\eta)$ is quantifier-free whenever $\eta \in \mathcal{L}_{\omega,\omega}(L_{1-j})$ is quantifier-free, and further,

$$\begin{aligned} T_{1-j} &\vdash \Psi_j \circ \Psi_{1-j}(\rho) \leftrightarrow \rho, \\ T_{1-j} &\vdash \Psi_j(x = y) \leftrightarrow (x = y), \\ T_{1-j} &\vdash \neg\Psi_j(\chi) \leftrightarrow \Psi_j(\neg\chi), \\ T_{1-j} &\vdash \Psi_j(\chi\wedge\varphi) \leftrightarrow (\Psi_j(\chi)\wedge\Psi_j(\varphi)), \qquad \text{and} \\ T_{1-j} &\vdash (\exists x)\Psi_j(\psi(x)) \leftrightarrow \Psi_j((\exists x)\psi(x)) \end{aligned}$$

for all $\rho \in \mathcal{L}_{\omega,\omega}(L_{1-j})$ and $\chi$, $\varphi$, $\psi(x) \in \mathcal{L}_{\omega,\omega}(L_j)$, and such that the free variables of $\Psi_j(\upsilon)$ are the same as those of $\upsilon$ for every $\upsilon \in \mathcal{L}_{\omega,\omega}(L_j)$.

For $\mathfrak{n} \in \mathbb{N} \cup \{\mathbb{N}\}$, the interdefinition $\Psi$ induces maps $\Psi^*_{j,\mathfrak{n}} : [\![T_j]\!]_{[\mathfrak{n}]} \to [\![T_{1-j}]\!]_{[\mathfrak{n}]}$ for $j \in \{0, 1\}$ satisfying, for any structure $\mathcal{N} \models T_j$, tuple $\mathbf{m} \in \mathfrak{n}$, and formula $\varphi \in \mathcal{L}_{\omega,\omega}(L_{1-j})$,

$$\Psi^*_{j,\mathfrak{n}}(\mathcal{N}) \models \varphi(\mathbf{m}) \qquad \text{if and only if} \qquad \mathcal{N} \models \Psi_{1-j}(\varphi)(\mathbf{m}).$$

It is immediate that each $\Psi^*_{j,\mathfrak{n}}$ is a bijection.

Quantifier-free interdefinitions between $\Pi_1$ theories preserve entropy functions, as we now show.

**Lemma 1.16.** *Let $\Psi = (\Psi_0, \Psi_1)$ be a quantifier-free interdefinition between a $\Pi_1$ $L_0$-theory $T_0$ and a $\Pi_1$ $L_1$-theory $T_1$. Suppose $\mu$ is an invariant measure on $Str_{L_0}(\mathbb{N})$ concentrated on $[\![T_0]\!]$ and let $\nu$ be the pushforward of $\mu$ along $\Psi^*_{0,\mathbb{N}}$. Then $\nu$ is an invariant measure on $Str_{L_1}(\mathbb{N})$ concentrated on $[\![T_1]\!]$, and* $\mathrm{Ent}(\nu) = \mathrm{Ent}(\mu)$.

**Proof.** It is immediate that $\nu$ is an invariant measure on $Str_{L_1}(\mathbb{N})$ concentrated on $[\![T_1]\!]$, by the definition of pushforward and the fact that $\Psi^*_{0,\mathbb{N}}$ is a bijection between $[\![T_0]\!]$ and $[\![T_1]\!]$.

We will next show, for each $n \in \mathbb{N}$, that there is a measure-preserving bijection between the atoms of $\mu_n$ and the atoms of $\nu_n$. By Definition 1.2, this will establish that $h(\mu_n) = h(\nu_n)$ for each $n \in \mathbb{N}$, and so $\mathrm{Ent}(\mu) = \mathrm{Ent}(\nu)$.

Let $n \in \mathbb{N}$. If $\mathcal{M} \in [\![T_0]\!]_{\mathbb{N}}$ then $\mathcal{M}|_{[n]} \in [\![T_0]\!]_{[n]}$, because $\mu$ concentrates on $[\![T_0]\!]_{\mathbb{N}}$ and $T_0$ is $\Pi_1$. Hence the measure $\mu_n$ concentrates on $[\![T_0]\!]_{[n]}$. Similarly, $\nu_n$ concentrates on $[\![T_1]\!]_{[n]}$. Therefore every atom of $\mu_n$ is in $[\![T_0]\!]_{[n]}$ and every atom of $\nu_n$ is in $[\![T_1]\!]_{[n]}$.

Write $\nu_n^*$ for the pushforward of $\mu_n$ along $\Psi^*_{0,[n]}$. It is clear that $\Psi^*_{0,[n]}$ is a measure-preserving bijection between the atoms of $\mu_n$ and the atoms of $\nu_n^*$. Let $\mathcal{B}$ be an atom of $\nu_n$; we have $\mathcal{B} \in [\![T_1]\!]_{[n]}$ because $\nu_n(\mathcal{B}) > 0$. We will show that $\nu_n(\mathcal{B}) = \nu_n^*(\mathcal{B})$.

Let $\mathcal{A} = \Psi^*_{1,[n]}(\mathcal{B})$. For $\mathcal{N} \in [\![T_1]\!]_{\mathbb{N}}$, by Definition 1.15 we have $\mathcal{N}|_{[n]} = \mathcal{B}$ if and only if $\Psi^*_{1,\mathbb{N}}(\mathcal{N})|_{[n]} = \mathcal{A}$. By this fact, the definitions of $\mu_n$, $\nu_n$, and pushforward, the surjectivity of $\Psi^*_{1,\mathbb{N}}$, and the fact that $\mu$ and $\nu$ are concentrated on $[\![T_0]\!]_{\mathbb{N}}$ and $[\![T_1]\!]_{\mathbb{N}}$ respectively, we have

$$\begin{aligned}
\nu_n(\mathcal{B}) &= \nu(\{\mathcal{N} \in [\![T_1]\!]_{\mathbb{N}} : \mathcal{N}|_{[n]} = \mathcal{B}\}) \\
&= \nu(\{\mathcal{N} \in [\![T_1]\!]_{\mathbb{N}} : \Psi^*_{1,\mathbb{N}}(\mathcal{N})|_{[n]} = \mathcal{A}\}) \\
&= \mu(\{\mathcal{M} \in [\![T_0]\!]_{\mathbb{N}} : \mathcal{M}|_{[n]} = \mathcal{A}\}) \\
&= \mu_n(\mathcal{A}).
\end{aligned}$$

But $\mu_n(\mathcal{A}) = \nu_n^*(\mathcal{B})$ by the definition of pushforward. Hence $\nu_n(\mathcal{B}) = \nu_n^*(\mathcal{B})$, as desired. □

Non-redundant invariant measures are defined only for relational languages. Hence towards showing, in Proposition 1.21, that every entropy function is the entropy function of a non-redundant invariant measure, we need a quantifier-free interdefinition between the empty theory in an arbitrary language and a particular $\Pi_1$ theory in a relational language. There is a standard such interdefinition that maps every function symbol to a relation symbol representing the graph of the function, but it entails an increase in arity, as each function symbol of arity $\ell$ is mapped to a relation symbol of arity $\ell + 1$.

While this standard interdefinition would suffice for our purposes in Proposition 1.21, here we provide a more parsimonious interdefinition, to highlight a connection with what is known about invariant measures for languages containing function symbols. Namely, using results from [AFP17], in Lemmas 1.18 and 1.19 we show how to avoid increasing the arity of symbols, by providing a quantifier-free interdefinition that replaces each function symbol of arity $\ell$ with finitely many relation symbols, each of arity $\ell$.

As we will see in Lemma 1.18, every invariant measure is concentrated on structures in which every function is a “selector”, sometimes called a “choice function”, i.e., a function for which the output is always one of the inputs. For example, the only unary selector is the identity function. We consider constant symbols to be 0-ary function symbols; observe that no constant is a selector.

**Definition 1.17.** Let $f \in L_0$ be a function symbol, and let $\ell$ be the arity of $f$. Define $\theta_f$ to be the sentence

$$(\forall x_0, \ldots, x_{\ell-1})\Big(\bigvee_{i\in[\ell]} f(x_0, \ldots, x_{\ell-1}) = x_i\Big),$$

asserting that $f$ is a *selector*, and define the $\Pi_1$ $L_0$-theory

$$\mathrm{Th}_{\mathrm{sel}}(L_0) := \{\theta_f : f \in L_0 \text{ is a function symbol}\}.$$

**Lemma 1.18.** *If $\mu$ is an invariant measure on $Str_{L_0}(\mathbb{N})$ then $\mu([\![\mathrm{Th}_{\mathrm{sel}}(L_0)]\!]) = 1$.*

**Proof.** Let $\nu$ be an ergodic invariant measure on $Str_{L_0}(\mathbb{N})$. By [AFP17] Lemma 2.4,

$$Th(\nu) := \{\varphi \in \mathcal{L}_{\omega,\omega}(L) : \nu([\![\varphi]\!]) = 1\},$$

is a complete deductively-closed first-order theory. For a function symbol $f \in L_0$, if $\neg\theta_f \in Th(\nu)$ then $Th(\nu)$ has non-trivial definable closure, contradicting [AFP17] Proposition 6.1, and so we must have $\theta_f \in Th(\nu)$.

Therefore $\nu([\![\mathrm{Th}_{\mathrm{sel}}(L_0)]\!]) = 1$. Because $\mu$ is a mixture of ergodic invariant measures $\nu$, we also have $\mu([\![\mathrm{Th}_{\mathrm{sel}}(L_0)]\!]) = 1$. □

We now provide the desired quantifier-free interdefinition.

**Lemma 1.19.** *Suppose $L_1$ is the relational language consisting of*

- *each relation symbol in $L_0$, along with*
- *a relation symbol $E_{f,i}$ for each function symbol $f \in L_0$ and $i \in [\ell]$, where $\ell$ is the arity of $f$.*

*Let $T$ be the $\Pi_1$ $L_1$-theory consisting of the sentences*

$$(\forall x_0, \ldots, x_{\ell-1}) \bigvee_{i\in[\ell]} E_{f,i}(x_0, \ldots, x_{\ell-1}),$$

*and*

$$(\forall x_0, \ldots, x_{\ell-1}) \bigwedge_{i<j\in[\ell]} \neg(E_{f,i}(x_0, \ldots, x_{\ell-1}) \wedge E_{f,j}(x_0, \ldots, x_{\ell-1})),$$

*for each function symbol $f \in L_0$, where $\ell$ is the arity of $f$.*

*Then there is a quantifier-free interdefinition $(\Psi_0, \Psi_1)$ between* $\mathrm{Th}_{\mathrm{sel}}(L_0)$ *and $T$.*

**Proof.** For each relation symbol $R \in L_0$, let $\Psi_1(R(x_0, \ldots, x_{k-1}))$ be the formula $R(x_0, \ldots, x_{k-1})$, where $k$ is the arity of $R$. For each function symbol $f \in L_0$ and $i \in [\ell]$, let $\Psi_1(E_{f,i}(x_0, \ldots, x_{\ell-1}))$ be the formula $f(x_0, \ldots, x_{\ell-1}) = x_i$, where $\ell$ is the arity of $f$. It is easy to define the analogous map $\Psi_0$ and check that $(\Psi_0, \Psi_1)$ is a quantifier-free interdefinition between $\mathrm{Th}_{\mathrm{sel}}(L_0)$ and $T$. □

Next we provide a quantifier-free interdefinition which, combined with the previous results, will allow us to prove Proposition 1.21.

**Lemma 1.20.** *Suppose $L_0$ is relational. Then there is a countable relational language $L_1$ and a quantifier-free interdefinition $(\Psi_0, \Psi_1)$ between the empty $L_0$-theory and the $\Pi_1$ $L_1$-theory* $\mathrm{Th}_{\mathrm{nr}}(L_1)$.

**Proof.** For each relation symbol $R \in L_0$ and equivalence relation $E$ on $[k]$, where $k \in \mathbb{N}$ is the arity of $R$, we define the following. Let $\ell$ be the number of $E$-equivalence classes. Let $f_E \colon [k] \to [k]$ send each $i \in [k]$ to the least element of its $E$-equivalence class, and let $\langle y_0^E, \ldots, y_{\ell-1}^E \rangle$ be the increasing enumeration of the image of $f_E$. Let $R_E$ be an $\ell$-ary relation symbol and let

$$\phi_E(x_0, \ldots, x_{k-1}) := R_E(x_{y_0^E}, \ldots, x_{y_{\ell-1}^E}) \wedge \bigwedge_{\substack{i,j \in [k] \\ iEj}} x_i = x_j \wedge \bigwedge_{\substack{i,j \in [k] \\ \neg(iEj)}} x_i \neq x_j.$$

Define $L_1$ to be the collection of all such symbols $R_E$. Recall the $\Pi_1$ theory $\mathrm{Th}_{\mathrm{nr}}(L_1)$, as defined in Definition 1.13.

Define the map $\Psi_0$ on atomic $L_0$-formulas by

$$\Psi_0(R(x_0, \ldots, x_{k-1})) := \bigvee_{\substack{E \text{ is an equivalence} \\ \text{relation on } [k]}} \phi_E(x_0, \ldots, x_{k-1}),$$

for each $R \in L_0$, where $k$ is the arity of $R$. Define the map $\Psi_1$ on atomic $L_1$-formulas by

$$\Psi_1(R_E(x_0, \ldots, x_{\ell-1})) := R(x_{f_E(0)}, \ldots, x_{f_E(k-1)})$$

for each $R_E \in L_1$, where $\ell$ is the arity of $R_E$. Extend $\Psi_0$ and $\Psi_1$ to all formulas in $\mathcal{L}_{\omega,\omega}(L_0)$ and $\mathcal{L}_{\omega,\omega}(L_1)$, respectively, in the natural way.

Observe that for any $\mathcal{M} \in Str_{L_0}(\mathbb{N})$, the $L_1$-structure $\Psi^*_{0,\mathbb{N}}(\mathcal{M})$ is non-redundant, and conversely, every non-redundant structure in $Str_{L_1}(\mathbb{N})$ is in the image of $\Psi^*_{0,\mathbb{N}}$. One can check that $(\Psi_0, \Psi_1)$ is a quantifier-free interdefinition between the empty $L_0$-theory and $\mathrm{Th}_{\mathrm{nr}}(L_1)$. □

We can now show that every entropy function is the entropy function of some non-redundant invariant measure.

**Proposition 1.21.** *Let $\mu$ be an invariant measure on $Str_{L_0}(\mathbb{N})$. Then there is a countable relational language $L_1$ and a non-redundant invariant measure $\nu$ on $Str_{L_1}(\mathbb{N})$ such that* $\mathrm{Ent}(\mu) = \mathrm{Ent}(\nu)$. *Further, if $L_0$ is finite and of maximum arity $k$, then so is $L_1$.*

**Proof.** Recall the $\Pi_1$ $L_0$-theory $\mathrm{Th}_{\mathrm{sel}}(L_0)$ from Definition 1.17. First observe by Lemma 1.18 that $\mu([\![\mathrm{Th}_{\mathrm{sel}}(L_0)]\!]) = 1$. By Lemma 1.19 there is a quantifier-free interdefinition $(\Psi_0, \Psi_1)$ between $\mathrm{Th}_{\mathrm{sel}}(L_0)$ and some $\Pi_1$ theory $T$ in a countable relational language $L'$. When $L_0$ is finite and of maximum arity $k$, so is $L'$.

By Lemma 1.20 there is a countable relational language $L_1$, and a quantifier-free interdefinition $\Theta = (\Theta_0, \Theta_1)$ between the empty $L'$-theory and the $\Pi_1$ $L_1$-theory $\mathrm{Th}_{\mathrm{nr}}$. It is easy to see from Definition 1.15 that $\Theta$ is also a quantifier-free interdefinition between $T$ and its image $\Theta_0(T)$, which is a $\Pi_1$ $L_1$-theory. Again, when $L'$ is finite and of maximum arity $k$, so is $L_1$.

By two applications of Lemma 1.16, there is an invariant measure $\nu$ on $Str_{L_1}(\mathbb{N})$ that is concentrated on $[\![\Theta_0(T)]\!]$ and that satisfies $\mathrm{Ent}(\mu) = \mathrm{Ent}(\nu)$. Finally, $\nu$ is non-redundant because $\Theta_0(T)$ contains $\mathrm{Th}_{\mathrm{nr}}$. □

## 2. Invariant measures arising from extended $L$-hypergraphons

In this section we study the growth of entropy functions of non-redundant invariant measures for a relational language $L$ whose relation symbols all have the same arity $k \geq 1$. By a variant of the Aldous–Hoover–Kallenberg theorem, these invariant measures are precisely the ones that arise as the distribution of a certain random $L$-structure $G(\mathbb{N}, W)$, where $W$ is a type of measurable function called an *extended L-hypergraphon*. In Theorem 2.7 we express the growth of the entropy function of $G(\mathbb{N}, W)$ in terms of $W$. This result generalizes a theorem of Aldous[Ald85] Chapter 15 and Janson[Jan13] Theorem D.5, and our argument mirrors that of Janson. As an immediate consequence of Theorem 2.7 and the Aldous–Hoover–Kallenberg theorem, we see in Corollary 2.8 that when $L$ is finite, the entropy function of any non-redundant invariant measure for $L$ is $O(n^k)$.

*For the rest of this section, fix $k \geq 1$ and let L be a countable relational language (possibly infinite) all of whose relations have arity k.*

We will define an extended $L$-hypergraphon to be a probability kernel from $[0,1]^{\mathfrak{P}_{<k}(k)}$ to a space of quantifier-free $k$-types, satisfying a specific coherence condition.

**Definition 2.1.** A **complete non-redundant quantifier-free $L$-type** with free variables $x_0, \ldots, x_{k-1}$ is a maximal consistent collection of atomic formulas or negations of atomic formulas containing $\{x_i \neq x_j : i < j < k\}$, and whose free variables are all contained in $\{x_0, \ldots, x_{k-1}\}$.

Let $\mathfrak{qf}^{\mathrm{nr}}{}_L$ be the space of complete non-redundant quantifier-free $L$-types with free variables $x_0, \ldots, x_{k-1}$, where a subbasic clopen set consists of complete non-redundant quantifier-free $L$-types containing a given atomic formula or negated atomic formula all of whose free variables are contained in $\{x_0, \ldots, x_{k-1}\}$.

Note that every complete non-redundant quantifier-free $L$-type in the above sense implies a complete quantifier-free $\mathcal{L}_{\omega_1,\omega}(L)$-type.

In the next definition, we introduce the notion of an extended $L$-hypergraphon. As we describe below, this generalizes the standard notion of a hypergraphon[Lov12] §23.3, which is a higher-arity version of a graphon.

The definition of extended $L$-hypergraphon involves a coherence condition, specified in terms of an action of Sym($k$). In Definition 2.5, we describe how an extended $L$-hypergraphon gives rise to a distribution on $L$-structures by determining, for each $k$-tuple of elements, the distribution on its quantifier-free type. The coherence condition ensures that the order in which the $k$-tuple is specified does not affect the resulting distribution on its quantifier-free type.

Consider the action of Sym($k$) on $\mathfrak{qf}^{\mathrm{nr}}{}_L$ given by

$$\sigma \cdot u = \{\varphi(x_{\sigma(i_0)}, \ldots, x_{\sigma(i_{\ell-1})}) : \varphi(x_{i_0}, \ldots, x_{i_{\ell-1}}) \in u \text{ and } \ell \leq k\}$$

for $\sigma \in \mathrm{Sym}(k)$ and $u \in \mathfrak{qf}^{\mathrm{nr}}{}_L$. Note that this action of Sym($k$) extends to $\mathcal{P}(\mathfrak{qf}^{\mathrm{nr}}{}_L)$ in the natural way, namely,

$$(\sigma \cdot \nu)(B) = \nu(\{\sigma \cdot u : u \in B\})$$

for $\nu \in \mathcal{P}(\mathfrak{qf}^{\mathrm{nr}}{}_L)$ and Borel $B \subseteq \mathfrak{qf}^{\mathrm{nr}}{}_L$.

**Definition 2.2.** An **extended $L$-hypergraphon** is a measurable map

$$W\colon [0,1]^{\mathfrak{P}_{<k}(k)} \to \mathcal{P}(\mathfrak{qf}^{\mathrm{nr}}{}_L)$$

such that for any $\langle x_F \rangle_{F \in \mathfrak{P}_{<k}(k)} \in [0,1]^{\mathfrak{P}_{<k}(k)}$ and $\sigma \in \mathrm{Sym}(k)$,

$$W(\langle x_{\sigma(F)} \rangle_{F \in \mathfrak{P}_{<k}(k)}) = \sigma \cdot W(\langle x_F \rangle_{F \in \mathfrak{P}_{<k}(k)}).$$

The next technical lemma and definitions show how an extended $L$-hypergraphon $W$ gives rise to a random non-redundant $L$-structure $G(\mathbb{N}, W)$ whose distribution is an invariant measure on $Str_L(\mathbb{N})$.

The following is a special case of a standard result from probability theory about the randomization of a kernel.

**Lemma 2.3** (see[Kal02] Lemma 3.22)**.** *Let $W\colon [0,1]^{\mathfrak{P}_{<k}(k)} \to \mathcal{P}(\mathfrak{qf}^{\mathrm{nr}}{}_L)$ be an extended L-hypergraphon. There is a measurable function $W^*\colon [0,1]^{\mathfrak{P}_{<k}(k)} \times [0,1] \to \mathfrak{qf}^{\mathrm{nr}}{}_L$ such that whenever $\zeta$ is a uniform random variable in $[0,1]$, then for all $t \in [0,1]^{\mathfrak{P}_{<k}(k)}$, the random variable $W^*(t,\zeta)$ has distribution $W(t)$.*

Definition 2.4 introduces notation that we will use throughout this section and the next.

**Definition 2.4.** Let $\mathfrak{n} \in \mathbb{N} \cup \{\mathbb{N}\}$, and suppose $\mathfrak{n} \geq k$ (when $\mathfrak{n} \in \mathbb{N}$). Let $J \in \mathfrak{P}_k(\mathfrak{n})$, and let $\tau_J \colon [k] \to J$ be the unique increasing bijection from $[k]$ to $J$. Define $\widehat{X_J}$ to be the sequence $\langle X_{\tau_J(F)} \rangle_{F \in \mathfrak{P}_{<k}(k)}$ consisting of terms of the form $X_I$, where $I$ is from $\mathfrak{P}_{<k}(\mathfrak{n})$.

Recall that $L$ consists of relation symbols all of arity $k$, and so in order to define a non-redundant $L$-structure, it suffices to describe the complete quantifier-free type of every strictly increasing $k$-tuple in the structure.

**Definition 2.5.** Let $W$ be an extended $L$-hypergraphon and $\mathfrak{n} \in \mathbb{N} \cup \{\mathbb{N}\}$, and suppose $\mathfrak{n} \geq k$ (when $\mathfrak{n} \in \mathbb{N}$).

- Define $M(\mathfrak{n}, W) \colon [0,1]^{\mathfrak{P}_{\leq k}(\mathfrak{n})} \to Str_L(\mathfrak{n})$ to be the map such that for all $\langle x_D \rangle_{D \in \mathfrak{P}_{\leq k}(\mathfrak{n})} \in [0,1]^{\mathfrak{P}_{\leq k}(\mathfrak{n})}$ and every $J \in \mathfrak{P}_k(\mathfrak{n})$, the quantifier-free type of the tuple $\langle \tau_J(0), \ldots, \tau_J(k-1) \rangle$ in the $L$-structure $M(\mathfrak{n}, W)(\langle x_D \rangle_{D \in \mathfrak{P}_{\leq k}(\mathfrak{n})})$ is $W^*(\widehat{x_J}, x_J)$.
- Let $\langle \zeta_D \rangle_{D \in \mathfrak{P}_{\leq k}(\mathfrak{n})}$ be an i.i.d. sequence of uniform random variables in $[0,1]$. Define $G(\mathfrak{n}, W)$ to be the random $L$-structure $M(\mathfrak{n}, W)(\langle \zeta_D \rangle_{D \in \mathfrak{P}_{\leq k}(\mathfrak{n})})$.
- For $J \in \mathfrak{P}_k(\mathfrak{n})$, define the random variable $E_J^W := W^*(\widehat{\zeta_J}, \zeta_J)$.

In summary, $G(\mathfrak{n}, W)$ is the random $L$-structure with underlying set $[\mathfrak{n}]$ whose quantifier-free $k$-types are given by the random variables $E_J^W$ for $J \in \mathfrak{P}_k(\mathfrak{n})$. It is easy to check that the distribution of $G(\mathfrak{n}, W)$ does not depend on the specific choice of i.i.d. uniform $\langle \zeta_D \rangle_{D \in \mathfrak{P}_{\leq k}(\mathfrak{n})}$ or on the specific function $W^*$ satisfying Lemma 2.3.

Observe that for an extended $L$-hypergraphon $W$, the distribution of $G(\mathbb{N}, W)$ is a probability measure on $Str_L(\mathbb{N})$ that is invariant because $\langle \zeta_D \rangle_{D \in \mathfrak{P}_{\leq k}(\mathbb{N})}$ is i.i.d., and is non-redundant because $W^*$ takes values in $\mathfrak{qf}^{\mathrm{nr}}{}_L$. In fact, as Theorem 2.6 asserts, every non-redundant invariant measure on $Str_L(\mathbb{N})$ arises from some $W$ in this way; this result is a variant of the usual Aldous–Hoover–Kallenberg theorem, and is a specialization of[Ack15] Theorem 2.37.

**Theorem 2.6 (Aldous–Hoover–Kallenberg theorem).** *A non-redundant probability measure $\mu$ on $Str_L(\mathbb{N})$ is invariant under the action of* Sym($\mathbb{N}$) *if and only if $\mu$ is the distribution of $G(\mathbb{N}, W)$ for some extended $L$-hypergraphon $W$.*

For more details on the usual Aldous–Hoover–Kallenberg theorem, which is stated in terms of exchangeable arrays, see[Kal05] Chapter 7 and the historical notes to that chapter. There is an essentially equivalent statement, in terms of hypergraphons, that provides a correspondence between hypergraphons (of arity $k$) and ergodic invariant measures on the space of $k$-uniform graphs (see[DJ08] and[Aus08]). The version that we have stated above in Theorem 2.6 provides an analogous correspondence between extended $L$-hypergraphons and non-redundant invariant measures on $Str_L(\mathbb{N})$. Note that, unlike those invariant measures that arise from hypergraphons, the ones arising in Theorem 2.6 can in general be mixtures of ergodic invariant measures, and can be concentrated on structures that include multiple relations (though finitely many, of the same arity) which need not be symmetric.

In fact, we could have defined extended $L$-hypergraphons, and then described the associated sampling procedure $G(\mathbb{N}, W)$ and stated the Aldous–Hoover–Kallenberg theorem, for the even more general case that does not stipulate non-redundance. However, the invariant measure resulting from the sampling procedure would then be sensitive to measure 0 changes in the extended $L$-hypergraphon, unlike the situation for standard hypergraphons or the usual Aldous–Hoover–Kallenberg representation, and so we have restricted our definitions and results to the non-redundant case.

Using the machinery we have developed above, we can now prove the main theorems of this section. By Theorem 2.6, in order to study the entropy function of a non-redundant invariant measure on $Str_L(\mathbb{N})$, we may ask for a suitable extended $L$-hypergraphon $W$ and then analyze the entropy function of $G(\mathbb{N}, W)$. In Theorem 2.7, we consider such a $G(\mathbb{N}, W)$, and express the leading term of its entropy function as a function of $W$. As a consequence, in Corollary 2.8, when $L$ is finite we obtain bounds on the entropy function of a non-redundant invariant measure for $L$.

The proof of Theorem 2.7 closely follows that of[Jan13] Theorem D.5.

**Theorem 2.7.**
*Fix an extended L-hypergraphon W. Suppose that* $C := \int h(W(\widehat{z_{[k]}}))\, d\lambda(\widehat{z_{[k]}})$ *is finite. Then*

$$\lim_{n\to\infty} \frac{\mathrm{Ent}(G(\mathbb{N}, W))(n)}{|\mathfrak{P}_k(n)|} = C.$$

*In particular,* $\mathrm{Ent}(G(\mathbb{N}, W))(n) = Cn^k + o(n^k)$.

**Proof.** Observe that for all $n \in \mathbb{N}$, we have $\mathrm{Ent}(G(\mathbb{N}, W))(n) = h(G(n, W))$. We first show a lower bound on $h(G(n, W))$ for all $n \geq k$. For $J \in \mathfrak{P}_k(n)$, the random variables $E^W_J$ are independent conditioned on $\langle \zeta_I \rangle_{I \in \mathfrak{P}_{<k}(n)}$ and so, as the conditional

entropy of conditionally independent random variables is additive, we have

$$\begin{aligned} H(G(n,W) \mid \langle \zeta_I \rangle_{I\in\mathfrak{P}_{<k}(n)}) &= \sum_{J\in\mathfrak{P}_k(n)} H(E_J^W \mid \langle \zeta_I \rangle_{I\in\mathfrak{P}_{<k}(n)}) \\ &= \sum_{J\in\mathfrak{P}_k(n)} H(\mathbb{E}(W(\widehat{\zeta_J}) \mid \langle \zeta_I \rangle_{I\in\mathfrak{P}_{<k}(n)})) \\ &= \sum_{J\in\mathfrak{P}_k(n)} H(W(\widehat{\zeta_J})) \end{aligned}$$

a.s., where the last equality follows from the fact that every random variable in the sequence $\widehat{\zeta_J}$ occurs within $\langle \zeta_I \rangle_{I\in\mathfrak{P}_{<k}(n)}$.

By Lemma 1.8, and then taking expectations of the first and last terms in the previous chain of equalities, we have

$$\begin{aligned} h(G(n,W)) &\geq \mathbb{E}(H(G(n,W) \mid \langle \zeta_I \rangle_{I\in\mathfrak{P}_{<k}(n)})) \\ &= \sum_{J\in\mathfrak{P}_k(n)} \mathbb{E}(H(W(\widehat{\zeta_J}))). \end{aligned}$$

Because the distribution of the random variable $\widehat{\zeta_J}$ is the same for all $J \in \mathfrak{P}_k(n)$, we have

$$\begin{aligned} \sum_{J\in\mathfrak{P}_k(n)} \mathbb{E}(H(W(\widehat{\zeta_J}))) &= |\mathfrak{P}_k(n)| \cdot \mathbb{E}(H(W(\widehat{\zeta_{[k]}}))) \\ &= |\mathfrak{P}_k(n)| \cdot \int h(W(\widehat{z_{[k]}}))\, d\lambda(\widehat{z_{[k]}}). \end{aligned}$$

This equation, together with the previous inequality, yields

$$\frac{h(G(n,W))}{|\mathfrak{P}_k(n)|} \geq \int h(W(\widehat{z_{[k]}}))\, d\lambda(\widehat{z_{[k]}}).$$

Now we show an upper bound on $h(G(n,W))$ for all $n \geq k$. Let $r \in \mathbb{N}$ be positive.

For $I \in \mathfrak{P}_{<k}(n)$, define $Y_I := \lfloor r \cdot \zeta_I \rfloor$, so that $Y_I = \ell$ precisely when $\frac{\ell}{r} \leq \zeta_I < \frac{\ell+1}{r}$. Then by Lemma 1.7, we have

$$\begin{aligned} h(G(n,W)) &\leq h(G(n,W), \langle Y_I \rangle_{I\in\mathfrak{P}_{<k}(n)}) \\ &= h(\langle Y_I \rangle_{I\in\mathfrak{P}_{<k}(n)}) + \mathbb{E}\Big(H(G(n,W) \mid \langle Y_I \rangle_{I\in\mathfrak{P}_{<k}(n)})\Big). \end{aligned}$$

The first term in this last expression is straightforward to calculate; we have

$$h(\langle Y_I \rangle_{I\in\mathfrak{P}_{<k}(n)}) = \left|\mathfrak{P}_{<k}(n)\right| \cdot \log_2(r),$$

as the $Y_I$ are uniformly distributed on $[r]$ and independent (because the $\zeta_I$ are i.i.d. uniform on $[0,1]$).

We next calculate the second term; we will show that

$$\mathbb{E}\Big(H(G(n,W)\mid \langle Y_I\rangle_{I\in\mathfrak{P}_{<k}(n)})\Big) \;=\; |\mathfrak{P}_k(n)|\cdot\int h(W_r(\widehat{z}_{[k]}))\,d\lambda(\widehat{z}_{[k]}).$$

By (the standard conditional extension of) Lemma 1.9, we have

$$\begin{aligned} H(G(n,W)\,\big|\,\langle Y_I\rangle_{I\in\mathfrak{P}_{<k}(n)}) \;&\le\; \sum_{J\in\mathfrak{P}_k(n)} H(E^W_J\,\big|\,\langle Y_I\rangle_{I\in\mathfrak{P}_{<k}(n)}) \\ &=\; \sum_{J\in\mathfrak{P}_k(n)} H(E^W_J\,\big|\,\widehat{Y}_J) \end{aligned}$$

a.s., where the last equality follows from the fact that the only random variables in $\langle\zeta_I\rangle_{I\in\mathfrak{P}_{<k}(n)}$ on which a given $E^W_J$ depends are those among $\widehat{\zeta}_J$.

Given a function $\alpha\colon \mathfrak{P}_{<k}(k)\to[r]$, define

$$w_r(\alpha) \;:=\; r^{|\mathfrak{P}_{<k}(k)|}\cdot\int W(\widehat{z}_{[k]})\cdot\prod_{F\in\mathfrak{P}_{<k}(k)}\mathbb{1}_{\left[\frac{\alpha(F)}{r},\frac{\alpha(F)+1}{r}\right)}(z_F)\,d\lambda(\widehat{z}_{[k]}),$$

where $\mathbb{1}_S$ denotes the characteristic function of a set $S$ (in this case, a half-open interval). Observe that

$$w_r(\alpha) \;=\; \mathbb{E}\Big(W(\widehat{\zeta}_{[k]})\,\Big|\,\bigwedge_{F\in\mathfrak{P}_{<k}(k)} Y_F=\alpha(F)\Big)$$

a.s. (We may form this conditional expectation because the set of signed measures on $\mathfrak{qf}^{\mathrm{nr}}{}_L$ is a Banach space, and hence integration on this space is well-defined.)

For $\widehat{z}_{[k]}\in[0,1]^{\mathfrak{P}_{<k}(k)}$, define $\beta_{\widehat{z}_{[k]}}\colon\mathfrak{P}_{<k}(k)\to[r]$ by $\beta_{\widehat{z}_{[k]}}(F)=\lfloor r\cdot z_F\rfloor$ for $F\in\mathfrak{P}_{<k}(k)$, and let $W_r(\widehat{z}_{[k]}):=w_r(\beta_{\widehat{z}_{[k]}})$. Observe that while we have defined continuum-many instances of $\beta_{\widehat{z}_{[k]}}$, they range over the merely finitely many functions from $\mathfrak{P}_{<k}(k)$ to $[r]$. Further note that $W_r$ is a step function, and that as $r\to\infty$, the function $W_r$ converges to $W$ pointwise a.e. We can think of $W_r$ as the result of discretizing $W$ along blocks of width $1/r$.

Recall the function $\tau_J$ defined in Definition 2.4. For any $J\in\mathfrak{P}_k(n)$, we have

$$\begin{aligned} \mathbb{P}\Big(E^W_J\in\cdot\,\Big|\,\bigwedge_{F\in\mathfrak{P}_{<k}(k)}Y_{\tau_J(F)}=\beta_{\widehat{z}_{[k]}}(F)\Big) &= \mathbb{E}\Big(W(\widehat{\zeta}_J)\,\Big|\,\bigwedge_{F\in\mathfrak{P}_{<k}(k)}Y_{\tau_J(F)}=\beta_{\widehat{z}_{[k]}}(F)\Big) \\ &= \mathbb{E}\Big(W(\widehat{\zeta}_{[k]})\,\Big|\,\bigwedge_{F\in\mathfrak{P}_{<k}(k)}Y_F=\beta_{\widehat{z}_{[k]}}(F)\Big) \\ &= w_r(\beta_{\widehat{z}_{[k]}}) \end{aligned}$$

a.s., because $\tau_J$ is a bijection from $[k]$ to $J$ and the distribution of the random variable $\widehat{\zeta}_J$ is the same as that of $\widehat{\zeta}_{[k]}$. Therefore

$$h\Big(E^W_J\in\cdot\,\Big|\,\bigwedge_{F\in\mathfrak{P}_{<k}(k)}Y_{\tau_J(F)}=\beta_{\widehat{z}_{[k]}}(F)\Big)=h(w_r(\beta_{\widehat{z}_{[k]}})).$$

Summing both sides over all possible choices of $\beta_{\widehat{z_{[k]}}}$, we have

$$\sum_{\alpha\colon \mathfrak{P}_{<k}(k)\to[r]} h\Big(E_J^W \in \cdot \,\Big|\, \bigwedge_{F\in\mathfrak{P}_{<k}(k)} Y_{\tau_J(F)} = \alpha(F)\Big) = \sum_{\alpha\colon \mathfrak{P}_{<k}(k)\to[r]} h(w_r(\alpha))).$$

As before,

$$H(E_J^W \mid \langle Y_I\rangle_{I\in\mathfrak{P}_{<k}(n)}) = H(E_J^W \mid \widehat{Y}_J).$$

One can directly show that

$$\mathbb{E}\Big(H(E_J^W \mid \widehat{Y}_J)\Big) = r^{-|\mathfrak{P}_{<k}(k)|} \sum_{\alpha\colon \mathfrak{P}_{<k}(k)\to[r]} h\Big(E_J^W \in \cdot \,\Big|\, \bigwedge_{F\in\mathfrak{P}_{<k}(k)} Y_{\tau_J(F)} = \alpha(F)\Big).$$

Hence

$$\begin{aligned}\mathbb{E}\Big(H(E_J^W \mid \langle Y_I\rangle_{I\in\mathfrak{P}_{<k}(n)})\Big) &= r^{-|\mathfrak{P}_{<k}(k)|}\cdot \sum_{\alpha\colon \mathfrak{P}_{<k}(k)\to[r]} h(w_r(\alpha))\\ &= \int h(W_r(\widehat{z_{[k]}}))\, d\lambda(\widehat{z_{[k]}}).\end{aligned}$$

But once again by (an extension of) Lemma 1.9, we have

$$\begin{aligned}\mathbb{E}\Big(H(G(n,W) \mid \langle Y_I\rangle_{I\in\mathfrak{P}_{<k}(n)})\Big) &\le \sum_{J\in\mathfrak{P}_k(n)} \mathbb{E}\Big(H(E_J^W \mid \langle Y_I\rangle_{I\in\mathfrak{P}_{<k}(n)})\Big)\\ &= |\mathfrak{P}_k(n)|\cdot \int h(W_r(\widehat{z_{[k]}}))\, d\lambda(\widehat{z_{[k]}}).\end{aligned}$$

Putting together our two calculations, we obtain

$$h(G(n,W)) \le |\mathfrak{P}_{<k}(n)|\cdot \log_2(r) + |\mathfrak{P}_k(n)|\cdot \int h(W_r(\widehat{z_{[k]}}))\, d\lambda(\widehat{z_{[k]}}).$$

Hence for each positive $r$, we have

$$\frac{h(G(n,W))}{|\mathfrak{P}_k(n)|} \le \frac{|\mathfrak{P}_{<k}(n)|}{|\mathfrak{P}_k(n)|}\cdot \log_2(r) + \int h(W_r(\widehat{z_{[k]}}))\, d\lambda(\widehat{z_{[k]}}),$$

and so

$$\limsup_{n\to\infty} \frac{h(G(n,W))}{|\mathfrak{P}_k(n)|} \le \int h(W_r(\widehat{z_{[k]}}))\, d\lambda(\widehat{z_{[k]}}).$$

Letting $r\to\infty$, recall that $W_r \to W$ a.e., and so by the dominated convergence theorem, we have

$$\limsup_{n\to\infty} \frac{h(G(n,W))}{|\mathfrak{P}_k(n)|} \le \int h(W(\widehat{z_{[k]}}))\, d\lambda(\widehat{z_{[k]}}).$$

Combining our lower and upper bounds, we have

$$\lim_{n\to\infty} \frac{h(G(\mathbb{N},W))}{|\mathfrak{P}_k(n)|} = \int h(W(\widehat{z_{[k]}}))\, d\lambda(\widehat{z_{[k]}}),$$

as desired.

Finally, because $\lim_{n\to\infty} \frac{|\mathfrak{P}_k(n)|}{n^k} = 1$, we have

$$\mathrm{Ent}(G(\mathbb{N}, W))(n) = Cn^k + o(n^k),$$

where $C = \int h(W(\widehat{z_{[k]}}))\, d\lambda(\widehat{z_{[k]}})$. □

As a corollary, when $L$ is finite, we obtain a bound on the growth of entropy functions of non-redundant invariant measures for $L$.

**Corollary 2.8.** *Suppose $L$ is finite, and let $\mu$ be a non-redundant invariant measure on $Str_L(\mathbb{N})$. Then* $\mathrm{Ent}(\mu)(n) = Cn^k + o(n^k)$ *for some constant $C$. In particular,* $\mathrm{Ent}(\mu)(n) = O(n^k)$.

**Proof.** By Theorem 2.6 there is some extended $L$-hypergraphon $W$ such that $\mu$ is the distribution of the random $L$-structure $G(\mathbb{N}, W)$. Because $L$ is finite, $W$ is bounded, and so $\int h(W(\widehat{z_{[k]}}))\, d\lambda(\widehat{z_{[k]}})$ is finite. Hence the entropy function $\mathrm{Ent}(\mu)$ is of the desired form by Theorem 2.7. □

## 3. Invariant measures sampled from a Borel hypergraph

We have seen in Corollary 2.8 that for a finite relational language $L$ all of whose relation symbols have the same arity $k \geq 1$, the entropy function of an invariant measure on $Str_L(\mathbb{N})$ is of the form $Cn^k + o(n^k)$, where $C$ is a constant depending on the invariant measure. In this section we consider the situation where $C = 0$.

For $k = 1$, consider an invariant measure $\mu$ on $Str_L$, and let $W$ be an extended $L$-hypergraphon such that $\mu$ is the distribution of $G(\mathbb{N}, W)$. By Theorem 2.7, we have $\mathrm{Ent}(\mu)(n) = Cn + o(n)$, where $C = \int h(W(z_\emptyset))\, d\lambda(z_\emptyset)$. Suppose $C = 0$. Then $h(W(z_\emptyset)) = 0$ for a.e. $z_\emptyset$, and so $W(z_\emptyset)$ is a point mass a.e. Hence $G(\mathbb{N}, W)$ is a random $L$-structure where every element of $\mathbb{N}$ has the same quantifier-free 1-type, a.s. Therefore $\mu$ is a mixture of finitely many point masses, and $\mathrm{Ent}(\mu)(n)$ is a constant that does not depend on $n$. In summary, for $k = 1$, the only possible entropy functions of sublinear growth are the constant functions.

Theorem 3.5, the main result of this section, states that in contrast, for $k > 1$, Corollary 2.8 is tight in the sense that for any given function $\gamma$ that is $o(n^k)$, there is some non-redundant invariant measure whose entropy function is $o(n^k)$ but grows faster than $\gamma$. As discussed in the introduction, this result is a generalization of[HN13] Theorem 1.1, and the arguments in this section closely follow their proof.

*For the rest of this section, fix $k \geq 2$ and let $L$ be the language of $k$-uniform hypergraphs, i.e., $L = \{E\}$ where $E$ is a $k$-ary relation symbol.*

A *k-uniform hypergraph* is a non-redundant $L$-structure satisfying

$$\bigwedge_{\sigma \in \mathrm{Sym}(k)} (\forall x_0, \ldots, x_{k-1}) \big(E(x_0, \ldots, x_{k-1}) \leftrightarrow E(x_{\sigma(0)}, \ldots, x_{\sigma(k-1)})\big),$$

and we call the instantiation of $E$ its *edge set*.

By a *Borel hypergraph*, we mean a $k$-uniform hypergraph $\mathcal{M}$ whose underlying set is $[0, 1]$ and such that for any atomic formula $\varphi$ in the language of hypergraphs, the set $\{\overline{a} \in \mathcal{M} : \mathcal{M} \models \varphi(\overline{a})\}$ of realizations of $\varphi$ in $\mathcal{M}$ is Borel.

Any extended $L$-hypergraphon $W$ *yields* a non-redundant invariant measure, namely the distribution of the random $L$-structure $\mathbb{G}(\mathbb{N}, W)$. We say that $W$ *induces a Borel hypergraph* when this invariant measure can be obtained by sampling a random subhypergraph of some Borel hypergraph, as we make precise in Definition 3.1. We will see, in Lemma 3.2, that when $W$ induces a Borel hypergraph, it yields an invariant measure whose entropy function is $o(n^k)$. The main construction of this section, in Theorem 3.5, builds non-redundant invariant measures whose entropy functions have arbitrarily high growth within $o(n^k)$ by sampling from certain extended $L$-hypergraphons that induce Borel hypergraphs.

Observe that there are only two non-redundant quantifier-free $k$-types in $L$ that are consistent with the theory of $k$-uniform hypergraphs. Let $u_\top, u_\bot \in \mathfrak{qf}^{\mathrm{nr}}{}_L$ be the unique non-redundant quantifier-free types containing

$$\bigwedge_{\sigma \in \mathrm{Sym}(k)} E(x_{\sigma(0)}, \ldots, x_{\sigma(k-1)})$$

and

$$\bigwedge_{\sigma \in \mathrm{Sym}(k)} \neg E(x_{\sigma(0)}, \ldots, x_{\sigma(k-1)}),$$

respectively, and let $\delta_\top, \delta_\bot \in \mathcal{P}(\mathfrak{qf}^{\mathrm{nr}}{}_L)$ be the respective point masses concentrated on them.

Recall from Definition 2.4 that $\widehat{x}_{[k]}$ denotes the tuple of variables $\langle x_I \rangle_{I \in \mathfrak{P}_{<k}(k)}$.

**Definition 3.1.** We say that an extended $L$-hypergraphon $W : [0, 1]^{\mathfrak{P}_{<k}(k)} \to \mathcal{P}(\mathfrak{qf}^{\mathrm{nr}}{}_L)$ **induces a Borel hypergraph** if

- for a.e. pair of sequences $\widehat{x}_{[k]}, \widehat{y}_{[k]}$ of elements of $[0, 1]$ with $x_{\{i\}} = y_{\{i\}}$ for all $i \in [k]$, we have $W(\widehat{x}_{[k]}) = W(\widehat{y}_{[k]})$, and
- for a.e. sequence $\widehat{x}_{[k]}$ of elements of $[0, 1]$, the distribution $W(\widehat{x}_{[k]})$ is either $\delta_\top$ or $\delta_\bot$.

It follows that an extended $L$-hypergraphon $W$ induces a Borel hypergraph precisely when there is a Borel hypergraph $\mathcal{B}$ such that $W(\widehat{x}_{[k]})$ is a point mass concentrated on the quantifier-free type of $\langle x_{\{0\}}, \dots, x_{\{k-1\}} \rangle$ in $\mathcal{B}$. In this case we say that $W$ *induces the Borel hypergraph* $\mathcal{B}$.

The notion of an extended $L$-hypergraphon inducing a Borel hypergraph is closely related, in the case $k = 2$, to that of a graphon being random-free. A *graphon* is a symmetric Borel function from $[0, 1]^2$ to $[0, 1]$; as described in [LS10] and [Jan13], it is called *random-free* when it is $\{0, 1\}$-valued a.e. Every graphon gives rise to a random undirected graph on $\mathbb{N}$ whose distribution is an invariant measure. For $k = 2$, an extended $L$-hypergraphon $W$ that yields an invariant measure concentrated on undirected graphs can be expressed as a mixture of invariant measures, each obtained via a graphon. In the case where such a $W$ corresponds to a single graphon, $W$ induces a Borel hypergraph precisely when the corresponding graphon is random-free.

The notion of a random-free graphon also essentially appeared, in the context of separate exchangeability, in work of Aldous in [Ald81] Proposition 3.6 and [Ald85] (14.15) and p. 133, and Diaconis–Freedman [DF81] (4.10). Further, Kallenberg [Kal99] describes, for all $k \geq 2$, the similar notion of a *simple array*; this corresponds to our notion of inducing a Borel hypergraph, in the case where the distribution of the simple array is ergodic. Random-free graphons arise as well in [PV10] and [AFP16] §6.1, which consider invariant measures that are concentrated on a given orbit of the logic action.

Theorem 2.7 implies that any extended $L$-hypergraphon $W$ yields an invariant measure whose entropy function is $O(n^k)$. Observe that there are extended $L$-hypergraphons achieving this upper bound, i.e., that yield an invariant measure whose entropy function is $\Omega(n^k)$. For example, one can directly calculate that the Erdős–Rényi extended $L$-hypergraphon given by the constant function

$$W_{\mathrm{ER}}(\widehat{x_{[k]}}) = \mathrm{Uniform}(\{u_\top, u_\bot\})$$

satisfies $\mathrm{Ent}(G(\mathbb{N}, W_{\mathrm{ER}}))(n) = \binom{n}{k}$.

However, we now show that this growth rate cannot be achieved for an extended $L$-hypergraphon that induces a Borel hypergraph.

**Lemma 3.2.** *Let $W$ be an extended $L$-hypergraphon, and suppose $W$ induces a Borel hypergraph. Then* $\mathrm{Ent}(G(\mathbb{N}, W))(n) = o(n^k)$.

**Proof.** Because $W$ induces a Borel hypergraph, it takes the value $\delta_\top$ or $\delta_\bot$ a.e. But $h(\delta_\top) = h(\delta_\bot) = 0$, and so $\int h(W(\widehat{z_{[k]}}))\, d\lambda(\widehat{z_{[k]}}) = 0$. Hence by Theorem 2.7, we have $\mathrm{Ent}(G(\mathbb{N}, W))(n) = o(n^k)$. □

As noted previously, Aldous [Ald85] and Janson [Jan13] have versions of Theorem 2.7 for $k = 2$. They also observe that their respective results immediately imply that a random-free graphon yields an invariant measure with entropy function that is $o(n^2)$; their proofs are similar to that of Lemma 3.2. Their setting involves working with graphons, which yield ergodic invariant measures. Under the restriction of ergodicity, it is easily seen that the converse of Lemma 3.2 holds for $k = 2$, as noted by Janson[Jan13] Theorem 10.16.

In contrast, the converse of Lemma 3.2 itself does not hold, as our notion of extended $L$-hypergraphon allows for ones that yield non-ergodic invariant measures. For example, consider the extended $L$-hypergraphon

$$W_{\blacksquare\square}(\widehat{x_{[k]}}) = \begin{cases} \delta_\top & \text{if } x_\emptyset < \frac{1}{2}, \\ \delta_\bot & \text{otherwise.} \end{cases}$$

Define the extended $L$-hypergraphons $W_{\blacksquare}(\widehat{x_{[k]}}) = \delta_\top$ and $W_{\square}(\widehat{x_{[k]}}) = \delta_\bot$, which each induce a Borel hypergraph. The random hypergraph $G(\mathbb{N}, W_{\blacksquare\square})$ is the complete hypergraph or the empty hypergraph, each with probability $\frac{1}{2}$, and so its distribution is a non-trivial mixture of the distributions of $G(\mathbb{N}, W_{\blacksquare})$ and $G(\mathbb{N}, W_{\square})$, hence a non-ergodic invariant measure. The extended $L$-hypergraphon $W_{\blacksquare\square}$ does not induce a Borel hypergraph as it depends on the variable $x_\emptyset$, yet the entropy function of $G(\mathbb{N}, W_{\blacksquare\square})$ is $o(n^k)$.

But in fact, for $k \geq 3$, there is a more interesting obstruction to a converse of Lemma 3.2, even among extended $L$-hypergraphons that yield ergodic invariant measures. The following example for $k = 3$ (which is easily generalized to larger values of $k$) makes fundamental use of the variables indexed by pairs from [3], and yet also yields an invariant measure whose entropy function is $o(n^3)$:

$$W_{\triangle}(\widehat{x_{[3]}}) = \begin{cases} \delta_\top & \text{if } x_{\{0,1\}} < \frac{1}{2} \text{ and } x_{\{0,2\}} < \frac{1}{2} \text{ and } x_{\{1,2\}} < \frac{1}{2}, \\ \delta_\bot & \text{otherwise.} \end{cases}$$

The random hypergraph $G(\mathbb{N}, W_{\triangle})$ can be thought of as first building a "virtual" Erdős–Rényi graph with independent 2-edge probabilities $\frac{1}{2}$, then adding a 3-edge for each triangle existing in the graph, and then throwing away the virtual 2-edges. (For more about this example, see[Aus08] p. 92 and[Lov12] Example 23.11.)

In Lemma 3.2, we established that any extended $L$-hypergraphon $W$ that induces a Borel hypergraph is such that the entropy function of $G(\mathbb{N}, W)$ is $o(n^k)$. We now proceed to show that there are such $W$ for which the growth of $\mathrm{Ent}(G(\mathbb{N}, W))(n)$ is arbitrarily close to $n^k$.

We first define a kind of "blow up" that creates an extended $L$-hypergraphon from a countably infinite $L$-structure. We will use this notion in Lemma 3.4. Blow ups are a standard technique for expanding a countable structure into a continuum-sized structure that has a positive-measure worth of "copies" of each point from the original. (See, e.g., the use of step function graphons in[Lov12].)

**Definition 3.3.** Let $\mathcal{M} \in Str_L(\mathbb{N})$ and let $\pi\colon [0,1] \to \mathbb{N}$ be a Borel map such that $\lambda(\pi^{-1}(i)) > 0$ for all $i \in \mathbb{N}$. The **$\pi$-blow up of $\mathcal{M}$** is defined to be the extended $L$-hypergraphon $W\colon [0,1]^{\mathfrak{P}_{<k}(k)} \to \{\delta_\top, \delta_\bot\}$ given by

$$W(\widehat{x_{[k]}}) = \begin{cases} \delta_\top & \text{if } \mathcal{M} \models E(\pi(x_{\{0\}}), \pi(x_{\{1\}}), \ldots, \pi(x_{\{k-1\}})), \\ \delta_\bot & \text{otherwise.} \end{cases}$$

Observe that such a $W$ induces a Borel hypergraph, as on every input it outputs either the value $\delta_\top$ or $\delta_\bot$, and it depends only on the variables $x_{\{0\}}, x_{\{1\}}, \ldots, x_{\{k-1\}}$. In particular, $W$ induces the Borel hypergraph $\mathcal{B}$ with underlying set $[0,1]$ given by:

$$\mathcal{B} \models E(x_0, x_1, \ldots, x_{k-1})$$

if and only if

$$\mathcal{M} \models E(\pi(x_0), \pi(x_1), \ldots, \pi(x_{k-1})).$$

Hence we can think of $\pi^{-1}$ as a Borel partition of the unit interval into positive measure pieces, such that each element of $\mathcal{M}$ is "blown up" into a piece of the partition in $\mathcal{B}$, and each piece of the partition arises in this way.

Next, we proceed to construct an extended $L$-hypergraphon $W$ that yields an invariant measure whose entropy function has the desired growth. Analogously to Hatami–Norine[HN13], we let $W$ be the blow up of a particular countably infinite structure, the *Rado k-hypergraph*, a well-known generalization of the *transversal-uniform graph* used in[HN13]. The Rado $k$-hypergraph is the countable homogeneous-universal $k$-uniform hypergraph, namely, the unique (up to isomorphism) countable $k$-uniform hypergraph satisfying the so-called "Alice's restaurant" axioms. These axioms state that for any possible way of extending a finite induced subhypergraph of the Rado $k$-hypergraph by one vertex to obtain a $k$-uniform hypergraph, there is some element of the Rado $k$-hypergraph that realizes this extension.

We now describe an inductive construction of an instantiation $\mathcal{R}_k$ of the Rado $k$-hypergraph, with underlying set $\mathbb{N}$. At each stage $\ell \in \mathbb{N}$ we define a finite set $A_\ell$ of new vertices, which we call *generation $\ell$*, and build a $k$-uniform hypergraph $G_\ell$ with underlying set $V_\ell := \bigcup_{j \le \ell} A_j$.

Stage 0: Let $A_0 := \{0\}$ consist of a single vertex, and let $G_0$ be the empty hypergraph with vertex set $V_0 = A_0$.

Stage $\ell > 0$: Let $A_\ell$ consist of one new vertex $a_X$ for each subset $X$ of unordered $(k-1)$-tuples from $V_{\ell-1}$, with the elements of $A_\ell$ chosen to be the consecutive least elements of $\mathbb{N}$ not yet used. Let $G_\ell$ be the $k$-uniform hypergraph on vertex set $V_\ell = V_{\ell-1} \cup A_\ell$ whose edges are those of $G_{\ell-1}$ along with, for each such $X$ and every unordered tuple $d \in X$, an edge consisting of $a_X$ and the $k-1$ vertices in $d$.

Define $\mathcal{R}_k$ to be the union of the hypergraphs $G_\ell$, i.e., the hypergraph with vertex set $\bigcup_{\ell\in\mathbb{N}} V_\ell$ and edge set $\bigcup_{\ell\in\mathbb{N}} E^{G_\ell}$. Observe that $\mathcal{R}_k$ is a $k$-uniform hypergraph with underlying set $\mathbb{N}$ that satisfies the Alice's restaurant axioms: Given a finite induced subhypergraph $D$ of $\mathcal{R}_k$, all one-vertex extensions of $D$ are realized in stage $\ell + 1$, for any $\ell$ such that $V_\ell$ contains the vertices of $D$.

The following lemma, which we will use in the proof of Theorem 3.5, provides a lower bound on the entropy function of a random hypergraph sampled from a blow up of $\mathcal{R}_k$ in the case where elements of $\mathcal{R}_k$ belonging to the same generation get blown up to sets of equal measure. Both the statement and proof of the lemma are directly analogous to those of[HN13] Lemma 2.1.

Recall from Definition 2.5 that $\langle \zeta_D \rangle_{D\in\mathfrak{P}_{\leq k}(\mathbb{N})}$ is the collection of i.i.d. uniform random variables in $[0, 1]$ in terms of which the random $L$-structure $G(\mathbb{N}, W)$ is defined.

**Lemma 3.4.** *Let $\pi\colon [0,1] \to \mathbb{N}$ be a Borel map such that $\lambda(\pi^{-1}(i)) > 0$ for all $i \in \mathbb{N}$, and let $W$ be a $\pi$-blow up of $\mathcal{R}_k$. Suppose that $\lambda(\pi^{-1}(a)) = \lambda(\pi^{-1}(b))$ for all $\ell \in \mathbb{N}$ and $a, b \in A_\ell$. Then for all $n \in \mathbb{N}$ and $\rho\colon [n] \to \mathbb{N}$, we have*

$$H\Big(G(n, W) \,\Big|\, \bigwedge_{j\in[n]} \pi(\zeta_{\{j\}}) \in A_{\rho(j)}\Big) \geq \binom{|\rho([n])|}{k},$$

*a.s.*

**Proof.** Let $S \subseteq [n]$ be maximal such that $\rho$ is injective on $S$, and write $G(S, W)$ to denote the random induced substructure of $G(n, W)$ with underlying set $S$. Then

$$H\Big(G(n, W) \,\Big|\, \bigwedge_{j\in[n]} \pi(\zeta_{\{j\}}) \in A_{\rho(j)}\Big) \geq H\Big(G(S, W) \,\Big|\, \bigwedge_{j\in S} \pi(\zeta_{\{j\}}) \in A_{\rho(j)}\Big).$$

Consider the random measure

$$\mathbb{E}\Big(G(S, W) \,\Big|\, \bigwedge_{j\in S} \pi(\zeta_{\{j\}}) \in A_{\rho(j)}\Big). \tag{$\dagger$}$$

Since $W$ is a blow up of $\mathcal{R}_k$, the random hypergraph $G(S, W)$ is a $|S|$-element sample with replacement from $\mathcal{R}_k$, with the vertices relabeled by $S$. By the injectivity of $\rho$, the condition in ($\dagger$) constrains the elements of $G(S, W)$ to be obtained from distinct generations of $\mathcal{R}_k$, which implies that the random distribution on hypergraphs given by ($\dagger$) is actually sampled from $\mathcal{R}_k$ without replacement a.s.

Further, each $\zeta_{\{j\}}$ is uniform, and $\pi^{-1}$ assigns sets of equal measure to vertices in $\mathcal{R}_k$ of the same generation. So ($\dagger$) is a.s. the distribution of the random hypergraph $Q$ with underlying set $S$ obtained by, for each $j \in S$, uniformly selecting a vertex of $\mathcal{R}_k$ from among those in generation $\rho(j)$, and taking the edges induced from $\mathcal{R}_k$.

Let $\ell \in \mathbb{N}$, and consider a subset $U \subseteq V_\ell$. Write $T$ for the set of unordered $(k-1)$-tuples from $U$. Suppose $\ell' > \ell$. For every subset $X \subseteq T$, exactly a $2^{-|T|}$-fraction of the vertices in $A_{\ell'}$ form an edge with every $(k-1)$-tuple in $X$ and with no $(k-1)$-tuple in $T \setminus X$. For a vertex $v$ selected uniformly at random from $A_{\ell'}$, let $G_{U,v}$ be the (not necessarily induced) random subhypergraph of $G_{\ell'}$ that has vertex set $U \cup \{v\}$ and whose edges are those in $G_{\ell'}$ that consist of $v$ along with $k-1$ vertices from $U$. Then $G_{U,v}$ is equally likely to be any of the hypergraphs with vertex set $U \cup \{v\}$ whose edges all include $v$.

For each $j \in S$, define $S_j := \{i \in S : \rho(i) < \rho(j)\}$, and let $Q_j$ be the (not necessarily induced) random subhypergraph of $Q$ that has vertex set $S_j \cup \{j\}$ and whose edges are those in $Q$ that consist of $j$ along with $k-1$ vertices from $S_j$. Then for each $j \in S$, the random hypergraph $Q_j$ is equally likely to be any given hypergraph $F_j$ with vertex set $S_j \cup \{j\}$ all of whose edges include $j$. Further, for any choice of hypergraphs $\{F_j : j \in S\}$ as above, the events $Q_j = F_j$ for $j \in S$ are independent because the random variables $\zeta_{\{j\}}$ are independent.

Now, for any hypergraph $F$ with underlying set $S$, we can write $F$ as the union, over $j \in S$, of the subhypergraph of $F$ that has vertex set $S_j \cup \{j\}$ and whose edges are those in $F$ consisting of $j$ along with $k-1$ vertices from $S_j$. We therefore see that $Q$ is equally likely to be any hypergraph on vertex set $S$.

In summary, the random distribution (†) is a.s. the uniform measure on $k$-uniform hypergraphs with underlying set $S$. Hence

$$H\Big(G(S,W)\ \Big|\ \bigwedge_{j\in S} \pi(\zeta_{\{j\}}) \in A_{\rho(j)}\Big) = \binom{|S|}{k}$$

a.s., establishing the lemma. □

We now prove the main result of this section, which asserts that the entropy function of an invariant measure can have arbitrarily large growth rate within $o(n^k)$. This result is a higher-arity version of[HN13] Theorem 1.1, and its proof proceeds via the same steps. We include the proof here (with appropriately modified parameters and notation) for completeness.

**Theorem 3.5.** *Suppose* $\gamma\colon \mathbb{N} \to [0,1]$ *is a function such that* $\lim_{n\to\infty} \gamma(n) = 0$. *Then there is an extended L-hypergraphon W that induces a Borel hypergraph and is such that* $\mathrm{Ent}(G(\mathbb{N},W))(n) = o(n^k)$ *and* $\mathrm{Ent}(G(\mathbb{N},W))(n) = \Omega(\gamma(n)\cdot n^k)$.

**Proof.** We will define a Borel map $\pi\colon [0,1] \to \mathbb{N}$ satisfying $\lambda(\pi^{-1}(i)) > 0$ for all $i \in \mathbb{N}$ in such a way that the $\pi$-blow up of $\mathcal{R}_k$, which we denote by $W$, has the desired properties.

By the observation that follows Definition 3.3, if $\lambda(\pi^{-1}(i)) > 0$ for all $i \in \mathbb{N}$, then the $\pi$-blow up $W$ induces a Borel hypergraph, and so by Lemma 3.2, we have $\mathrm{Ent}(G(\mathbb{N}, W))(n) = o(n^k)$. Hence it suffices to construct $\pi$ satisfying $\lambda(\pi^{-1}(i)) > 0$ for all $i \in \mathbb{N}$ in such a way that $\mathrm{Ent}(G(\mathbb{N}, W))(n) = \Omega(\gamma(n) \cdot n^k)$.

For positive $r \in \mathbb{N}$, define

$$g_r := \max\{\{2^{r+3}k\} \cup \{n \in \mathbb{N} : \gamma(n) > 2^{-(r+1)k-3k-1}k^{-k}\}\}.$$

Note that $\lim_{n\to\infty} \gamma(n) = 0$ and so for each $r$ there are only finitely many $n$ such that $\gamma(n) > 2^{-(r+1)k-3k-1}k^{-k}$; hence $g_r$ is well-defined.

Observe that for all $n \geq g_1 + 1$, the inequalities

$$n > 2^{r+2}k \qquad \text{and} \qquad \gamma(n) \leq 2^{-rk-3k-1}k^{-k} \tag{$\star$}$$

hold when $r = 1$. The remainder of the proof establishes that for all such $n$, we have $\mathrm{Ent}(G(\mathbb{N}, W))(n) \geq n^k \cdot \gamma(n)$.

Fix $n \geq g_1 + 1$. We have seen that there is at least one $r$ satisfying ($\star$); on the other hand, there are only finitely many choices of $r$ for which ($\star$) holds. Let $q$ be the largest such $r$. Then either $n < 2^{q+3}k$ or $\gamma(n) > 2^{-(q+1)k-3k-1}k^{-k}$, and so $n \leq g_q$ by the definition of $g_q$.

For each $r \geq 1$, define

$$\Gamma_r := \{\ell \in \mathbb{N} : \textstyle\sum_{i=1}^{r-1} g_i \leq \ell < \sum_{i=1}^{r} g_i\},$$

so that $\{\Gamma_r\}_{r\geq 1}$ is a partition of $\mathbb{N}$. For every $\ell \in \Gamma_r$, let $\alpha_\ell := \frac{1}{g_r 2^r}$. Observe that for each $r \geq 1$, we have $|\Gamma_r| = g_r$, and so $\sum_{\ell\in\mathbb{N}} \alpha_\ell = 1$.

As a consequence, there is a Borel map $\pi\colon [0, 1] \to \mathbb{N}$ such that for all $a \in \mathcal{R}_k$, we have

$$\lambda(\pi^{-1}(a)) = \frac{\alpha_\ell}{|A_\ell|}$$

where $\ell$ is such that $a \in A_\ell$. In other words, vertices in $\mathcal{R}_k$ of the same generation $A_\ell$ are blown up to sets of the same positive measure, and the entire generation $A_\ell$ is blown up to a set of measure $\alpha_\ell$. We may therefore apply Lemma 3.4. Hence for any $\rho\colon [n] \to \mathbb{N}$, we have

$$H\Big(G(n, W) \Big| \bigwedge_{j\in[n]} \pi(\zeta_{\{j\}}) \in A_{\rho(j)}\Big) \geq \binom{|\rho([n])|}{k}$$

a.s.

By Lemma 1.8, we have

$$\begin{aligned}
\mathrm{Ent}(G(\mathbb{N}, W))(n) &\geq \mathbb{E}\Big(H\Big(G(n, W) \Big| \bigvee_{\rho\colon [n]\to\mathbb{N}} \Big(\bigwedge_{j\in[n]} \pi(\zeta_{\{j\}}) \in A_{\rho(j)}\Big)\Big)\Big) \\
&\geq \sum_{\rho\colon [n]\to\mathbb{N}} \mathbb{P}\Big(\bigwedge_{j\in[n]} \pi(\zeta_{\{j\}}) \in A_{\rho(j)}\Big) \cdot \binom{|\rho([n])|}{k} \\
&\geq \mathbb{P}\Big(|Z| \geq n \cdot 2^{-q-2}\Big) \cdot \binom{\lceil n \cdot 2^{-q-2}\rceil}{k}, \qquad (\mathbf{d}ag)
\end{aligned}$$

where we define the random set $Z := \bigcup_{j\in[n]}\{\ell \in \mathbb{N} : \pi(\zeta_{\{j\}}) \in A_\ell\}$.

Now define the random quantity $X := |\{Z \cap \Gamma_q]\}|$, and note that we always have $X \leq |Z|$. Because $\langle \zeta_{\{j\}} \rangle_{j\in[n]}$ is an i.i.d. uniform sequence, we have

$$\begin{aligned}
\mathbb{E}(X) &= \sum_{\ell\in\Gamma_q} \mathbb{P}\Big(\bigvee_{j\in[n]} \pi(\zeta_{\{j\}}) \in A_\ell\Big) \\
&= \sum_{\ell\in\Gamma_q} (1 - (1 - \alpha_\ell)^n) \\
&= g_q \cdot \left(1 - \left(1 - \frac{1}{g_q 2^q}\right)^n\right).
\end{aligned}$$

Observe that $(1 - x)^n \leq 1 - nx + n^2x^2 \leq 1 - \frac{nx}{2}$ holds for all $x \in [0, \frac{1}{2n}]$. Since $n \leq g_q$ and $q \geq 1$, we have $\frac{1}{g_q 2^q} \in [0, \frac{1}{2n}]$, and so

$$1 - \left(1 - \frac{1}{g_q 2^q}\right)^n \geq \frac{n}{g_q 2^{q+1}}.$$

Putting these together, we get

$$\mathbb{E}(X) \geq n \cdot 2^{-q-1}.$$

By Chebyshev's inequality, we have

$$\mathbb{P}\left[|X - \mathbb{E}(X)| \geq \frac{\mathbb{E}(X)}{2}\right] \leq \frac{4Var(X)}{(\mathbb{E}(X))^2}.$$

Hence,

$$\begin{aligned}
1 - \frac{4Var(X)}{(\mathbb{E}(X))^2} &\leq \mathbb{P}\left[|X - \mathbb{E}(X)| < \frac{\mathbb{E}(X)}{2}\right] \\
&\leq \mathbb{P}\left[X > \frac{\mathbb{E}(X)}{2}\right] \\
&\leq \mathbb{P}(X > n \cdot 2^{-q-2}) \\
&\leq \mathbb{P}(|Z| > n \cdot 2^{-q-2}).
\end{aligned}$$

For distinct $\ell, \ell' \in \Gamma_q$, the events $\ell \in Z$ and $\ell' \in Z$ have negative correlation, which implies that $Var(X) \leq \mathbb{E}(X)$. Hence

$$\mathbb{P}(|Z| > n \cdot 2^{-q-2}) \geq 1 - \frac{4}{\mathbb{E}(X)} \geq 1 - \frac{4}{n \cdot 2^{-q-1}} \geq \frac{1}{2}.$$

Finally, substituting in (**d**$ag$) and recalling that, by ($\star$) for $r = q$, we have $n \cdot 2^{-q-2} > k$ and $\gamma(n) \leq 2^{-qk-3k-1}k^{-k}$, we obtain

$$\begin{aligned}
\mathrm{Ent}(G(\mathbb{N}, W))(n) &\geq \frac{1}{2} \cdot \binom{\lceil n \cdot 2^{-q-2} \rceil}{k} \\
&\geq \frac{1}{2} \cdot \frac{(n \cdot 2^{-q-2} - k)^k}{k^k} \\
&\geq \frac{1}{2} \cdot (n \cdot 2^{-q-3})^k k^{-k} \\
&= n^k \cdot 2^{-qk-3k-1} k^{-k} \\
&\geq n^k \cdot \gamma(n),
\end{aligned}$$

as desired. □

## 4. Non-redundant invariant measures

In this section we consider entropy functions of invariant measures for countable languages that may be of unbounded arity. If an invariant measure fails to be non-redundant, then its entropy function may take the value $\infty$ even when there are only finitely many relation symbols of each arity in the language. Hence we restrict to the case of non-redundant invariant measures for relational languages, and provide an upper bound on the entropy function in terms of the number of relation symbols of each arity.

*For the rest of this section, let $L$ be a countable relational language (possibly infinite).*

We show, in Proposition 4.3, that the entropy function of a non-redundant invariant measure for $L$ is dominated by that of a particular random $L$-structure that generalizes the Erdős–Rényi random graph having edge probability $\frac{1}{2}$. We also calculate, in Lemma 4.2, the entropy function of such a maximal entropy structure explicitly in terms of the number of relation symbols of each arity in $L$. In the case where $L$ has finitely many relation symbols of each arity, this provides a more precise version of Lemma 1.14, which states that such an entropy function takes values in $\mathbb{R}$. Moreover, this calculation shows that there are $\mathbb{R}$-valued entropy functions that grow arbitrarily fast, in contrast to the situation for finite languages, where the growth is at most polynomial.

We now define the *uniform non-redundant measure* for $L$, which is the distribution of a random structure obtained by independently flipping a fair coin to decide every relation on a tuple of distinct elements, and setting all relations on tuples with repeated elements to false.

For a relation symbol $R \in L$, write $\mathrm{arity}(R)$ to denote its arity.

**Definition 4.1.** Given a set $X$, define

$$\mathrm{NR}_{L,X} := \{\langle R, \overline{x}\rangle : R \in L,\ \overline{x} \in X \text{ has distinct entries, and } |\overline{x}| = \mathrm{arity}(R)\}.$$

Let $\{\xi_{R(\overline{x})} : \langle R, \overline{x}\rangle \in \mathrm{NR}_{L,\mathbb{N}}\}$ be a collection of i.i.d. uniform $\{\top, \bot\}$-valued random variables and let $\Xi$ be the $Str_L(\mathbb{N})$-valued random variable given by

$$\Xi \models R(\overline{x}) \qquad \text{if and only if} \qquad \langle R, \overline{x}\rangle \in \mathrm{NR}_{L,\mathbb{N}} \quad \text{and} \quad \xi_{R(\overline{x})} = \top$$

for every $R \in L$ and $\mathrm{arity}(R)$-tuple $\overline{x}$ of elements from $\mathbb{N}$. Define $\mu_{\mathrm{un},L}$ to be the distribution of $\Xi$, and call it the **uniform non-redundant measure** for $L$.

It is easy to see that $\mu_{\mathrm{un},L}$ is both invariant and non-redundant. Note that for $n \in \mathbb{N}$, the measure $(\mu_{\mathrm{un},L})_n$ is the uniform distribution on non-redundant structures in $Str_L(n)$.

Let $\mathfrak{a}_L \colon \mathbb{N} \to \mathbb{N} \cup \{\infty\}$ be the function sending each $n \in \mathbb{N}$ to the number of relation symbols in $L$ having arity $n$. In the following lemma we calculate the entropy function of $\mu_{\mathrm{un},L}$ in terms of the function $\mathfrak{a}_L$.

**Lemma 4.2.** *For any $n \in \mathbb{N}$, we have*

$$\mathrm{Ent}(\mu_{\mathrm{un},L})(n) = \sum_{r \leq n} \binom{n}{r} \cdot r! \cdot \mathfrak{a}_L(r).$$

**Proof.** The invariant measure $\mu_{\mathrm{un},L}$ is non-redundant, and so for each $n \in \mathbb{N}$, the invariant measure $(\mu_{\mathrm{un},L})_n$ is concentrated on elements of $Str_L(n)$ in which no relations of arity greater than $n$ hold. Hence $(\mu_{\mathrm{un},L})_n$ is determined by the set $\{\xi_{R(\overline{x})} : \langle R, \overline{x}\rangle \in \mathrm{NR}_{L,[n]}\}$ of random variables. Because these random variables are independent, we have

$$\begin{aligned}\mathrm{Ent}(\mu_{\mathrm{un},L})(n) &= \sum_{(R,\overline{x}) \in \mathrm{NR}_{L,[n]}} h(\xi_{R(\overline{x})}) \\ &= \left|\mathrm{NR}_{L,[n]}\right| \\ &= \sum_{r \leq n} \binom{n}{r} \cdot r! \cdot \mathfrak{a}_L(r),\end{aligned}$$

as the number of $r$-tuples from $[n]$ consisting of distinct elements is $\binom{n}{r} \cdot r!$, and there are $\mathfrak{a}_L(r)$-many relation symbols of arity $r$ in $L$. □

Observe that when $L$ has only finitely many relation symbols of each arity, Lemma 4.2 shows that $\mathrm{Ent}(\mu_{\mathrm{un},L})$ is $\mathbb{R}$-valued. Hence, by varying the choice of such an $L$ (and hence the function $\mathfrak{a}_L$), we can obtain $\mathbb{R}$-valued entropy functions that grow arbitrarily fast.

We now show that $\mu_{\mathrm{un},L}$ has the fastest growing entropy function among non-redundant invariant measures on $Str_L(\mathbb{N})$.

**Proposition 4.3.** *Let $\nu$ be a non-redundant invariant measure on $Str_L(\mathbb{N})$. Then*

$$\mathrm{Ent}(\nu)(n) \ \leq \ \mathrm{Ent}(\mu_{\mathrm{un},L})(n)$$

*for all $n \in \mathbb{N}$.*

**Proof.** Suppose $\mathcal{N}$ is a random $Str_L(\mathbb{N})$-structure with distribution $\nu$. For $R \in L$ and $\overline{x} \in \mathbb{N}$ with $|\overline{x}| = \mathrm{arity}(R)$, recall that the random instantiation $R^{\mathcal{N}}$ satisfies

$$R^{\mathcal{N}}(\overline{x}) = \top \qquad \text{if and only if} \qquad \mathcal{N} \models R(\overline{x}).$$

By Lemma 1.9 we then have, for $n \in \mathbb{N}$,

$$\mathrm{Ent}(\nu)(n) \ = \ \mathrm{Ent}(\mathcal{N})(n) \ \leq \sum_{\substack{R\in L,\ \overline{x}\in[n],\text{ and}\\ |\overline{x}|=\mathrm{arity}(R)}} h(R^{\mathcal{N}}(\overline{x})).$$

If $\overline{x}$ has duplicate entries then we know that $h(R^{\mathcal{N}}(\overline{x})) = 0$ as $\mathcal{N}$ is non-redundant a.s. Further, if $\overline{x}$ has no duplicate entries then $h(R^{\mathcal{N}}(\overline{x})) \leq h(\xi_{R(\overline{x})})$ as the distribution of $\xi_{R(\overline{x})}$ is uniform on $\{\top, \bot\}$, and this is the distribution with maximal entropy on $\{\top, \bot\}$.

Therefore, we have

$$\begin{aligned}\sum_{\substack{R\in L,\ \overline{x}\in[n],\text{ and}\\ |\overline{x}|=\mathrm{arity}(R)}} h(R^{\mathcal{N}}(\overline{x})) \ &\leq \sum_{(R,\overline{x})\in\mathrm{NR}_{L,[n]}} h(\xi_{R(\overline{x})})\\ &= \ \mathrm{Ent}(\Xi)(n)\\ &= \ \mathrm{Ent}(\mu_{\mathrm{un},L})(n),\end{aligned}$$

and so $\mathrm{Ent}(\nu)(n) \ \leq \ \mathrm{Ent}(\mu_{\mathrm{un},L})(n)$, as desired. □

Putting together the previous lemma and proposition we immediately obtain the following bound.

**Corollary 4.4.** *Let $\nu$ be a non-redundant invariant measure on $Str_L(\mathbb{N})$. Then*

$$\mathrm{Ent}(\nu)(n) \leq \sum_{r\leq n} \binom{n}{r} \cdot r! \cdot \mathfrak{a}_L(r)$$

*for all $n \in \mathbb{N}$. In particular, when $L$ is finite with maximum arity $k$, we have* $\mathrm{Ent}(\nu)(n) = O(n^k)$.

Recall from Lemma 1.14 that when $L$ has only finitely many relation symbols of each arity, any non-redundant invariant measure on $Str_L(\mathbb{N})$ has an $\mathbb{R}$-valued entropy function. Corollary 4.4 improves this by providing an explicit upper bound.

In fact, we do not need the invariant measure to be non-redundant, nor the language to be relational, to obtain the final line of Corollary 4.4; by Proposition 1.21 we see that the polynomial bound $O(n^k)$ holds for the entropy function of an arbitrary invariant measure for a finite language (possibly with constant or function symbols) of maximum arity $k$, thereby strengthening Lemma 1.12.

An interesting question for future work is to characterize precisely those functions from $\mathbb{N}$ to $\mathbb{R} \cup \{\infty\}$ that can be the entropy function of an invariant measure. By Proposition 1.21, it suffices to consider entropy functions of non-redundant invariant measures for relational languages.

**Acknowledgements**

The authors thank Jan Reimann and Daniel M. Roy for helpful conversations, and the anonymous referee for their comments.

**References**

Ack15. N. Ackerman, *Representations of Aut(M)-invariant measures: Part 1*, arXiv e-print 1509.06170v1 (2015).

AFKrP17. N. Ackerman, C. Freer, A. Kruckman, and R. Patel, *Properly ergodic structures*, ArXiv e-print 1710.09336 (2017).

AFKwP17. N. Ackerman, C. Freer, A. Kwiatkowska, and R. Patel, *A classification of orbits admitting a unique invariant measure*, Ann. Pure Appl. Logic **168** (2017), no. 1, 19–36.

AFNP16. N. Ackerman, C. Freer, J. Nešetřil, and R. Patel, *Invariant measures via inverse limits of finite structures*, European J. Combin. **52** (2016), 248–289.

AFP16. N. Ackerman, C. Freer, and R. Patel, *Invariant measures concentrated on countable structures*, Forum Math. Sigma **4** (2016), e17, 59pp.

AFP17. ______, *Countable infinitary theories admitting an invariant measure*, arXiv e-print 1710.06128v1 (2017).

Ald81. D. J. Aldous, *Representations for partially exchangeable arrays of random variables*, J. Multivariate Anal. **11** (1981), no. 4, 581–598.

Ald85. ______, *Exchangeability and related topics*, École d'été de probabilités de Saint-Flour, XIII (1983), Lecture Notes in Math., vol. 1117, Springer, Berlin, 1985, pp. 1–198.

Aus08. T. Austin, *On exchangeable random variables and the statistics of large graphs and hypergraphs*, Probab. Surv. **5** (2008), 80–145.

AZ86. G. Ahlbrandt and M. Ziegler, *Quasi-finitely axiomatizable totally categorical theories*, Ann. Pure Appl. Logic **30** (1986), no. 1, 63–82.

CD13. S. Chatterjee and P. Diaconis, *Estimating and understanding exponential random graph models*, Ann. Statist. **41** (2013), no. 5, 2428–2461.

CT06. T. M. Cover and J. A. Thomas, *Elements of information theory*, 2nd ed., Wiley, Hoboken, NJ, 2006.

CV11. S. Chatterjee and S. R. S. Varadhan, *The large deviation principle for the Erdős-Rényi random graph*, European J. Combin. **32** (2011), no. 7, 1000–1017.

DF81. P. Diaconis and D. Freedman, *On the statistics of vision: the Julesz conjecture*, J. Math. Psych. **24** (1981), no. 2, 112–138.

DJ08. P. Diaconis and S. Janson, *Graph limits and exchangeable random graphs*, Rend. Mat. Appl. (7) **28** (2008), no. 1, 33–61.

HJS18. H. Hatami, S. Janson, and B. Szegedy, *Graph properties, graph limits, and entropy*, J. Graph Theory **87** (2018), no. 2, 208–229.

HN13. H. Hatami and S. Norine, *The entropy of random-free graphons and properties*, Combin. Probab. Comput. **22** (2013), no. 4, 517–526.

Jan13. S. Janson, *Graphons, cut norm and distance, couplings and rearrangements*, New York Journal of Mathematics (NYJM) Monographs, vol. 4, State University of New York, University at Albany, 2013.

Kal99. O. Kallenberg, *Multivariate sampling and the estimation problem for exchangeable arrays*, J. Theoret. Probab. **12** (1999), no. 3, 859–883.

Kal02. ______, *Foundations of modern probability*, 2nd ed., Probability and its Applications, Springer, New York, 2002.

Kal05. ______, *Probabilistic symmetries and invariance principles*, Probability and its Applications, Springer, New York, 2005.

Kec95. A. S. Kechris, *Classical descriptive set theory*, Graduate Texts in Mathematics, vol. 156, Springer-Verlag, New York, 1995.

Lov12. L. Lovász, *Large networks and graph limits*, American Mathematical Society Colloquium Publications, vol. 60, American Mathematical Society, Providence, RI, 2012.

LS10. L. Lovász and B. Szegedy, *Regularity partitions and the topology of graphons*, An irregular mind, Bolyai Soc. Math. Stud., vol. 21, János Bolyai Math. Soc., Budapest, 2010, pp. 415–446.

PV10. F. Petrov and A. Vershik, *Uncountable graphs and invariant measures on the set of universal countable graphs*, Random Structures & Algorithms **37** (2010), no. 3, 389–406.

RS13. C. Radin and L. Sadun, *Phase transitions in a complex network*, J. Phys. A **46** (2013), no. 30, 305002.

# Linear Orders and Categoricity Spectra

Nikolay Bazhenov

*Sobolev Institute of Mathematics,*
*4 Acad. Koptyug Avenue,*
*Novosibirsk, 630090, Russia*

*Department of Mathematics,*
*School of Science and Technology,*
*Nazarbayev University,*
*53 Qabanbaybatyr Avenue,*
*Astana, 010000, Kazakhstan*
*E-mail: bazhenov@math.nsc.ru*

We study effective categoricity for computable linear orders. The categoricity spectrum of a computable structure $S$ is the set of all Turing degrees capable of computing isomorphisms among arbitrary computable presentations of $S$. The degree of categoricity for $S$ is the least degree in this spectrum. The degree of categoricity **d** is strong if there are two computable copies of $S$ such that any isomorphism between the copies computes **d**.

We give a new series of computable linear orders with no degree of categoricity: for every computable successor ordinal $\alpha \geq 4$, the set of PA degrees over $\mathbf{0}^{(\alpha)}$ is the categoricity spectrum for a scattered linear order. We also build the first examples of linear orders with non-strong degrees of categoricity: If $\alpha$ is a computable infinite ordinal, then there is a scattered linear order such that it is $\Delta^0_{\alpha+2}$ categorical, not $\Delta^0_\alpha$ categorical, and has non-strong degree of categoricity.

*Keywords*: Computable categoricity, linear order, categoricity spectrum, degree of categoricity, PA degrees, decidable categoricity.

## 1. Introduction

We study algorithmic complexity of isomorphisms between different computable presentations of a linear order.

Let **d** be a Turing degree. A computable structure $\mathcal{S}$ is **d**-*computably categorical* if for any computable structure $\mathcal{A}$ isomorphic to $\mathcal{S}$, there is a **d**-computable isomorphism from $\mathcal{A}$ onto $\mathcal{S}$. If $\mathbf{d} = \mathbf{0}$, then $\mathcal{S}$ is usually called *computably categorical*. The study of computable categoricity goes back to the works of Fröhlich and Shepherdson [1], and Mal'tsev [2,3]. For many familiar classes of structures $K$, there is a nice description of computably categorical members of $K$: e.g., a computable linear order is computably categorical if and only if it has only finitely many pairs of adjacent elements [4,5].

The *categoricity spectrum* of a computable structure $\mathcal{S}$ (denoted by $CatSpec(\mathcal{S})$) is the set of all degrees $\mathbf{d}$ such that $\mathcal{S}$ is $\mathbf{d}$-computably categorical. If $\mathbf{c}$ is the least degree in the spectrum $CatSpec(\mathcal{S})$, then $\mathbf{c}$ is called the *degree of categoricity* of $\mathcal{S}$.

Categoricity spectra and degrees of categoricity were introduced by Fokina, Kalimullin, and Miller[6]. In Refs. 6,7, it was proved that for any computable non-limit ordinal $\alpha$, every Turing degree $\mathbf{d}$ which is d.c.e. in and above $\mathbf{0}^{(\alpha)}$ is a degree of categoricity for some computable structure. Recently, Csima and Ng proved that every $\Delta^0_2$ degree is a degree of categoricity.

Some known examples of degrees of categoricity can be realized in the class of linear orders: Frolov[8] showed that any degree d.c.e. in and above $\mathbf{0}^{(2)}$ is a degree of categoricity for a linear order. In Refs. 9,10, it was proved that for any computable successor ordinal $\alpha \geq 3$, every degree c.e. in and above $\mathbf{0}^{(\alpha)}$ is a degree of categoricity for a linear order.

Miller[11] built the first example of a structure with no degree of categoricity. Miller and Shlapentokh[12] constructed an algebraic field $\mathcal{F}$ such that its categoricity spectrum contains precisely the *PA* degrees (in particular, this implies that $\mathcal{F}$ has no degree of categoricity). In Ref. 13, it was shown that for every computable successor ordinal $\alpha \geq 2$, there is a distributive lattice such that its categoricity spectrum consists of *PA* degrees over $\mathbf{0}^{(\alpha)}$.

Note that an example of a computable linear order with no degree of categoricity can be easily obtained as follows. Csima, Franklin, and Shore[7] proved that any degree of categoricity is hyperarithmetic. It is well-known (see Ref. 14, or Ref. 9 for a discussion) that the Harrison linear order $\mathcal{H} = \omega_1^{CK}(1+\eta)$ has two computable copies which are not hyperarithmetically isomorphic. Hence, $\mathcal{H}$ has no degree of categoricity.

The first main result of the paper gives new examples of linear orders with no degree of categoricity: For every computable successor ordinal $\alpha \geq 4$, there is a scattered linear order such that its categoricity spectrum contains precisely the *PA* degrees over $\mathbf{0}^{(\alpha)}$ (Theorem 3.1).

Suppose that $\mathbf{d}$ is the degree of categoricity for a computable structure $\mathcal{S}$. The degree $\mathbf{d}$ is called the *strong degree of categoricity* for $\mathcal{S}$ if it satisfies the following:

(∗) There are two computable copies $\mathcal{A}$ and $\mathcal{B}$ of $\mathcal{S}$ such that every isomorphism from $\mathcal{A}$ onto $\mathcal{B}$ computes $\mathbf{d}$.

We say that $\mathbf{d}$ is the *non-strong degree of categoricity* for $\mathcal{S}$ if it does not satisfy (∗).

In most cases, if a structure $\mathcal{S}$ has a degree of categoricity $\mathbf{d}$, then $\mathbf{d}$ is the strong degree of categoricity for $\mathcal{S}$ (see, e.g., Refs. 6,7,15 for a discussion). The first examples of computable structures with non-strong degrees of categoricity were independently constructed by Bazhenov, Kalimullin, Yamaleev [16], and Csima, Stephenson [17]. In Ref. 16, it was proved that there is a rigid structure $\mathcal{S}$ such that $\mathbf{0}'$ is the non-strong degree of categoricity for $\mathcal{S}$ (see also Ref. 15 which is a significantly extended version of the communication [16]). In Ref. 17, the authors prove that there is a rigid structure with computable dimension 3 and non-strong degree of categoricity $\mathbf{d} \leq \mathbf{0}^{(2)}$.

Our second main result obtains first examples of linear orders with non-strong degrees of categoricity: For every computable successor ordinal $\alpha \geq 4$, we build a Turing degree $\mathbf{d}$ and a computable scattered linear order $\mathcal{L}$ such that $\mathbf{0}^{(\alpha)} \leq \mathbf{d} \leq \mathbf{0}^{(\alpha+1)}$ and $\mathbf{d}$ is the non-strong degree of categoricity for $\mathcal{L}$ (Theorem 4.1). Note that if $\alpha = \beta + 1 > \omega$, then the order $\mathcal{L}$ is $\Delta^0_{\beta+2}$ categorical and not $\Delta^0_\beta$ categorical.

The outline of the paper is as follows. Section 2 contains the necessary preliminaries. In Sec. 3, we give a proof of Theorem 3.1. We also discuss a consequence which deals with decidable structures. Section 4 proves Theorem 4.1.

## 2. Preliminaries

We consider only computable languages, and structures with domain contained in the set of natural numbers $\omega$. Linear orders are treated as structures in the language $\{\leq\}$. The reader is referred to the books [14,18] for the background on computable structures.

For a language $L$, *infinitary L-formulas* are formulas of the logic $L_{\omega_1,\omega}$. For a countable ordinal $\alpha$, infinitary $\Sigma_\alpha$ and $\Pi_\alpha$ formulas are defined in a standard way (see Chap. 6 in Ref. 14). *Computable infinitary L-formulas* were introduced in Ref. 19. Here we give an informal description of these formulas:

(1) Computable $\Sigma_0$ and $\Pi_0$ formulas are quantifier-free first-order formulas of $L$.
(2) For a non-zero computable ordinal $\alpha$, a computable $\Sigma_\alpha$ formula (or $\Sigma^c_\alpha$ formula) is a computably enumerable (c.e.) disjunction $\bigvee\!\!\!\bigvee_i \exists \bar{y}_i \psi_i(\bar{x}, \bar{y}_i)$, where $\psi_i$ is a computable $\Pi_{\beta_i}$ formula for some $\beta_i < \alpha$.
(3) A computable $\Pi_\alpha$ formula (or $\Pi^c_\alpha$ formula) is a c.e. conjunction $\bigwedge\!\!\!\bigwedge_i \forall \bar{y}_i \psi_i(\bar{x}, \bar{y}_i)$, where $\psi_i$ is a $\Sigma^c_{\beta_i}$ formula for some $\beta_i < \alpha$.

For a formal definition of $\Sigma^c_\alpha$ and $\Pi^c_\alpha$ formulas, the reader is referred to Chap. 7 in Ref. 14.

For a computable ordinal $\alpha$, one can define the following infinitary formulas in the language of linear orders (see Ref. 10 and Sec. 6.2. in Ref. 14 for details):

(a) A $\Sigma^c_{2\alpha}$ formula $F_\alpha(x, y)$ such that for a well-order $\mathcal{L}$, the condition $\mathcal{L} \models F_\alpha(a, b)$ holds iff the interval $[a; b)_\mathcal{L}$ is isomorphic to some $\beta < \omega^\alpha$.

(b) A $\Pi^c_{2\alpha+1}$ formula $S_\alpha(x, y)$ such that the condition $\mathcal{L} \models S_\alpha(a, b)$ is equivalent to $[a; b)_\mathcal{L} \cong \omega^\alpha$.

A linear order $\mathcal{L}$ is *scattered* if the ordering of rationals $\eta$ is not isomorphically embeddable into $\mathcal{L}$. For the background on countable linear orders, we refer the reader to Refs. 20,21.

Recall that in a linear order $\mathcal{L}$, elements $a$ and $b$ belong to the same *block* (denoted by $Block(a, b)$) if there are only finitely many elements between $\min_\mathcal{L}(a, b)$ and $\max_\mathcal{L}(a, b)$. Note that the relation $Block(x, y)$ is definable by a $\Sigma^c_2$ formula $F_1(x, y) \vee F_1(y, x)$.

By $\zeta$ we denote the standard ordering of integers. We say that an element $a \in \mathcal{L}$ belongs to a *$\zeta$-block* if the interval $\{x : Block^\mathcal{L}(x, a)\}$ is isomorphic to $\zeta$.

Trees are treated as subtrees of $\omega^{<\omega}$. For a tree $T$, the *branching function* $b_T : T \to \omega \cup \{\omega\}$ is defined as follows:

$$b_T(\sigma) = card(\{n \in \omega : \sigma^\frown\langle n\rangle \in T\}).$$

In other words, $b_T(\sigma)$ is the number of children of $\sigma$ in $T$.

Suppose that $X \subseteq \omega$. A Turing degree $\mathbf{d}$ is called a *PA degree over X* if for any infinite $X$-computable, finite-branching tree $T$ with an $X$-computable branching function $b_T$, there is a $\mathbf{d}$-computable (infinite) path through $T$. Note that the notion of a *PA* degree over $X$ depends only on the choice of $\deg_T(X)$.

We will use the following characterization of *PA* degrees (due to Scott[22], Jockusch and Soare[23], and Solovay):

**Proposition 2.1 (see Theorem 6.6 in Ref. 24).** *A Turing degree* $\mathbf{d}$ *is a PA degree over X if and only if there is a* $\mathbf{d}$*-computable set A with the following properties:*

$$\{e : \varphi^X_e(e)\downarrow = 1\} \subseteq A, \quad \{e : \varphi^X_e(e)\downarrow = 0\} \subseteq \overline{A}.$$

It is well-known (see, e.g., Theorem 6.2 in Ref. 24) that the set of *PA* degrees over $X$ is upwards closed. In addition (Theorem 6.5.i in Ref. 24), if $\mathbf{d}$ is a *PA* degree over $X$, then there is a degree $\mathbf{c}$ such that $\mathbf{c} < \mathbf{d}$ and $\mathbf{c}$ is also a *PA* degree over $X$. Thus, the set of *PA* degrees over $X$ does not have minimal elements.

### 2.1. *Pairs of computable structures*

Suppose that $\alpha$ is a countable ordinal, and $\mathcal{A}$ and $\mathcal{B}$ are $L$-structures. We say that $\mathcal{A} \leq_\alpha \mathcal{B}$ if every infinitary $\Pi_\alpha$ sentence true in $\mathcal{A}$ is also true in $\mathcal{B}$. The relations $\leq_\alpha$ are called *standard back-and-forth relations.*

Ash[25] obtained a complete description of standard back-and-forth relations for well-orders. Here we give only an excerpt from the description:

**Lemma 2.1 (Lemma 7 in Ref. 25, see also Lemma 15.10 in Ref. 14).** *Suppose that $\alpha$ is a countable ordinal. Then we have:*

*(i)* $\omega^\alpha \cdot 2 \leq_{2\alpha+1} \omega^\alpha$;
*(ii)* $\omega^{\alpha+1} + \omega^\alpha \leq_{2\alpha+2} \omega^{\alpha+1}$.

Let $\alpha$ be a computable ordinal. A family of $L$-structures $K = \{\mathcal{A}_i\}_{i\in I}$ is $\alpha$-*friendly* if the structures $\mathcal{A}_i$ are computable uniformly in $i \in I$, and the relations

$$BF_\beta = \{(i, \bar{a}, j, \bar{b}) : i, j \in I,\ \bar{a} \in \mathcal{A}_i,\ \bar{b} \in \mathcal{A}_j,\ (\mathcal{A}_i, \bar{a}) \leq_\beta (\mathcal{A}_j, \bar{b})\}$$

are computably enumerable, uniformly in $\beta < \alpha$.

The technique of pairs of computable structures was developed by Ash and Knight[14,26]. This technique is useful for encoding $\Sigma^0_\beta$ sets into structures. In this paper, we use the following result which is based on the technique:

**Theorem 2.1 (Theorem 3.1 from Ref. 26).** *Let $\alpha$ be a non-zero computable ordinal. Suppose that $\mathcal{A}$ and $\mathcal{B}$ are L-structures such that $\mathcal{B} \leq_\alpha \mathcal{A}$ and the family $\{\mathcal{A}, \mathcal{B}\}$ is $\alpha$-friendly. Then for any $\Sigma^0_\alpha$ set X, there is a uniformly computable sequence of structures $(C_n)_{n\in\omega}$ such that*

$$C_n \cong \begin{cases} \mathcal{A}, & \text{if } n \notin X, \\ \mathcal{B}, & \text{if } n \in X. \end{cases}$$

It is well-known (see, e.g., Proposition 15.11 in Ref. 14) that for any computable ordinals $\alpha, \beta_0, \dots, \beta_n$, there is an $\alpha$-friendly family $\{\mathcal{B}_0, \dots, \mathcal{B}_n\}$ such that for each $i \leq n$, $\mathcal{B}_i$ is isomorphic to $\beta_i$. Therefore, Lemma 2.1 and Theorem 2.1 together imply the following:

**Corollary 2.1.** *Suppose that $\alpha$ is a computable ordinal.*

*(i) For any $\Sigma^0_{2\alpha+1}$ set X, there is a computable sequence of structures $(C_n)_{n\in\omega}$ such that*

$$C_n \cong \begin{cases} \omega^\alpha, & \text{if } n \notin X, \\ \omega^\alpha \cdot 2, & \text{if } n \in X. \end{cases}$$

*(ii) For any $\Sigma^0_{2\alpha+2}$ set X, there is a computable sequence of structures $(C_n)_{n\in\omega}$ such that*

$$C_n \cong \begin{cases} \omega^{\alpha+1}, & \text{if } n \notin X, \\ \omega^{\alpha+1} + \omega^\alpha, & \text{if } n \in X. \end{cases}$$

### 2.2. $\Delta^0_\alpha$ *categoricity*

We say that a family of computable $L$-structures $K = \{\mathcal{S}_0, \mathcal{S}_1, \dots, \mathcal{S}_n\}$ is *uniformly* $\Delta^0_\alpha$ *categorical* if given computable indices of computable structures $\mathcal{A}$ and $\mathcal{B}$ such that $\mathcal{A} \cong \mathcal{B} \cong \mathcal{S} \in K$, one can effectively compute a $\Delta^0_\alpha$ index of an isomorphism $f$ from $\mathcal{A}$ onto $\mathcal{B}$.

In the proof of Theorem 4 from Ref. 13, the following result was obtained:

**Proposition 2.2.** *Suppose that $\alpha$ is a computable ordinal.*

*(a) The family $\{\omega^\alpha, \omega^\alpha \cdot 2\}$ is uniformly $\Delta^0_{2\alpha+1}$ categorical (in Ref. 13, see Proposition 2 and p. 609).*
*(b) The family $\{\omega^{\alpha+1}, \omega^{\alpha+1} + \omega^\alpha\}$ is uniformly $\Delta^0_{2\alpha+2}$ categorical (see p. 610 in Ref. 13).*

## 3. PA degrees over $\mathbf{0}^{(\alpha)}$

In the section, we build new examples of linear orders which do not have degree of categoricity:

**Theorem 3.1.** *Suppose that $\alpha$ is a computable successor ordinal such that $\alpha \geq 4$. Then there is a computable scattered linear order $\mathcal{L}$ such that the categoricity spectrum of $\mathcal{L}$ contains precisely the PA degrees over $\mathbf{0}^{(\alpha)}$. In particular, $\mathcal{L}$ has no degree of categoricity.*

**Proof.** For a computable non-zero ordinal $\beta$, we introduce the following notation:

$$\emptyset^{\langle\beta\rangle} := \begin{cases} \emptyset^{(\beta-1)}, & \text{if } \beta < \omega, \\ \emptyset^{(\beta)}, & \text{if } \beta \geq \omega. \end{cases} \tag{1}$$

The notation will be useful in our proofs: e.g., note that a set $A$ is $\Sigma^0_\beta$ if and only if $A$ is c.e. in $\emptyset^{\langle\beta\rangle}$.

For a computable ordinal $\gamma \neq 0$, let

$$PA_\gamma := \{\mathbf{d} : \mathbf{d} \text{ is a } PA \text{ degree over } \emptyset^{\langle\gamma\rangle}\}.$$

The proof uses ideas from Ref. 13. We consider two separate cases: the first case deals with $PA_{2\beta+1}$, and the second one works with $PA_{2\beta+2}$.

**Case I.** Suppose that $\beta$ is a computable ordinal such that $\beta \geq 2$. We construct a computable scattered linear order $\mathcal{L}$ with $CatSpec(\mathcal{L}) = PA_{2\beta+1}$.

By Corollary 2.1, we can build computable sequences of linear orders $(\mathcal{A}_e)_{e\in\omega}$ and $(\mathcal{B}_e)_{e\in\omega}$ such that for any $e$, we have:

$$\mathcal{A}_e \cong \begin{cases} \omega^\beta \cdot 2, & \text{if } \varphi_e^{\emptyset^{\langle 2\beta+1\rangle}}(e)\downarrow = 1, \\ \omega^\beta, & \text{otherwise;} \end{cases} \qquad \mathcal{B}_e \cong \begin{cases} \omega^\beta \cdot 2, & \text{if } \varphi_e^{\emptyset^{\langle 2\beta+1\rangle}}(e)\downarrow = 0, \\ \omega^\beta, & \text{otherwise.} \end{cases}$$

Our computable order $\mathcal{L}$ is obtained as a sum

$$\sum_{e\in\omega}(e+2+(\cdots+\mathcal{A}_e+1+\zeta+\mathcal{B}_e+1+\zeta+\mathcal{A}_e+1+\zeta+\mathcal{B}_e+1+\zeta+ \\ \mathcal{A}_e+1+\zeta+\mathcal{B}_e+1+\zeta+\dots)).$$

**Lemma 3.1.** *Suppose that $\mathcal{M}$ is a computable copy of $\mathcal{L}$, and $\mathbf{d}\in PA_{2\beta+1}$. Then there is a $\mathbf{d}$-computable isomorphism from $\mathcal{M}$ onto $\mathcal{L}$.*

**Proof.** Using the oracle $\mathbf{0}^{(4)}$, one can find computable indices of linear orders $C_{e,k}$, where $e\in\omega$ and $k\in\mathbb{Z}$, such that

$$\mathcal{M}=\sum_{e\in\omega}(e+2+(\cdots+C_{e,-2}+1+\zeta+C_{e,-1}+1+\zeta+C_{e,0}+1+\zeta+C_{e,1}+1+\zeta+ \\ C_{e,2}+1+\zeta+\dots),$$

and $C_{e,k}$ is isomorphic either to $\omega^\beta$, or to $\omega^\beta\cdot 2$. A procedure of finding $C_{e,k}$ can be arranged as follows:

(1) With the oracle $\mathbf{0}^{(2)}$, we find intervals $\mathcal{F}_e=[p_e;q_e]_{\mathcal{M}}$, $e\in\omega$, such that $card(\mathcal{F}_e)=e+2$, $p_e$ has no immediate predecessor in $\mathcal{M}$, and $q_e$ has no immediate successor in $\mathcal{M}$.
(2) Using $\mathbf{0}^{(4)}$, for each $e\in\omega$, we find a sequence $\{r_{e,k}\}_{k\in\mathbb{Z}}$ such that for every $k$, we have: $q_e<_{\mathcal{M}} r_{e,k}<_{\mathcal{M}} r_{e,k+1}<_{\mathcal{M}} p_{e+1}$, and $r_{e,k}$ and $r_{e,k+1}$ belong to adjacent $\zeta$-blocks (i.e. they lie in different $\zeta$-blocks $B_k$ and $B_{k+1}$, and there are no $\zeta$-blocks between $B_k$ and $B_{k+1}$).
(3) Effectively in $\mathbf{0}^{(3)}$, we find elements $a_{e,k}$ and $b_{e,k}$ such that $a_{e,k}$ is the least element with $a_{e,k}>_{\mathcal{M}} r_{e,k}$ and $\neg Block(a_{e,k},r_{e,k})$; and $b_{e,k}$ is the greatest element with $b_{e,k}<_{\mathcal{M}} r_{e,k+1}$ and $\neg Block(b_{e,k},r_{e,k+1})$. Then the domain of $C_{e,k}$ is given by the interval $[a_{e,k};b_{e,k})_{\mathcal{M}}$.

Consider a $\Sigma^c_{2\beta+1}$ formula

$$\Psi(x):=\exists y[(x<y)\&\neg F_\beta(x,y)]. \tag{2}$$

Note that the element $a_{e,k}$ from the procedure above is the least element in $C_{e,k}$. If $C_{e,k}$ is isomorphic to $\omega^\beta$, then $C_{e,k}\not\models\Psi(a_{e,k})$. If $C_{e,k}\cong\omega^\beta\cdot 2$, then $C_{e,k}\models\Psi(a_{e,k})$ (since there are infinitely many $y\in C_{e,k}$ such that the interval $[a_{e,k};y)$ is isomorphic to an ordinal $\geq\omega^\beta$).

The sets

$$U(\mathcal{M}):=\{e\in\omega : C_{e,0}\models\Psi(a_{e,0})\},$$
$$V(\mathcal{M}):=\{e\in\omega : C_{e,1}\models\Psi(a_{e,1})\}$$

are disjoint $\Sigma^0_{2\beta+1}$ sets. For $W \in \{U, V\}$, we fix a strongly $\mathbf{0}^{\langle 2\beta+1\rangle}$-computable sequence of finite sets $\{W^s\}_{s\in\omega}$ such that $W(\mathcal{M}) = \bigcup_{s\in\omega} W^s$ and $W^s \subseteq W^{s+1}$ for all $s$.

We construct a tree $T \subseteq 2^{<\omega}$ as follows: for $\sigma \in 2^{<\omega}$, $\sigma \in T$ if and only if for every $e < |\sigma|$, we have

$$(e \in U^{|\sigma|} \;\Rightarrow\; \varphi_e^{\emptyset^{\langle 2\beta+1\rangle}}(e)\downarrow = \sigma(e)), \text{ and}$$
$$(e \in V^{|\sigma|} \;\Rightarrow\; \varphi_e^{\emptyset^{\langle 2\beta+1\rangle}}(e)\downarrow = 1 - \sigma(e)).$$

The tree $T$ is well-defined: Indeed, for $W \in \{U, V\}$, if $e \in W^{|\sigma|} \subseteq W(\mathcal{M})$, then some $C_{e,k}$ is isomorphic to $\omega^\beta \cdot 2$, and by the definition of $\mathcal{L}$, we have $\varphi_e^{\emptyset^{\langle 2\beta+1\rangle}}(e)\downarrow \in \{0, 1\}$.

It is straightforward to show that both $T$ and its branching function $b_T$ are $\mathbf{0}^{\langle 2\beta+1\rangle}$-computable. Therefore, there is a **d**-computable path $P$ through the tree $T$. Define

$$\mathcal{D}_e := \begin{cases} \mathcal{A}_e, & \text{if } P(e) = 1, \\ \mathcal{B}_e, & \text{if } P(e) = 0; \end{cases} \qquad \mathcal{E}_e := \begin{cases} \mathcal{A}_e, & \text{if } P(e) = 0, \\ \mathcal{B}_e, & \text{if } P(e) = 1. \end{cases}$$

Consider three disjoint cases:

(1) If $e \notin U(\mathcal{M}) \cup V(\mathcal{M})$, then $C_{e,0} \cong C_{e,1} \cong \mathcal{A}_e \cong \mathcal{B}_e \cong \omega^\beta$.
(2) If $e \in U(\mathcal{M})$, then $P(e) = \varphi_e^{\emptyset^{\langle 2\beta+1\rangle}}(e)$ and $C_{e,0} \cong \mathcal{D}_e \cong \omega^\beta \cdot 2$.
(3) If $e \in V(\mathcal{M})$, then $P(e) = 1 - \varphi_e^{\emptyset^{\langle 2\beta+1\rangle}}(e)$, $C_{e,1} \cong \mathcal{E}_e \cong \omega^\beta \cdot 2$, and $\mathcal{D}_e \cong \omega^\beta$. Hence, $C_{e,0} \cong \mathcal{D}_e$.

Thus, in each of the cases, $C_{e,0} \cong \mathcal{D}_e$ and $C_{e,1} \cong \mathcal{E}_e$. By Proposition 2.2, one can recover (uniformly in $e$) $\Delta^0_{2\beta+1}$ indices for $\mathbf{0}^{\langle 2\beta+1\rangle}$-computable isomorphisms $f_{e,0}\colon C_{e,0} \to \mathcal{D}_e$ and $f_{e,1} : C_{e,1} \to \mathcal{E}_e$.

Since **d** is a *PA*-degree over $\emptyset^{\langle 2\beta+1\rangle}$ and $\beta \geq 2$, we have $\mathbf{d} > \mathbf{0}^{\langle 2\beta+1\rangle} \geq \mathbf{0}^{(4)}$. Therefore, one can extend the **d**-computable map $\bigcup_{e\in\omega}(f_{e,0} \cup f_{e,1})$ to a **d**-computable isomorphism from $\mathcal{M}$ onto $\mathcal{L}$. Lemma 3.1 is proved. □

Now we build a "bad" computable copy $\mathcal{S}$ of the structure $\mathcal{L}$. Using Corollary 2.1 again, we obtain computable sequences of linear orders $(\widehat{\mathcal{A}}_e)_{e\in\omega}$ and $(\widehat{\mathcal{B}}_e)_{e\in\omega}$ such that for any $e$,

$$\widehat{\mathcal{A}}_e \cong \begin{cases} \omega^\beta \cdot 2, & \text{if } \varphi_e^{\emptyset^{\langle 2\beta+1\rangle}}(e)\downarrow \in \{0, 1\}, \\ \omega^\beta, & \text{otherwise}; \end{cases} \qquad \widehat{\mathcal{B}}_e \cong \omega^\beta.$$

The order $\mathcal{S}$ is given by a sum

$$\sum_{e\in\omega} (e + 2 + (\cdots + \widehat{\mathcal{A}}_e + 1 + \zeta + \widehat{\mathcal{B}}_e + 1 + \zeta + \widehat{\mathcal{A}}_e + 1 + \zeta + \widehat{\mathcal{B}}_e + 1 + \zeta + \widehat{\mathcal{A}}_e + 1 + \zeta + \widehat{\mathcal{B}}_e + 1 + \zeta + \ldots)).$$

We choose a computable sequence of elements $(m_e)_{e\in\omega}$ such that every $m_e$ belongs to a copy of $\widehat{\mathcal{A}}_e$ inside $\mathcal{S}$. Suppose that $f$ is an isomorphism from $\mathcal{S}$ onto $\mathcal{L}$. We define a set

$$Y := \{e : f(m_e) \text{ belongs to a copy of } \mathcal{A}_e \text{ inside } \mathcal{L}\}.$$

Then we have $Y \leq_T f$, and:

- If $\varphi_e^{\emptyset^{\langle 2\beta+1\rangle}}(e) = 1$, then $e \in Y$.
- If $\varphi_e^{\emptyset^{\langle 2\beta+1\rangle}}(e) = 0$, then $e \notin Y$.

By Proposition 2.1, the degree $\deg_T(f)$ is a $PA$ degree over $\emptyset^{\langle 2\beta+1\rangle}$. Therefore, every degree from the categoricity spectrum of $\mathcal{L}$ must be a $PA$ degree over $\emptyset^{\langle 2\beta+1\rangle}$. This fact and Lemma 3.1 together prove the first case of the theorem.

**Case II.** Suppose again that $\beta$ is a computable ordinal such that $\beta \geq 2$. We construct a computable linear order $\mathcal{L}$ with categoricity spectrum equal to $PA_{2\beta+2}$. The proof is essentially the same as in Case I, modulo the following key modification: We need to use the orders $\omega^{\beta+1}$ and $\omega^{\beta+1} + \omega^{\beta}$ in place of $\omega^{\beta}$ and $\omega^{\beta} \cdot 2$, respectively.

Note also that the formula $\Psi$ from Eq. (2) should be replaced by a $\Sigma^c_{2\beta+2}$ formula

$$\Psi_1(x) := \exists y[(x < y) \& \forall z(y \leq z \to F_{\beta}(y, z))].$$

Theorem 3.1 is proved. □

Recall that a computable structure $\mathcal{S}$ is *decidable* if given a tuple $\bar{a} = a_0, a_1, \ldots, a_n$ from $\mathcal{S}$ and a first-order formula $\psi(x_0, x_1, \ldots, x_n)$, one can effectively check whether $\psi(\bar{a})$ is true in $\mathcal{S}$.

Let $\mathbf{d}$ be a Turing degree. A decidable structure $\mathcal{S}$ is *decidably* $\mathbf{d}$*-categorical* if for any decidable copy $\mathcal{A}$ of $\mathcal{S}$, there is a $\mathbf{d}$-computable isomorphism from $\mathcal{A}$ onto $\mathcal{S}$. The *decidable categoricity spectrum* of $\mathcal{S}$ is the set of all degrees $\mathbf{d}$ such that $\mathcal{S}$ is decidably $\mathbf{d}$-categorical. Decidable categoricity spectra were introduced by Goncharov[27] (note that in Ref. 27, they are called autostability spectra relative to strong constructivizations).

Using the proof of Theorem 3.1 and the methods from Sec. 5 in Ref. 28, it is not difficult to obtain the following:

**Corollary 3.1.** *Suppose that $\alpha$ is a computable successor ordinal such that $\alpha > \omega$. Then there is a decidable discrete linear order $\mathcal{M}$ such that the decidable categoricity spectrum of $\mathcal{M}$ contains precisely the PA degrees over $\mathbf{0}^{(\alpha)}$.*

## 4. Non-strong degrees of categoricity

**Theorem 4.1.** *Suppose that $\alpha$ is a computable successor ordinal such that $\alpha \geq 4$. Then there exist a Turing degree* $\mathbf{d}$ *and a computable scattered linear order* $\mathcal{L}$ *such that* $\mathbf{0}^{(\alpha)} \leq \mathbf{d} \leq \mathbf{0}^{(\alpha+1)}$ *and* $\mathbf{d}$ *is the non-strong degree of categoricity for* $\mathcal{L}$.

The proof consists of two parts. In the first part (Sec. 4.1), we discuss the setup of our construction: we introduce an encoding of a $\mathbf{0}^{(\alpha)}$-limitwise monotonic function $F(x)$ possessing some special properties into a computable scattered linear order $\mathcal{L}[F]$. In the second part (Sec. 4.2), we build a specific $\mathbf{0}^{(\alpha)}$-limitwise monotonic function $G$ and show that the degree $\mathbf{d} = \deg_T(G)$ and the order $\mathcal{L}[G]$ satisfy the theorem. The proof uses a result from Ref. 15 (see Proposition 4.1), but we tried to make exposition as self-contained as possible.

We use the same notation $\emptyset^{\langle\beta\rangle}$ as in Eq. (1). Here we give a detailed proof of the theorem for the case when we obtain a degree $\mathbf{d}$ such that

$$\mathbf{0}^{\langle 2\beta+1\rangle} \leq \mathbf{d} \leq \mathbf{0}^{\langle 2\beta+2\rangle}, \text{ where } \beta \geq 2. \quad (3)$$

The proof for the case $\mathbf{0}^{\langle 2\beta+2\rangle} \leq \mathbf{d} \leq \mathbf{0}^{\langle 2\beta+3\rangle}$ can be easily recovered from the proofs for Eq. (3) and Theorem 3.1.

### 4.1. *Setup of the construction*

Let $\mathbf{c}$ be a Turing degree. A function $F\colon \omega \to \omega$ is called $\mathbf{c}$-*limitwise monotonic* if there is a total $\mathbf{c}$-computable function $f(x, s)$ such that for any $x$ and $s$, we have:

(a) $f(x, s) \leq f(x, s+1)$;
(b) $\lim_s f(x, s) = F(x)$.

We call such a function $f(x, s)$ a $\mathbf{c}$-*limitwise monotonic approximation* (or a $\mathbf{c}$-*l.m. approximation* for short) of the function $F$.

Suppose that $\beta$ is a computable ordinal such that $\beta \geq 2$. Let $F(x)$ be a $\mathbf{0}^{\langle 2\beta+1\rangle}$-limitwise monotonic function with the following property:

$$\forall x(F(x) \leq 2). \quad (4)$$

Note that the exposition in this subsection (including, in particular, construction of the order $\mathcal{L}[F]$) can be easily generalized to the case when Eq. (4) is replaced by the condition $\forall x(F(x) \leq N+1)$, where $N$ is a fixed natural number. Nevertheless, for simplicity, here we work only with functions $F$ satisfying Eq. (4).

We choose a $\mathbf{0}^{\langle 2\beta+1\rangle}$-l.m. approximation $f(x, s)$ for $F$. Using Corollary 2.1, we build a computable sequence of linear orders $(C_{m,k})_{m\in\omega,\, k\in\{1,2\}}$ such that

$$C_{m,k} \cong \begin{cases} \omega^{\beta} \cdot 2, & \text{if } \exists s(f(m, s) \geq k), \\ \omega^{\beta}, & \text{otherwise.} \end{cases}$$

It is clear that the isomorphism types of $C_{m,k}$ do not depend on the choice of an approximation $f$.

**Notation 4.1.** Suppose that $\sigma \in 2^{<\omega}$ and $\sigma$ is non-empty. We say that a linear order $\mathcal{M}$ is a *$\beta$-code* of $\sigma$ (denoted by $\mathcal{M} \in Code(\beta;\sigma)$) if

$$\mathcal{M} = \mathcal{S}_0 + 1 + \zeta + \mathcal{S}_1 + 1 + \zeta + \cdots + \mathcal{S}_{|\sigma|-2} + 1 + \zeta + \mathcal{S}_{|\sigma|-1},$$
$$\text{where } \mathcal{S}_i \cong \begin{cases} \omega^\beta \cdot 2, & \text{if } \sigma(i) = 1, \\ \omega^\beta, & \text{if } \sigma(i) = 0. \end{cases}$$

Note that in the theorem, we will use $\beta$-codes only for strings of the form $1^n 0^m$, $n, m \in \omega$.

Suppose that $m$ is a natural number. We say that a string $\sigma$ of length 2 is the *$m$-change string* for the function $F$ (denoted by $\sigma = chan(F;m)$) if for $i \in \{0,1\}$, we have

$$\sigma(i) = \begin{cases} 1, & \text{if } F(m) \geq i+1, \\ 0, & \text{otherwise.} \end{cases}$$

It is easy to see that $\sigma \in \{00, 10, 11\}$.

We define orders:

$$\widehat{C_m} := C_{m,1} + 1 + \zeta + C_{m,2},$$
$$\mathcal{A}_m := \omega^\beta \cdot 2 + 1 + \zeta + \omega^\beta \cdot 2 + 1 + \zeta + \widehat{C_m},$$
$$\mathcal{B}_m := \omega^\beta \cdot 2 + 1 + \zeta + \widehat{C_m} + 1 + \zeta + \omega^\beta.$$

If $\sigma_m = chan(F;m)$, then it is clear that $\mathcal{A}_m$ is a $\beta$-code of the string $1\ulcorner\sigma_m$, and $\mathcal{B}_m \in Code(\beta; \ulcorner\sigma_m\frown 0)$

The function $F(x)$ will be *encoded* by a computable scattered linear order $\mathcal{L}[F]$ which is defined as follows:

$$\mathcal{L}[F] := \sum_{m\in\omega} (m + 3 + (\cdots + \mathcal{A}_m + 2 + \zeta + \mathcal{B}_m + 2 + \zeta + \mathcal{A}_m + 2 + \zeta + \mathcal{B}_m + 2 + \zeta + \mathcal{A}_m + 2 + \zeta + \mathcal{B}_m + 2 + \zeta + \ldots)). \quad (5)$$

The next three lemmas contain some useful properties of the structure $\mathcal{L}[F]$.

**Lemma 4.1.** *Suppose that* $\mathbf{d} = \deg_T(F)$ *and* $\mathbf{x} = \mathbf{d} \cup \mathbf{0}^{\langle 2\beta+1\rangle}$. *Then the order* $\mathcal{L}[F]$ *is* $\mathbf{x}$*-computably categorical.*

**Proof.** Let $\mathcal{M}$ be a computable copy of $\mathcal{L}[F]$. Using an argument similar to that one from the proof of Lemma 3.1, we can show that effectively in $\mathbf{0}^{(4)}$, one can recover computable indices of linear orders $\mathcal{D}_{m,k}$, where $m \in \omega$ and $k \in \mathbb{Z}$, such that

$$\mathcal{M} = \sum_{m\in\omega} (m + 3 + (\cdots + \mathcal{D}_{m,-1} + 2 + \zeta + \mathcal{D}_{m,0} + 2 + \zeta + \mathcal{D}_{m,1} + 2 + \zeta + \ldots), \quad (6)$$

where each $\mathcal{D}_{m,k}$ is isomorphic either to $\mathcal{A}_m$ or to $\mathcal{B}_m$.

Furthermore, effectively in $\mathbf{0}^{(4)}$ (and uniformly in $m, k$), we can find a decomposition

$$\mathcal{D}_{m,k} = \mathcal{E}_{m,k,1} + 1 + \zeta + \mathcal{E}_{m,k,2} + 1 + \zeta + \mathcal{E}_{m,k,3} + 1 + \zeta + \mathcal{E}_{m,k,4}, \quad (7)$$

where each $\mathcal{E}_{m,k,j}$ is isomorphic either to $\omega^\beta$ or to $\omega^\beta \cdot 2$.

Since $\mathbf{x}$ computes $F$, we can $\mathbf{x}$-effectively find the $m$-change string $\sigma_m := chan(F; m)$. Notice that the number of 1-s in the string $\tau = 1 \frown \sigma_m$ is strictly greater than the number of 1-s inside $\xi = \frown \sigma_m \frown 0$. Thus, using the $\Sigma^c_{2\beta+1}$ formula $\Psi$ from Eq. (2) (apply it to the orders $\mathcal{E}_{m,k,j}$), we $\mathbf{0}^{\langle 2\beta+1\rangle}$-effectively determine whether $\mathcal{D}_{m,0}$ is a $\beta$-code for $\tau$ or for $\xi$. In other words, we find out which one of the structures $\mathcal{A}_m$ or $\mathcal{B}_m$ is isomorphic to $\mathcal{D}_{m,0}$.

After this, we proceed similarly to Lemma 3.1: We recover $\Delta^0_{2\beta+1}$ indices for isomorphisms between appropriate computable copies of $\mathcal{S} \in \{\omega^\beta, \omega^\beta \cdot 2\}$ (e.g., if $\mathcal{D}_{m,0}$ and $\mathcal{B}_m$ are isomorphic, then, in particular, we need to find a $\Delta^0_{2\beta+1}$ computable isomorphism from $\mathcal{E}_{m,0,2}$ onto $C_{m,1}$). Using these indices, we build an $\mathbf{x}$-computable isomorphism from $\mathcal{M}$ onto $\mathcal{L}[F]$. Lemma 4.1 is proved. □

**Notation 4.2 (Ref. 15).** *Suppose that $g(x)$ and $\psi(x, y)$ are partial functions. Then $\psi \diamond g$ denotes the partial function*

$$\psi \diamond g(x) := \psi(x, g(x)).$$

**Lemma 4.2.** *Suppose that $\mathcal{M}$ and $\mathcal{N}$ are computable copies of $\mathcal{L}[F]$. Then there exist an isomorphism $h$ from $\mathcal{M}$ onto $\mathcal{N}$ and a partial $\mathbf{0}^{\langle 2\beta+1\rangle}$-computable function $\psi(x, y)$ with the following properties: $range(\psi) \subseteq \{0, 1\}$, $\psi \diamond F$ is total, and*

$$h \le_T (\psi \diamond F) \oplus \mathbf{0}^{\langle 2\beta+1\rangle}.$$

**Proof.** As in the previous lemma, we fix a decomposition from Eqs. (6)–(7) for the order $\mathcal{M}$. We choose a similar decomposition for the structure $\mathcal{N}$, this decomposition will use notations $\widehat{\mathcal{D}}_{m,k}$ and $\widehat{\mathcal{E}}_{m,k,j}$ in place of $\mathcal{D}_{m,k}$ and $\mathcal{E}_{m,k,j}$, respectively. We emphasize that both decompositions can be obtained effectively in $\mathbf{0}^{(4)} \le \mathbf{0}^{\langle 2\beta+1\rangle}$.

Consider $\Sigma^0_{2\beta+1}$ sets:

$$W := \{(m,k,j) : m \in \omega,\ k \in \{0,1\},\ j \in \{2,3,4\},\ \mathcal{E}_{m,k,j} \cong \omega^\beta \cdot 2\},$$
$$\widehat{W} := \{(m,k,j) : m \in \omega,\ k \in \{0,1\},\ j \in \{2,3,4\},\ \widehat{\mathcal{E}}_{m,k,j} \cong \omega^\beta \cdot 2\}.$$

Then a partial function $\psi(x,y)$ is defined as follows:

(a) If $y \geq 3$, then set $\psi(x,y) = 0$.
(b) If $F(x) < y \leq 2$, then the value $\psi(x,y)$ is undefined.
(c) Suppose that $F(x) \geq y$. Then we find some (e.g., first under a fixed enumeration) elements $(x,k,2+y) \in W$ and $(x,\widehat{k},2+y) \in \widehat{W}$. If $k = \widehat{k}$, then set $\psi(x,y) := 1$. Otherwise, define $\psi(x,y) := 0$.

Note that the function $\psi$ is well-defined: indeed, the definition of $\mathcal{L}[F]$ shows that if $F(x) \geq y$, then at least one of the orders $\mathcal{E}_{x,0,2+y}$ or $\mathcal{E}_{x,1,2+y}$ is isomorphic to $\omega^\beta \cdot 2$. Furthermore, the function $\psi \diamond F(x)$ is total.

Recall that $F$ is a $\mathbf{0}^{\langle 2\beta+1\rangle}$-limitwise monotonic function. Therefore, it is easy to prove that $\psi$ can be chosen partial $\mathbf{0}^{\langle 2k+1\rangle}$-computable. Notice that $\psi$ is $\{0,1\}$-valued.

We sketch a construction of an isomorphism $h$ from $\mathcal{M}$ onto $\mathcal{N}$.

For $m \in \omega$, the definition of $\mathcal{L}$ implies that the number $j_0 := 2 + F(m)$ is the maximal $j$ such that $(m,0,j) \in W$ or $(m,1,j) \in W$. Moreover, both elements $(m,0,j_0)$ and $(m,1,j_0)$ cannot simultaneously belong to $W$. A similar fact is true for $\widehat{W}$.

We consider two cases depending on the value $\psi \diamond F(m)$:

*Case 1.* Assume that $\psi \diamond F(m) = 1$. Then exactly one of the following two conditions holds: either $(m,0,j_0) \in W \cap \widehat{W}$, or $(m,1,j_0) \in W \cap \widehat{W}$. W.l.o.g., we may assume that $(m,0,j_0) \in W \cap \widehat{W}$. Then $\mathcal{E}_{m,0,j_0} \cong \widehat{\mathcal{E}}_{m,0,j_0} \cong \omega^\beta \cdot 2$ and $\mathcal{E}_{m,1,j_0} \cong \widehat{\mathcal{E}}_{m,1,j_0} \cong \omega^\beta$. This implies that $\mathcal{D}_{m,0} \cong \widehat{\mathcal{D}}_{m,0} \ncong \widehat{\mathcal{D}}_{m,1} \cong \mathcal{D}_{m,1}$. Using this fact, similarly to the previous lemma, we can build a $\mathbf{0}^{\langle 2\beta+1\rangle}$-computable isomorphism

$$h_m \colon \sum_{k\in\mathbb{Z}} (2 + \zeta + \mathcal{D}_{m,k}) \to \sum_{k\in\mathbb{Z}} (2 + \zeta + \widehat{\mathcal{D}}_{m,k}). \tag{8}$$

*Case 2.* Suppose that $\psi \diamond F(m) = 0$. Then again, there are two disjoint conditions: either $(m,0,j_0) \in W$ and $(m,1,j_0) \in \widehat{W}$, or $(m,1,j_0) \in W$ and $(m,0,j_0) \in \widehat{W}$. Assume that $(m,0,j_0) \in W$ and $(m,1,j_0) \in \widehat{W}$. Then $\mathcal{E}_{m,0,j_0} \cong \widehat{\mathcal{E}}_{m,1,j_0} \cong \omega^\beta \cdot 2$ and $\mathcal{D}_{m,0} \cong \widehat{\mathcal{D}}_{m,1} \ncong \widehat{\mathcal{D}}_{m,0} \cong \mathcal{D}_{m,1}$. We can construct a $\mathbf{0}^{\langle 2\beta+1\rangle}$-computable isomorphism $h_m$ satisfying Eq. (8).

The map $\bigcup_{m\in\omega} h_m$ can be easily extended to an isomorphism $h \colon \mathcal{M} \cong \mathcal{N}$, computable in $(\psi \diamond F) \oplus \mathbf{0}^{\langle 2\beta+1\rangle}$. Lemma 4.2 is proved. □

**Lemma 4.3.** *Suppose that* $\mathbf{d} = \deg_T(F)$ *and* $\mathbf{d} \geq \mathbf{0}^{\langle 2\beta+1\rangle}$. *Then* $\mathbf{d}$ *is the degree of categoricity for* $\mathcal{L}[F]$.

**Proof.** Since $\mathbf{d} \geq \mathbf{0}^{\langle 2\beta+1\rangle}$, by Lemma 4.1, the structure $\mathcal{L}[F]$ is $\mathbf{d}$-computably categorical. We build two computable copies $\mathcal{L}^0$ and $\mathcal{L}^1$ of $\mathcal{L}[F]$. Define

$$\mathcal{A}^0_m := \omega^\beta \cdot 2 + 1 + \zeta + \omega^\beta \cdot 2 + 1 + \zeta + C_{m,1} + 1 + \zeta + \omega^\beta,$$
$$\mathcal{B}^0_m := \omega^\beta \cdot 2 + 1 + \zeta + C_{m,1} + 1 + \zeta + C_{m,2} + 1 + \zeta + C_{m,2};$$
$$\mathcal{A}^1_m := \omega^\beta \cdot 2 + 1 + \zeta + \omega^\beta \cdot 2 + 1 + \zeta + C_{m,2} + 1 + \zeta + \omega^\beta,$$
$$\mathcal{B}^1_m := \omega^\beta \cdot 2 + 1 + \zeta + C_{m,1} + 1 + \zeta + C_{m,1} + 1 + \zeta + C_{m,2}.$$

For a better intuition, we give a table of $\beta$-codes (Table 1). The table is read as follows: if $F(m) = 2$, then the structure $\mathcal{A}^1_m$ is a $\beta$-code of the string 1110.

Table 1. Our linear orders treated as $\beta$-codes.

| Structure | $F(m) = 0$ | $F(m) = 1$ | $F(m) = 2$ |
|---|---|---|---|
| $\mathcal{A}_m$ | 1100 | 1110 | 1111 |
| $\mathcal{B}_m$ | 1000 | 1100 | 1110 |
| $\mathcal{A}^0_m$ | 1100 | 1110 | 1110 |
| $\mathcal{B}^0_m$ | 1000 | 1100 | 1111 |
| $\mathcal{A}^1_m$ | 1100 | 1100 | 1110 |
| $\mathcal{B}^0_m$ | 1000 | 1110 | 1111 |

The structure $\mathcal{L}^0$ is defined using Eq. (5), where we replace $\mathcal{A}_m$ and $\mathcal{B}_m$ by $\mathcal{A}^0_m$ and $\mathcal{B}^0_m$, respectively. The order $\mathcal{L}^1$ is defined in a similar way. It is straightforward to show (e.g., consider the table of $\beta$-codes above) that $\mathcal{L}^0$ and $\mathcal{L}^1$ are computable copies of $\mathcal{L}[F]$.

Choose a computable sequence $(w_m)_{m\in\omega}$ such that every $w_m$ belongs to a copy of $\mathcal{A}_m$ inside $\mathcal{L}[F]$. Suppose that $g\colon \mathcal{L}[F] \cong \mathcal{L}^0$ and $h\colon \mathcal{L}[F] \cong \mathcal{L}^1$. Then the following holds:

(1) If $g(w_m)$ belongs to a copy of $\mathcal{A}^0_m$ inside $\mathcal{L}^0$, then $F(m) \leq 1$.
(2) If $g(w_m)$ does not belong to a copy of $\mathcal{A}^0_m$, then it lies in a copy of $\mathcal{B}^0_m$, and $F(m) = 2$.
(3) If $h(w_m)$ belongs to a copy of $\mathcal{A}^1_m$ in $\mathcal{L}^1$, then $F(m) = 0$.
(4) If $h(w_m)$ is not in a copy of $\mathcal{A}^1_m$, then it belongs to a copy of $\mathcal{B}^1_m$, and $F(m) \geq 1$.

Therefore, we have $F \leq_T (g \oplus h)$, and every degree from the spectrum $CatSpec(\mathcal{L}[F])$ must compute $\mathbf{d}$. This fact and $\mathbf{d}$-computable categoricity of $\mathcal{L}[F]$ together imply that $\mathbf{d}$ is the degree of categoricity for $\mathcal{L}[F]$. □

### 4.2. *Building a special function G*

The goal of this subsection is to build a $\mathbf{0}^{\langle 2\beta+1\rangle}$-limitwise monotonic function $G(x)$ with the following properties:

(i) $G \geq_T \mathbf{0}^{\langle 2\beta+1\rangle}$. By Lemma 4.3, this implies that $\mathbf{d} := \deg_T(G)$ is the degree of categoricity for the order $\mathcal{L}[G]$. Moreover, the $\mathbf{0}^{\langle 2\beta+1\rangle}$-limitwise monotonicity of $G$ shows that $\mathbf{d} \leq \mathbf{0}^{\langle 2\beta+2\rangle}$.

(ii) The degree $\mathbf{d}$ cannot be strong degree of categoricity for $\mathcal{L}[G]$. This will be obtained by using Lemma 4.2 and Proposition 4.1 (see below).

First, we recall a result obtained in Ref. 15:

**Proposition 4.1 (simple version of Theorem 4.5 from Ref. 15).** *There is a limitwise monotonic function $F(x)$ such that $F(x) \leq 2$ for all $x$, and $F \not\leq_T \psi \diamond F$, for any $\{0,1\}$-valued partial computable function $\psi$ with $\psi \diamond F$ total.*

**Proof sketch.** Fix an effective list $\{(\Phi_e, \psi_e)\}_{e\in\omega}$ which enumerates all pairs containing a Turing functional $\Phi_e$ and a $\{0,1\}$-valued partial computable function $\psi_e(x)$. We build a l.m. approximation $F(x)[s]$ and satisfy the following series of requirements:

$R_e$: If $\psi \diamond F$ is total, then $F \neq \Phi_e(\psi_e \diamond F)$.

We describe the strategy for $R_e$ in isolation:

(1) Choose a fresh witness $x_e$ and define $F(x_e)[s] := 0$.
(2) Wait until we see a computation $\Phi_e(\psi_e \diamond F; x_e)[s_1] = 0$.
(3) Initialize all strategies of lower priority and set $F(x_e)[s_1] := 1$.
(4) Wait until we see $\Phi_e(\psi_e \diamond F; x_e)[s_2] = 1$.
(5) Initialize all lower priority strategies and define $F(x_e)[s_2] := 2$.

The intuition behind the strategy is as follows: Suppose that an $R_e$-strategy eventually reaches the fifth stage. Since we initialize the lower priority strategies at stages (2) and (4), the values $F(y)$, for $y \neq x_e$, used in the $\Phi_e$-computation will not be changed. This implies that $\psi_e \diamond F(x_e)[s_1] \neq \psi_e \diamond F(x_e)[s_2]$ (w.l.o.g., one can assume that these values are defined). Since $\psi_e \diamond F(x_e) \in \{0,1\}$, if the value $v_e := \Phi_e(\psi_e \diamond F; x_e)$ is defined, then either $v_e = \Phi_e(\psi_e \diamond F; x_e)[s_1] = 0$ or $v_e = \Phi_e(\psi_e \diamond F; x_e)[s_2] = 1$. Therefore, if our strategy reaches its fifth stage, then it is satisfied. If it never reaches stage (5), then it is satisfied in a trivial way.

As per usual, we arrange a finite injury argument for $R_e$-s and build the desired limitwise monotonic function $F$. □

As a consequence of Proposition 4.1, we obtain:

**Corollary 4.1.** *Suppose that* $X$ *is an oracle and* $\mathbf{x} = \deg_T(X)$. *Then there is an* $\mathbf{x}$*-limitwise monotonic function* $G(x)$ *such that* $G(x) \leq 2$ *for all* $x$, *and*

$$X \leq_T G \nleq_T (\psi \diamond G) \oplus X,$$

*for any* $\{0, 1\}$*-valued partial* $\mathbf{x}$*-computable function* $\psi$ *with* $\psi \diamond G$ *total.*

**Proof.** Proceed with a straightforward relativization of Proposition 4.1, modulo the following important modification: Beforehand, for all $x \in \omega$, we define $G(2x) := \chi_X(x)$, where $\chi_X$ is the characteristic function of $X$. After that, when we build our $G$, we will choose witnesses $x_e$ as odd numbers. □

We apply the corollary to the oracle $X = \mathbf{0}^{\langle 2\beta+1 \rangle}$ and obtain a $\mathbf{0}^{\langle 2\beta+1 \rangle}$-limitwise monotonic function $G(x)$. Note that by Lemma 4.3, the degree $\mathbf{d} := \deg_T(G)$ is the degree of categoricity for $\mathcal{L}[G]$.

Assume that $\mathbf{d}$ is a strong degree of categoricity for $\mathcal{L}[G]$. Then one can choose two computable copies $\mathcal{M}$ and $\mathcal{N}$ of $\mathcal{L}[G]$ such that any isomorphism $g$ from $\mathcal{M}$ onto $\mathcal{N}$ computes $\mathbf{d}$.

By Lemma 4.2, we can find an isomorphism $h$ from $\mathcal{M}$ onto $\mathcal{N}$ and a partial $\mathbf{0}^{\langle 2\beta+1 \rangle}$-computable, $\{0, 1\}$-valued function $\psi_0$ such that $\psi_0 \diamond G$ is total and $h \leq_T (\psi_0 \diamond G) \oplus \mathbf{0}^{\langle 2\beta+1 \rangle}$. Therefore,

$$G \leq_T h \leq_T (\psi_0 \diamond G) \oplus \mathbf{0}^{\langle 2\beta+1 \rangle}.$$

This contradicts Corollary 4.1. Theorem 4.1 is proved.

## Acknowledgements

The work was supported by Nazarbayev University Faculty Development Competitive Research Grants N090118FD5342.

## References

1. A. Fröhlich and J. C. Shepherdson. Effective procedures in field theory. *Philos. Trans. Roy. Soc. London Ser. A*, 248(950):407–432, 1956.
2. A. I. Mal'tsev. Constructive algebras I. *Russ. Math. Surv.*, 16(3):77–129, 1961.
3. A. I. Mal'tsev. On recursive abelian groups. *Sov. Math. Dokl.*, 32:1431–1434, 1962.
4. S. S. Goncharov and V. D. Dzgoev. Autostability of models. *Algebra Logic*, 19(1):28–37, 1980.
5. J. B. Remmel. Recursively categorical linear orderings. *Proc. Am. Math. Soc.*, 83(2):387–391, 1981.
6. E. B. Fokina, I. Kalimullin, and R. Miller. Degrees of categoricity of computable structures. *Arch. Math. Logic*, 49(1):51–67, 2010.

7. B. F. Csima, J. N. Y. Franklin, and R. A. Shore. Degrees of categoricity and the hyperarithmetic hierarchy. *Notre Dame J. Formal Logic*, 54(2):215–231, 2013.
8. A. N. Frolov. Effective categoricity of computable linear orderings. *Algebra Logic*, 54(5):415–417, 2015.
9. N. A. Bazhenov. Degrees of autostability for linear orders and linearly ordered abelian groups. *Algebra Logic*, 55(4):257–273, 2016.
10. N. Bazhenov. A note on effective categoricity for linear orderings. In T. V. Gopal, G. Jäger, and S. Steila, editors, *Theory and Applications of Models of Computation*, volume 10185 of *Lect. Notes Comput. Sci.*, pages 85–96. Springer, Cham, 2017.
11. R. Miller. **d**-computable categoricity for algebraic fields. *J. Symb. Log.*, 74(4):1325–1351, 2009.
12. R. Miller and A. Shlapentokh. Computable categoricity for algebraic fields with splitting algorithms. *Trans. Am. Math. Soc.*, 367(6):3955–3980, 2015.
13. N. A. Bazhenov. Effective categoricity for distributive lattices and Heyting algebras. *Lobachevskii J. Math.*, 38(4):600–614, 2017.
14. C. J. Ash and J. F. Knight. *Computable Structures and the Hyperarithmetical Hierarchy*. Elsevier, Amsterdam, 2000.
15. N. A. Bazhenov, I. Sh. Kalimullin, and M. M. Yamaleev. Degrees of categoricity and spectral dimension. *J. Symb. Log.*, 83(1):103–116, 2018.
16. N. A. Bazhenov, I. Sh. Kalimullin, and M. M. Yamaleev. Degrees of categoricity vs. strong degrees of categoricity. *Algebra Logic*, 55(2):173–177, 2016.
17. B. F. Csima and J. Stephenson. Finite computable dimension and degrees of categoricity. *Ann. Pure Appl. Logic*. In press. Available at https://doi.org/10.1016/j.apal.2018.08.012.
18. Yu. L. Ershov and S. S. Goncharov. *Constructive models*. Kluwer Academic/Plenum Publishers, New York, 2000.
19. C. J. Ash. Stability of recursive structures in arithmetical degrees. *Ann. Pure Appl. Logic*, 32(2):113–135, 1986.
20. R. G. Downey. Computability theory and linear orderings. In Yu. L. Ershov, S. S. Goncharov, A. Nerode, and J. B. Remmel, editors, *Handbook of Recursive Mathematics, vol. 2*, volume 139 of *Stud. Logic Found. Math.*, pages 823–976. Elsevier Science B.V., Amsterdam, 1998.
21. J. G. Rosenstein. *Linear orderings*, volume 98 of *Pure Appl. Math.* Academic Press, New York, 1982.
22. D. Scott. Algebras of sets binumerable in complete extensions of arithmetic. In J. Dekker, editor, *Recursive Function Theory*, volume 5 of *Proc. Sympos. Pure Math.*, pages 117–121. American Mathematical Society, Providence, 1962.
23. C. G. Jockusch and R. I. Soare. $\Pi^0_1$ classes and degrees of theories. *Trans. Am. Math. Soc.*, 173:33–56, 1972.
24. S. G. Simpson. Degrees of unsolvability: a survey of results. In J. Barwise, editor, *Handbook of Mathematical Logic*, volume 90 of *Stud. Logic Found. Math.*, pages 631–652. Elsevier, Amsterdam, 1977.
25. C. J. Ash. Recursive labelling systems and stability of recursive structures in hyperarithmetical degrees. *Trans. Am. Math. Soc.*, 298(2):497–514, 1986.
26. C. J. Ash and J. F. Knight. Pairs of recursive structures. *Ann. Pure Appl. Logic*, 46(3):211–234, 1990.

27. S. S. Goncharov. Degrees of autostability relative to strong constructivizations. *Proc. Steklov Inst. Math.*, 274:105–115, 2011.
28. N. Bazhenov. Autostability spectra for decidable structures. *Math. Struct. Comput. Sci.*, 28(3):392–411, 2018.

# FPC and the Symmetry Gap in Combinatorial Optimization

Anuj Dawar

*Department of Computer Science and Technology,*
*University of Cambridge,*
*Cambridge CB3 0FD, UK*
*E-mail: anuj.dawar@cl.cam.ac.uk*

Fixed-point logic with counting is a well-studied formalism in finite model theory whose expressive power defines a natural and powerful fragment of the polynomial-time decidable properties. It can be shown that it determines a class of properties that can be solved by *symmetric* polynomial-time algorithms. We review results that demonstrate that powerful combinatorial optimization techniques can be formulated in FPC. Inexpressibility results for the logic then yield strong lower bound results on the power of such algorithms, such as semidefinite programming techniques for constraint optimization problems. We argue that this describes a *symmetry gap* that any putative efficient algorithms for these problems must overcome.

*Keywords*: Logic, Combinatorial Optimization, Descriptive Complexity, Semidefinite Programming, Symmetry.

## 1. Introduction

Fixed-point logic with counting (FPC) is the extension of first-order logic with (1) a mechanism for defining relations recursively, and (2) an ability to count. It was originally proposed by Neil Immerman [1] as a logic in which one might be able to express all and only the polynomial-time decidable properties of finite structures. While it is straightforward to show that for every sentence $\varphi$ of FPC the collection of finite models of $\varphi$ is decidable in polynomial-time, the converse turns out to be false. A first counterexample was constructed by Cai, Fürer and Immerman [2] and by now many natural examples are known. Still, FPC defines a natural and diverse class of problems inside the complexity class P and has been the object of sustained research into its expressive power [3]. There are a number of reasons for this, which I explore further in this article, which will take us through a tour of a number of recent results that lead us beyond considerations in logic into an insight into what I call the *symmetry gap* in combinatorial optimization.

One reason for the interest in FPC is that it is very expressive. Most problems that are computationally feasible, in the sense of being decidable in P, can be expressed in FPC. It is not really possible to quantify the sense of "most" in the previous sentence, but it seems that if the obvious algorithm for solving a problem runs in polynomial-time, then it is likely that it can be expressed in FPC. Indeed, it is quite a challenge to come up with examples of problems in P that are not in FPC. And yet, the second attraction of FPC is that we *can* come up with such examples. While we have no good methods for proving lower bounds against P—the holy grail of complexity theory—we do have means of proving many natural NP-complete problems are not in FPC (though there are others for which the question of whether or not they are in FPC is equivalent to whether P = NP). Moreover, recent results have revealed that these inexpressibility results can be understood as lower bounds against a natural *symmetric* fragment of P. That is, FPC captures a natural notion of *symmetric* polynomial time algorithms, in a way I make precise below.

What makes FPC even more interesting is a number of discoveries that have shown that many natural problems, which are not *obviously* in P turn out ultimately to be in FPC. The problem of finding a maximum matching in a graph was shown to be in P by Edmonds[4]. The problem of solving linear programs was shown to be in P by Khachiyan[5]. Deep investigations in structural graph theory by Robertson and Seymour lead to the conclusion that any proper minor-closed class of graphs is in P[6]. Each of these results was a major breakthrough in its time and certainly none of the problems can be described as being obviously in P. We now know that all of them are, in fact, in FPC. That is to say, the efficient algorithms for solving them can be made symmetric in a precise sense.

The flip side of this is that when we know that a certain algorithmic technique respects symmetry, we can use the methods for proving inexpressibility in FPC to show its limitiations. A striking recent example of this is a dichotomy established on the power of semidefinite programming to solve finite-valued constraint satisfication problems[7], where it is shown that every such problem is either solved by its basic linear programming relaxation, or not solved by any sublinear number of levels of the Lasserre hierarchy of semidefinite programs (precise definitions are to follow, leading up to Theorem 7.1 below). Thus, we obtain a fundamental gap in the field of combinatorial optimization from logical inexpressibility results. This tells us that some of our most powerful algorithmic techniques cannot be used to efficiently solve certain NP-complete problems. If we are able to solve such problems, we would need symmetry-breaking algorithms.

In this paper, we lead up to this result by taking a tour that touches on topics from logic, combinatorics, optimization and other areas. The aim is to give a coherent account which allows us to state the results in a common language. No detailed proofs are given. For these I refer the reader to the original sources. The first half of the paper brings together the necessary definitions and results about fixed-point logic with counting. We define the logic, give its characterization in terms of circuit complexity, its relationship to the Weisfeiler-Leman approximations of graph isomorphism and lead up to a definition of counting width. The second half of the paper investigates the relationship with combinatorial optimization. We see how problems in this area can be formulated in FPC, and specifcally how the ellipsoid method can be expressed, leading up to the definability dichotomy result for finite-valued constraint satisfaction problems.

## 2. Fixed-point Logic with Counting

We write FPC to denote the *fixed-point logic with counting*. This is an extension of first-order logic with a mechanism for recursion and counting. We also write FPC to denote the collection of decision problems that can be defined in FPC. That is, we think of a sentence $\varphi$ of FPC as defining the decision problem of determining, given a *finite* structure $\mathbb{A}$, whether or not $\mathbb{A} \models \varphi$. Thus, the decision problems are isomorphism-closed classes of finite structures (which we also refer to as *properties* of finite structures). We consider decision problems of structures such as graphs, Boolean formulas and systems of equations. Where it is not obvious, I make clear how such objects are treated as finite relational structures.

Fix a relational vocabulary $\sigma$, and let FO[$\sigma$] denote first-order logic over this vocabulary. By this we may mean the collection of first-order $\sigma$-formulas or we may also mean the collection of $\sigma$-classes of finite structures that are definable in first-order logic. In general, we may drop the mention of the vocabulary $\sigma$ if it is implicit and just write FO. FP denotes the extension of FO with a fixed-point operator. Fixed-point operators come in different varieties, such as the least-fixed-point operator and the inflationary-fixed-point operator (see[8] for a discussion of these varieties of operators). For our purposes here, it is most convenient to just consider inflationary fixed-points. For details of the different logics one can obtain with such fixed-point operators and their expressive power, the interested reader may consult the textbook by Ebbinghaus and Flum[9]. By Immerman[1] and Vardi[10], we know that on ordered structures, FP expresses exactly those properties that are decidable in polynomial-time. However, in the absence of order, simple counting properties, such as saying that the number of elements in a structure is even, are not definable.

Fixed-point logic with counting (FPC) extends FP with the ability to express the cardinality of definable sets. The logic has two sorts of variables: $x_1, x_2, \ldots$ ranging over the domain of the structure, and $\nu_1, \nu_2, \ldots$ ranging over the non-negative integers. If we allow unrestricted quantification over non-negative integers, the logic would be powerful enough to express undecidable properties. In Immerman's original proposal, number variables were restricted to taking values in the set $\{0, \ldots, n\}$ where $n$ is the number of elements in the domain of the structure of interpretation. In the formal definition below, we adopt another convention (suggested by Grohe) of requiring quantification of number variables to be bounded by a term. In addition, we also have second order variables $X_1, X_2, \ldots$, each of which has a type which is a finite string in $\{\text{element}, \text{number}\}^*$. Thus, if $X$ is a variable of type (element, number), it is to be interpreted by a binary relation relating elements to numbers. The logic allows us to build up *counting terms* according to the following rule:

If $\varphi$ is a formula and $x$ is a variable, then $\#x\varphi$ is a term.

The intended semantics is that $\#x\varphi$ denotes the number (i.e. the non-negative integer) of elements that satisfy the formula $\varphi$. The formulas of FPC are now described by the following set of rules:

- all atomic formulas of first-order logic are formulas of FPC;
- if $\tau_1$ and $\tau_2$ are counting terms (that is each one is either a number variable or a term of the form $\#x\varphi$) then each of $\tau_1 < \tau_2$ and $\tau_1 = \tau_2$ is a formula;
- if $\varphi$ and $\psi$ are formulas then so are $\varphi \wedge \psi$, $\varphi \vee \psi$ and $\neg\psi$;
- if $\varphi$ is a formula, $x$ is an element variable, $\nu$ is a number variable, and $\eta$ is a counting term, then $\exists x\, \varphi$ and $\exists \nu \leq \eta\, \varphi$ are formulas; and
- if $X$ is a relation symbol of type $\alpha$, $\mathbf{z}$ is a tuple of variables whose sorts match the type $\alpha$, $\eta$ is a tuple of counting terms, one for each counting variable in $\mathbf{z}$ and $\mathbf{t}$ is a tuple of terms of type $\alpha$, then $[\mathbf{fp}_{X,\mathbf{z},\eta}\varphi](\mathbf{t})$ is a formula.
  The intended semantics here is that the tuple of elements denoted by $\mathbf{t}$ is in the inflationary fixed-point of the operator defined by $\varphi$, binding the variables in $X, \mathbf{z}$.

For details of the semantics and a lot more about the logic I refer the reader to Otto's excellent monograph[11]. Here, I illustrate the use of this logic with some examples. In what follows, we use $\mathbb{A}$ and $\mathbb{B}$ to denote finite structures over a vocabulary $\sigma$ and $A$ and $B$ to denote their respective universes.

A standard example of the power of inductive definitions is that we can use them to say a graph is connected, a property that is not expressible in FO.

**Example 2.1.** The following formula:

$$\forall u \forall v [\mathbf{fp}_{T,xy}(x = y \vee \exists z(E(x,z) \wedge T(z,y)))](u,v)$$

is satisfied in a graph $(V, E)$ if, and only if, the graph is connected. Indeed, the least relation $T$ that satisfies the equivalence $T(x,y) \equiv x = y \vee \exists z(E(x,z) \wedge T(z,y)))$ is the reflexive and transitive closure of $E$. This is, therefore the fixed point defined by the operator $\mathbf{fp}$. The formula can now be read as saying that every pair $u, v$ is in this reflexive-transitive closure.

A somewhat more involved example is obtained by considering the *circuit value problem* (CVP):

**Example 2.2 (CVP).** *A* circuit *is a labelled directed acyclic graph in which all sources (i.e. nodes with no incoming edges) are labelled either* 0 *or* 1 *and all internal nodes have labels from the set* $\{\wedge, \vee, \neg\}$, *with the proviso that any node labelled* $\neg$ *has exactly one incoming edge. A single node is designated as the output node. The* circuit value problem *(CVP) is the problem of deciding whether the output node evaluates to* 1 *under the standard evaluation. This can be expressed by the following formula, where we assume that the labels are given by unary predicates (*zero, one, and, or *and* not*) and the output node by a constant* out*.*

$$\begin{aligned}\mathbf{fp}_{V,xv\leq 1}[(&\mathsf{one}(x) \wedge v = 1 \vee \mathsf{zero}(x) \wedge v = 0)\vee\\ &(\mathsf{or}(x) \wedge v = 1 \wedge \exists y\, E(y,x) \wedge V(y,1))\vee\\ &(\mathsf{or}(x) \wedge v = 0 \wedge \forall y\, E(y,x) \rightarrow V(y,0))\vee\\ &(\mathsf{and}(x) \wedge v = 0 \wedge \exists y\, E(y,x) \wedge V(y,0))\vee\\ &(\mathsf{and}(x) \wedge v = 1 \wedge \forall y\, E(y,x) \rightarrow V(y,1))\vee\\ &(\mathsf{not}(x) \wedge v = 1 \wedge \forall y\, E(y,x) \rightarrow V(y,0))]\vee\\ &(\mathsf{not}(x) \wedge v = 0 \wedge \forall y\, E(y,x) \rightarrow V(y,1))](\mathsf{out}, 1).\end{aligned}$$

*In this formula, we use constants* 0 *and* 1 *for the natural numbers for convenience. They are easily definable and we do not have to assume they are present in the language.*

There are a couple of points to note about this example. First, CVP is a problem that is known to be P-complete under logarithmic-space reductions[12] and this means that the failure of FP to express all problems in P is not on the grounds of complexity. It is able to express the hardest problems, but the restrictions are different. Secondly, the use of counting variable and terms in the above example was inessential. All we need is two elements in the structure that can be distinguished. However, we could consider circuits with a richer basis of gates than just the Boolean functions used above. For example, we can allow *threshold* or *majority* gates. In this case we cannot use FP alone to express the corresponding circuit value problem. The use of the counting mechanism is then essential, but the problem is still expressible in FPC.

The standard example of a property not definable in FP is to say that the number of elements in a structure is even.

**Example 2.3.** The following sentence is satisfied in a structure $\mathbb{A}$ if, and only if, the number of elements of $\mathbb{A}$ that satisfy the formula $\varphi(x)$ is even.

$$\exists v_1 \leq [\#x\varphi]\exists v_2 \leq v_1(v_1 = [\#x\varphi] \wedge (v_2 + v_2 = v_1)).$$

In particular, taking $\varphi$ to be a universally true formula such as $x = x$, we get a sentence that defines evenness. Here we have used the addition symbol in the subformula $v_2 + v_2 = v_1$. It should be noted that this denotes a relation that is easily definable by induction (and therefore using a fixed-point operator) on the domain of numbers.

A key tool in analysing the expressive power of FPC is to look at equivalence in a weaker logic—first-order logic with counting quantifiers and a bounded number of variables. For each natural number $i$, we have a quantifier $\exists^i$ where $\mathbb{A} \models \exists^i x \varphi$ if, and only if, there are at least $i$ distinct elements $a \in A$ such that $\mathbb{A} \models \varphi[a/x]$. While the extension of first-order logic with counting quantifiers is no more expressive than FO itself (in contrast to the situation with counting terms), the presence of these quantifiers does affect the number of variables that are necessary to express a query. Let $C^k$ denote the $k$-variable fragment of first-order logic with counting quantifiers. That is, $C^k$ consists of those formulas in which no more than $k$ variables appear, free or bound. For two structures $\mathbb{A}$ and $\mathbb{B}$, we write $\mathbb{A} \equiv^k \mathbb{B}$ to denote that the two structures are not distinguished by any sentence of $C^k$. The link between this and FPC is the following fact, established by Immerman and Lander[13]:

**Theorem 2.1.** *For every sentence $\varphi$ of* FPC, *there is a $k$ such that if $\mathbb{A} \equiv^k \mathbb{B}$, then $\mathbb{A} \models \varphi$ if, and only if, $\mathbb{B} \models \varphi$.*

Indeed, this theorem follows from the fact that for any formula $\varphi$ of FPC, there is a $k$ so that on structures with at most $n$ elements, $\varphi$ is equivalent to a formula $\theta_n$ of $C^k$. Additionally, it can be shown that the quantifier depth of $\theta_n$ is bounded by a polynomial function of $n$, but the important bound for us is that the number of variables $k$ is bounded by a constant that only depends on $\varphi$.

The equivalence relations $\equiv^k$ have many different characterisations, some of which arose in contexts removed from the connection with logic, such as the Weisfeiler-Leman family of equivalences arising in the study of graph isomorphism (some of these characterizations and their connections are explored further in Section 4). Particularly interesting are the characterizations of $\equiv^k$ in terms of two-player games, including the counting game of Immerman and Lander[13] and the bijection game of Hella[14]. These provide a means of arguing that two structures $\mathbb{A}$ and $\mathbb{B}$ are $\equiv^k$-equivalent. By Theorem 2.1, this can then be used to show that some property is not definable in FPC, by showing that it is not closed under $\equiv^k$ for any fixed $k$.

Immerman's suggestion that FPC might be sufficient to express all properties in P arose from the intuition that it addresses the two obvious shortcomings of FO by providing a means to express inductive definitions and a means of counting. In a sense, all algorithms that are "obviously" polynomial-time can be translated into the logic. This has been reinforced over the years by results that establish definability in FPC for many non-obvious polynomial-time problems. One particular line of investigation has sought to examine classes of structures based on *sparse graphs*. An early result in this vein was due to Immerman and Lander[13], who showed that FPC captures P on trees. This was generalized in two distinct directions by Grohe[15], who showed that FPC captures P on planar graphs and Grohe and Mariño[16] who show that FPC captures P on graphs of bounded treewidth. These results were the beginning of a long progression which leads to Grohe's theorem[17] establishing that FPC captures P on any class of structures whose adjacency graphs form a proper minor-closed class. This monumental result generalizes all of the above and encapsulates in logical form a long development in structural graph theory.

On the other hand, it was already known, by an ingenious construction by Cai et al.[2] that there are graph properties decidable in polynomial time that are not in FPC. We have more on this in Section 4 below. For now, we note that a number of other conclusions on the limitations of the expressive power of FPC can be drawn from the result of Cai et al. Such results typically fall into one of two classes. In the first category are non-expressibility results that follow from the result of Cai et al. by means of reductions. For instance, there is no FPC sentence that defines the graphs with a Hamiltonian cycle. This is because, by a result of Dahlhaus[18], this problem is NP-complete under first-order reductions, and FPC is closed under such reductions. Hence, as long as we have some problem in NP that is not definable in FPC, it follows that Hamiltonicity is not definable. By an older result of Lovász and Gács[19] we also know that satisfiability of Boolean formulas in CNF (suitably encoded as a relational structure) is NP-complete under first-order reductions and therefore this is also not definable in FPC. Interestingly, it remains an open question whether 3-SAT is NP-complete under first-order reductions. In the second category of non-expressibility results that follow on from Cai et al. are those proved by adapting their methods. For instance, it is known that graph 3-colourability is *not* NP-complete under first-order reductions (indeed, the class of problems that reduce to it obeys a 0-1-law[20]). Still, it has been shown[21] that 3-colourability is not definable in FPC by a construction of graphs adapting that of Cai et al. Also, Atserias et al.[22] show that the problem of determining whether a system of linear equations (modulo 2) is solvable cannot be expressed in FPC, though this is a problem in P. By means of first-order reductions, they then show that this implies that a host of constraint satisfaction problems (characterised by algebraic properties) are not definable in FPC, including 3-colourability and 3-SAT. We have more information on this in Section 5.

## 3. FPC and Circuit Complexity

We want to justify the contention that FPC is a natural subclass of P which includes those problems that can be solved by efficient, *symmetric* algorithms. We make this notion of symmetry precise in the language of circuit complexity.

A language $L \subseteq \{0,1\}^*$ can be described by a family of *Boolean functions*:

$$(f_n)_{n\in\omega} : \{0,1\}^n \to \{0,1\}.$$

Each $f_n$ can be represented by a *circuit* $C_n$ which is a directed acyclic graph where we think of the vertices as gates suitably labeled by Boolean operators $\wedge, \vee, \neg$ for the internal gates and by inputs $x_1, \dots, x_n$ for the gates without incoming edges. One gate is distinguished as determining the output. If there is a polynomial $p(n)$ bounding the size of $C_n$ (i.e. the number of gates in $C_n$), then the language $L$ is said to be in the complexity class P/poly. If, in addition, the family of circuits is *uniform*, meaning that the function that takes $n$ to $C_n$ is itself computable in polynomial time, then $L$ is in P. For the definition of either of these classes, it does not make a difference if we expand the class of gates that we can use in the circuit beyond the Boolean basis to include, for instance, *threshold* or *majority* gates. The presence of such gates can make a difference for more restricted circuit complexity classes, for instance when we limit the depth of the circuit to be bounded by a constant, but not when we allow arbitrary polynomial-size circuits. Also, in the circuit characterization of P, it does not make a difference if we replace the uniformity condition with a stronger requirement. Say, we might require that the function taking $n$ to $C_n$ is computable in much weaker complexity classes such as DLogTime.

We are interested in languages that represent properties of relational structures such as graphs. For simplicity, in what follows, let us restrict attention to directed graphs, i.e. structures in a vocabulary with one binary relation. A property of such graphs that is in P can be recognised by a family $(C_n)_{n\in\omega}$ of Boolean circuits of polynomial size and uniformity, as before, where now the inputs to $C_n$ are labelled by the $n^2$ *potential edges* of an $n$-vertex graph, each taking a value of 0 or 1. Of course, given an $n$-vertex graph $G$, there are many ways that it can be mapped onto the inputs of the circuit $C_n$. So, to ensure that the family of circuits is really defining a property of graphs, we require it to be *invariant* under the choice of this mapping. That is, each input of $C_n$ carries a label of the form $(i, j)$ for $i, j \in [n]$ and we require the output to be unchanged under any permutation $\pi \in S_n$ acting on the inputs by the action $(i, j) \mapsto (\pi(i), \pi(j))$. It is clear that any property of graphs that is invariant under isomorphisms of graphs and is in P is decided by such a family of circuits.

It turns out that the properties of graphs that are definable in logics such as FPC are decided by circuits with a stronger invariance condition. Say that a circuit $C_n$ is *symmetric* if any permutation $\pi \in S_n$ can be extended to an *automorphism* of $C_n$ which takes each input $(i, j)$ to $(\pi(i), \pi(j))$. It is clear that symmetric circuits are necessarily invariant and it is not difficult to come up with examples that show that the converse is not true.

It is also straightforward to show we can translate a formula of first-order logic, say, in the language of graphs into a family of symmetric circuits. Given a first-order sentence $\varphi$ and a positive integer $n$, we define the circuit $C_n^{\varphi}$ by taking as gates the set of pairs $(\psi, \mathbf{a})$ where $\psi$ is a subformula of $\varphi$ and $\mathbf{a}$ is a tuple of values from $[n]$, one for each free variable occurring in $\psi$. The input gates correspond to the atomic formulas and the output gate is the one labelled by $\varphi$. If the outermost connective in $\psi$ is a Boolean operation, the gate $(\psi, \mathbf{a})$ is labelled by that operator and it is connected to the gates corresponding to subformulas in the obvious way: if $\psi$ is $\exists x\theta$, then $(\psi, \mathbf{a})$ would be an OR gate with inputs from $(\theta, \mathbf{a}a)$ for all values $a \in [n]$ and similarly a universal formula attaches to a large AND gate. The circuits constructed are symmetric as for any $\pi \in S_n$, the map that takes $(\psi, \mathbf{a})$ to $(\psi, \pi(\mathbf{a}))$ is an automorphism of the circuit. Moreover, the *depth* of the circuit is a function of $\varphi$, so a fixed formula yields a family of circuits of constant depth. On the other hand, we can take a formula $\varphi$ of fixed-point logic, FP, and for each $n$, obtain a first-order formula $\theta_n$ with $k$ (independent of $n$) variables and quantifier depth bounded by a polynomial in $n$ so that $\varphi$ and $\theta_n$ are equivalent on structures with at most $n$ elements. Converting this into a circuit by the translation above yields a polynomial-size family of symmetric Boolean circuits (the size is bounded by $c \cdot n^k$ where $c$ is the number of sub-formulas of $\varphi$) that is equivelent to $\varphi$. If we start with a sentence $\varphi$ of FPC instead, we can use the translation to $C^k$ (see Theorem 2.1) to obtain a family of symmetric circuits with threshold (or majority) gates. We need these additional kinds of gates to translate the counting quantifiers that appear in the $C^k$ formula and Boolean gates will no longer suffice. Symmetric circuits have been studied for the repesentation of properties defined in logic under different names in the literature[23,24].

Anderson and Dawar[25] establish that the correspondence between decidability by a family of symmetric circuits and definability in logic is tight by giving translations in the other direction. In particular, the following is shown.

**Theorem 3.1.** *A class of graphs is accepted by a polynomially uniform family of symmetric threshold circuits* if, and only if, *it is definable in* FPC.

Theorem 3.1 gives a natural and purely circuit-based characterization of FPC definability, justifying the claim that this is a natural and robust class. It also shows that inexpressibility results for FPC can be understood as lower bound results against a natural class of circuits. The holy grail of circuit complexity is to prove lower bounds against the class of polynomial-size circuit families. So far, lower bounds have been proved by imposing various restrictions on the families, such as monotonicity (where a famous result of Razborov[26] showed that no polynomial size family of *monotone* circuits can decide the clique problem) or bounded depth. We can now add symmetric circuits to the catalogue of circuit classes against which we are able to prove lower bounds.

Indeed, we can say more. Say that a Boolean function $f : \{0,1\}^n \to \{0,1\}$ is *symmetric* if it is invariant under all permutations of its input. This is to say that the value of $f$ depends solely on the number of its inputs that are 1 and 0. Note that in particular all the standard Boolean functions, as well as threshold or majority functions are symmetric in this sense. It has been observed[27] that we can strengthen Theorem 3.1 by allowing arbitrary symmetric functions as gates in our circuits. In other words, FPC can be characterized as the collection of all properties decided by uniform families of *symmetric* circuits using only *symmetric* functions as gates (note the two distinct meanings of the word "symmetric" in this formulation). This is the basis for the claim that FPC represents a natural notion of symmetric computation inside P. In short, we can say that FPC consists of those problems that are solvable in polynomial time by a symmetric algorithm.

## 4. Weisfeiler-Leman Equivalences

In Section 2 above, we introduced the equivalence relation $\equiv^k$ (for each $k \in \mathbb{N}$) on finite structures, denoting indistinguishability in the logic $C^k$. This family of equivalence relations has a large number of different characterizations arising in logic, combinatorics and algebra among others. They are commonly known as the Weisfeiler-Leman equivalences and we begin this section with a quick review of the various characterizations.

The origins of the Weisfeiler-Leman equivalences lie in attempts at finding an efficient test for graph isomorphism. The problem of testing whether two graphs $G$ and $H$ are isomorphic is algorithmically equivalent (in the sense that there are easy reductions in either direction) to the problem of determining for two vertices $u$ and $v$ in a graph $G$ whether there is an automorphism of $G$ that takes $u$ to $v$. A very easy to implement but incomplete test, classifies the vertices of $G$ by the number of neighbours they have, and then iteratively refines the classification by counting the number of neighbour in each class. This iteration converges to the coarsest partition of the vertices of $G$ such that for any pair $u, v$ of vertices in the same part $P$, and for any part $Q$ of the partition, $u$ and $v$ have the same number of neighbours in $Q$. This is a rough approximation of the partition of $G$ into the orbits of its automorphism group. It is easy to come up with examples of graphs on which the partition is strictly coarser than the orbit partition but nonetheless, this naïve vertex partition algorithm works on *almost all* graphs in a precise sense[28].

Weisfeiler and Leman[29] proposed a finer test, based on classification of *pairs* of vertices rather than individual vertices in a graph. The test is defined in terms of algebras of complex matrices that they call *cellular algebras*. Babai later generalized this test to the classification of $k$-tuples of vertices, gave it a purely combinatorial characterization (stripping out the complex algebra) and gave his test the name *k-dimensional Weisfeiler-Leman test*. In this classification, the naïve vertex classification algorithm is the 1-dimensional test and the original test of Weisfeiler and Leman the 2-dimensional test. The $k$-dimensional test yields, on a graph $G$ the coarsest partition of $k$-tuples of vertices such that if two tuples $\mathbf{u}, \mathbf{v}$ are in the same class $P$, then for each $k$-tuple $Q_1, \ldots, Q_k$ of classes, the sets $\{u \mid \mathbf{u}[u/i] \in Q_i\}$ and $\{v \mid \mathbf{v}[v/i] \in Q_i\}$ have the same number of elements. Here $\mathbf{u}[u/i]$ denotes the $k$-tuple obtained by replacing the $i$th element of $\mathbf{u}$ with $u$. Cai et al.[2] give a brief history of the Weisfeiler-Leman equivalences.

The connection with logic comes from the work of Immerman and Lander[13] who show, essentially, that the $k$-dimensional Weisfeiler-Leman equivalence is the same as indistingushability in the logic $C^{k+1}$. In this paper, we write $\equiv^k$ to denote the $k$-dimensional equivalence and it does not matter for our purposes whether we define it in combinatorial or logical terms as the additive constant 1 is never significant. Also, we abuse notation and treat $\equiv^k$ both as an equivalence relation on tuples of elements and an equivalence relation on graphs. That is, we write $G \equiv^k H$ to denote that $G$ and $H$ are not distinguished by the corresponding isomorphism test, and we also extend this to finite structures over an arbitrary relational vocabulary $\sigma$. Another recent characterization of this family of equivalences that is worth mentioning is as the graded isomorphisms in a co-Kleisli category defined by the pebbling comonad[30].

Since one of the reason for examining these equivalence relations is as efficient tests for isomorphism, we should have a look at their complexity. A simple argument shows that $\equiv^k$ can be decided in time $n^{O(k)}$. It is also easy to see that as we let $k$ grow, the equivalences cannot get coarser. That is to say, for all $k$, $G \equiv^{k+1} H$ implies $G \equiv^k H$. Also, if $k$ is as large as $n$, the number of vertices in $G$ or $H$, then $G \equiv^k H$ if, and only if, $G$ and $H$ are isomorphic. When Babai first proposed the $k$-dimensional Weisfeiler-Leman test, he asked whether there was a constant $k$ such that $\equiv^k$ was the same as isomorphism on all finite graphs. It was this question that was answered in the negative by an ingenious construction by Cai et al.[2]. They construct a sequence of pairs of graphs $(G_k, H_k)$, one for each $k \in \mathbb{N}$ such that:

- for each $k$, $G_k \equiv^k H_k$;
- for each $k$, $G_k \not\cong H_k$, i.e. the graphs are not isomorphic; and
- $G_k$ and $H_k$ have $O(k)$ vertices.

It is a consequence that not only does a constant value of $k$ not suffice, but any value of $k$ that is $o(n)$, where $n$ is the number of vertices is not sufficient.

**Linear Programming.** In the rest of this section, we examine one other surprising characterization of this family of equivalences as it provides an interesting connection with the second half of the paper. This is in terms of linear programming relaxations of the isomorphism problem. Suppose $G$ and $H$ are graphs on $n$ vertices given by $n \times n$ adjacency matrices $A$ and $B$ respectively. Then, $G \cong H$ if, and only if, there is a permutation matrix $P$ such that $PAP^{-1} = B$ or, equivalently $PA = BP$. The value of the second formulation is that, as the entries of $P$ are unknown, we can take them to be variables $x_{ij}$ and $PA = BP$ is a system of *linear* equations in the $n^2$ unknowns. If we add the equations $\sum_i x_{ij} = 1$ (for each $j$) and $\sum_j x_{ij} = 1$ (for each $i$), we obtain a system of equations that has a 0-1 solution (i.e. a solution in which all variables take the value 0 or 1) if, and only if, $G$ and $H$ are isomorphic. Of course, determining whether a system of equations has a solution in $\{0, 1\}$ is an NP-complete problem, so this reduction is not an advance. However, we can ask for non-integer solutions, putting in the additional requirement that $x_{ij} \geq 0$ for all $i, j$. Say that $G$ and $H$ are *fractionally isomorphic*, written $G \cong^f H$, if the resulting system of inequalities has any real solution. It turns out that $G$ and $H$ are fractionally isomorphic precisely if they are not distinguished by the 1-dimensional Weisfeiler-Leman test, i.e. the naïve vertex classification method[31].

This is, of course, an illustration of a standard method in the area of approximation algorithms. We can express a problem as an integer programming problem (i.e. a linear program for which we seek integer solutions) and consider the relaxed problem of finding any real solutions. Since linear programming is solvable in polynomial time, this gives us an efficient way of finding an approixmation to the solution of the original problem. Work in approximation algorithms also provides us with many systematic techniques for tightening the relaxation.

To make this precise, consider a linear program, i.e. a set $C$ of constraints over a set $V$ of variables. Each constraint $c$ consists of a vector $\mathbf{a}_c \in \mathbb{Q}^V$ and a rational $b_c \in \mathbb{Q}$. The *feasibility problem* is to determine for such an instance if there exists a vector $\mathbf{x} \in \mathbb{Q}^V$ such that $\mathbf{a}_c^T\mathbf{x} \leq b_c$ for all $c \in C$. Together, the collection of constraints define a convex region in $n$-dimensional Euclidean space (where $n$ is the size of $V$) and we aim to determine if the region is non-empty. In the integer programming problem, we are interested in knowing whether there is an integer point in this region. The aim of *lift-and-project* tightenings of the linear program is to introduce new variables and add constraints systematically to define a region in a higher-dimensional space whose projection onto the space defined by the original variables reduces the feasible region, without eliminating any integer points. One such widely-studied hierarchy of lift-and-project tightenings is defined by Sherali and Adams[32].

For each positive integer $k$, we define the $k$th Sherali-Adams lift of the set of linear constraints $C$ by taking, for each set $I$ of at most $k$ variables, and each $J \subseteq I$, the product of the constraint $\mathbf{a}^T\mathbf{x} \leq b$ with $\prod_{i \in I \setminus J} x_i \prod_{j \in J}(1 - x_j)$. We then *linearize* the constraints by replacing all occurrences of $x^2$ (for any variable $x$) by $x$ and replacing all occurrences of a monomial $\prod_{j \in I} x_j$ with a new variable $y_I$ for each $I$. We then add the constraints $y_\emptyset = 1$, $y_{\{x\}} = x$ and $y_I \leq y_{I'}$ whenever $I' \subseteq I$. Two things are immediate from this definition. First, the lift-and-project linear program is satisfied by all 0-1 solutions to the original program. This is because 0 and 1 satisfiy the equation $x^2 = x$, and if all variables take values either 0 or 1 then so does $\prod_{i \in I \setminus J} x_i \prod_{j \in J}(1 - x_j)$. Secondly, if the size of the original program is $n$, then the $k$th lift has size $n^{O(k)}$. A third point, which may not be as obvious, is that if we let $k$ be as large as $n$, then the projection of the resulting linear program really defines the convex hull of the integer solutions to the original program. Thus, solving the $n$th lift gives us an exact answer to our original progam but this lift is, in general, of size exponential in $n$.

Let $G \cong^{f,k} H$ denote that the $k$th Sherali-Adams lift of the graph isomorphism linear program on the graphs $G$ and $H$ admits a real solution. We call this equivalence relation the $k$th Sherali-Adams approximation of graph isomorphism. It turns out that this family of approximations is closely linked to the family of Weisfeiler-Leman equivalences. Essentially, we can establish the following implications: (i) $G \equiv^k H$ implies $G \cong^{f,k} H$ and (ii) $G \cong^{f,k} H$ implies $G \equiv^{k-1} H$[33,34]. We also know that the reverse implications are false[35] for any $k > 2$ and so the two families of equivalences strictly interleave.

Thus, as a family of approximations to the graph isomorphism relation, the Sherali-Adams lifts are essentially the same as the Weisfeiler-Leman relations. It turns out that the equivalence relations $\equiv^k$ can, in fact, tell us much more about the power of lift-and-project hierarchies, beyond just graph isomorphism and this is the topic we turn to next.

## 5. Counting Width and Constraint Satisfaction

We noted in Section 2 that Cai et al.[2] construct a class of graphs that is decidable in polynomial time but not definable in FPC. This is established by constructing a sequence of pairs of non-isomorphic graphs $G_k, H_k$ which are not distinguished by $\equiv^k$. Moreover, $G_k$ and $H_k$ can be constructed to have $O(k)$ vertices. One consequence of this is the following: if $k(n)$ is the least value such that any pair of non-isomorphic graphs on $n$ vertices is distinguished by $\equiv^{k(n)}$, then, the construction of Cai et al. shows that $k(n) = \Omega(n)$. This motivates the following definition, which provides a very useful measure of the complexity *in terms of symmetry* of a class of structures.

**Definition 5.1.** For any isomorphism-closed class of finite structures $C$, let $C_n$ denote the collection of structures in $C$ with at most $n$ elements. We write $\nu_C : \mathbb{N} \to \mathbb{N}$ for the function such that $\nu_C(n)$ is the least $k$ for which $C_n$ is closed under $\equiv^k$. We call $\nu_C$ the *counting width* of $C$.

It is clear that $\nu_C = O(n)$ for any class $C$. It is also a consequence of Theorem 2.1 that if $C$ is definable in FPC , then $\nu_C = O(1)$.

Atserias et al.[22] use a construction based on that of Cai et al. to show that the class of satisfiable 3-XOR formulas is not definable in FPC. A 3-XOR formula over a set $X$ of Booelan variables is a conjunction of clauses, each of which is the XOR of three literals. To think of such a Boolean formula as a relational structure, we consider structures whose universe is $X$ and which have two ternary relations $R_0$ and $R_1$: $R_0$ contains exactly those triples $(x_1, x_2, x_3)$ for which there is a clause in the formula with exactly these three variables, and an even number of them negated, while $R_1$ contains those triples $(x_1, x_2, x_3)$ for which there is a clause in the formula with exactly these three variables, and an odd number of them negated. We write 3XOR-SAT for the class of such structures that represent satisfiable formulas. The following result was derived[7] from the construction of Atserias et al.

**Theorem 5.1.** $\nu_{\text{3XOR-SAT}}(n) = \Omega(n)$.

This then enables us to prove lower bounds on the counting width of a variety of other decision problems, particularly through the means of reductions definable in FPC. Formally, this is done by means of FPC interpretations.

**FPC Interpretations.** Consider two signatures $\sigma$ and $\tau$. An *m-ary* FPC-*interpretation of* $\tau$ *in* $\sigma$ is a sequence of formulas of FPC in vocabulary $\sigma$ consisting of: (i) a formula $\delta(\mathbf{x})$; (ii) a formula $\varepsilon(\mathbf{x}, \mathbf{y})$; (iii) for each relation symbol $R \in \tau$ of arity $k$, a formula $\varphi_R(\mathbf{x}_1, \dots, \mathbf{x}_k)$; and (iv) for each constant symbol $c \in \tau$, a formula $\gamma_c(\mathbf{x})$, where each $\mathbf{x}$, $\mathbf{y}$ or $\mathbf{x}_i$ is an $m$-tuple of variables. We call $m$ the *width* of the interpretation. If $m = 1$, we say that the interpretaion is *linear*. We say that an interpretation $\Theta$ associates a $\tau$-structure $\mathbb{B}$ to a $\sigma$-structure $\mathbb{A}$ if there is a surjective map $h$ from the $m$-tuples $\{\mathbf{a} \in A^m \mid \mathbb{A} \models \delta[\mathbf{a}]\}$ to $\mathbb{B}$ such that:

- $h(\mathbf{a}_1) = h(\mathbf{a}_2)$ if, and only if, $\mathbb{A} \models \varepsilon[\mathbf{a}_1, \mathbf{a}_2]$;
- $R^{\mathbb{B}}(h(\mathbf{a}_1), \dots, h(\mathbf{a}_k))$ if, and only if, $\mathbb{A} \models \varphi_R[\mathbf{a}_1, \dots, \mathbf{a}_k]$;
- $h(a) = c^{\mathbb{B}}$ if, and only if, $\mathbb{A} \models \gamma_c[\mathbf{a}]$.

Note that an interpretation $\Theta$ associates a $\tau$-structure with $\mathbb{A}$ only if $\varepsilon$ defines an equivalence relation on $A^m$ that is a congruence with respect to the relations defined by the formulas $\varphi_R$ and $\gamma_c$. In such cases, however, $\mathbb{B}$ is uniquely defined up to isomorphism and we write $\Theta(\mathbb{A}) = \mathbb{B}$. It is also worth noting that the size of $\mathbb{B}$ is at most $n^m$, if $\mathbb{A}$ is of size $m$. But, it may in fact be smaller. We call an interpretation $d$-bounded, if $|\mathbb{B}| \leq |\mathbb{A}|^d$, and say the interpretation is *linearly bounded* if $d = 1$. Every linear interpretation is linearly bounded, but the converse is not necessarily the case.

The notion of interpretations is used to define logical reductions. Let $C$ and $\mathcal{D}$ be classes of $\sigma$- and $\tau$-structures respectively. We say that $C$ is FPC-*reducible* to $\mathcal{D}$ if there is an FPC-interpretation $\Theta$ of $\tau$ in $\sigma$, such that $\Theta(\mathbb{A}) \in \mathcal{D}$ if, and only if, $\mathbb{A} \in C$, and we write $C \leq_{\text{FPC}} \mathcal{D}$.

We can compose FPC formulas with interpretations, and so if $C \leq_{\text{FPC}} \mathcal{D}$ and $\mathcal{D}$ is definable in FPC, then so is $C$. Moreover, this also allows us to bound the counting width of $C$ in terms of the counting width of $\mathcal{D}$[7].

**Theorem 5.2.** *If $C \leq_{\text{FPC}} \mathcal{D}$ by a d-bounded interpretation of width m, then $\nu_C \leq (m\nu_{\mathcal{D}})^d$.*

An immediate consequence of this is the following corollary which enables us to establish linear lower bounds on classes of structures, using Theorem 5.1 and suitable linearly bounded interpretations.

**Corollary 5.1.** *If $C \leq_{\text{FPC}} \mathcal{D}$ by a linearly bounded interpretation, then $\nu_{\mathcal{D}} = \Omega(\nu_C)$.*

In particular, this implies a linear lower bound on the counting width of 3-SAT and the class of 3-colourable graphs as it turns out that we can construct linearly bounded reductions from 3XOR-SAT to these classes. Indeed, these are special cases of a rather more sweeping result about constraint satisfaction problems, which we review next.

**Constraint Satisfaction Problems.** Constraint satisfaction problems (CSP) are usually defined as decision problems where we are given a collection $V$ of variables, a domain $D$ of values and *constraints*. A constraint is a pair $(\mathbf{x}, R)$, where $\mathbf{x} \in V^k$ is a tuple of variables and $R \subseteq D^k$ a relation on $D$. An assignment $h : V \to D$ satisfies the constraint $(\mathbf{x}, R)$ if $h(\mathbf{x}) \in R$. The problem is to decide if all the constraints can be simultaneously satisfied. If we fix the domain $D$ in advance as well as the set $\Gamma$ of relations on $D$ that can appear in the constraints, we can see $\mathcal{D} = (D, \Gamma)$ as a finite relational structure. The instance is then a structure $\mathbb{A}$ in the same vocabulary and we define CSP($\mathcal{D}$) to be the class of structures $\mathbb{A}$ such that there is a homomorphism from $\mathbb{A}$ to $\mathcal{D}$. This view of CSP as essentially homomorphism problems is due to Feder and Vardi[36].

**Example 5.1.** Let $D = \{0, 1, 2\}$, and $\Gamma = \{\neq\}$, that is, $\Gamma$ contains the inequality relation over $D$. We obtain the 3-colourability problem as $\mathrm{CSP}(D, \Gamma)$.

By results of Atserias et al.[22] and Barto and Kozik[37] we can get the following classification of CSP in terms of their definability in FPC.

**Theorem 5.3.** *For any finite structure $\mathcal{D}$, either* $\mathrm{CSP}(\mathcal{D})$ *is definable in* FPC*, or* 3XOR-SAT $\leq_{\mathrm{FPC}}$ $\mathrm{CSP}(\mathcal{D})$.

Indeed, both the definability and the reductions can be formulated in logics weaker than FPC. For our purpose here, it is significant that this has been improved[7] to show that the reductions in this theorem can be made linearly bounded. As a consequence, we get the following corollary.

**Corollary 5.2.** *For any finite structure $\mathcal{D}$, either $\nu_{\mathrm{CSP}(\mathcal{D})}(n) = O(1)$ or $\nu_{\mathrm{CSP}(\mathcal{D})}(n) = \Omega(n)$.*

This establishes a sharp dichotomy on the counting width of constraint satisfaction problems. This should be compared with the complexity-theoretic dichotomy for such problems, originally conjectured by Feder and Vardi[36] and recently proved by Bulatov[38] and Zhuk[39]. This states that every $\mathrm{CSP}(\mathcal{D})$ is either in P or NP-complete. This dichotomy does not line up exactly with the one in Corollary 5.2. While it is the case that if $\mathrm{CSP}(\mathcal{D})$ has bounded width, it is in P, there are also examples, such as 3XOR-SAT of constraint satisfaction problems that are in P but have unbounded width. This means that they are not solvable by algorithms that are simultaneously polynomial-time *and* symmetric.

## 6. The Ellipsoid Method in FPC

Linear programming is a widely used approach to solving combinatorial optimization problems. It provides a powerful framework within which optimization problems can be represented, as well as efficient methods for solving the resulting programs. In particular, it is known since the work of Khachiyan[5] that there are polynomial-time algorithms that solve linear programs.

**Linear Programming.** Recall that a linear program consists of a set $C$ of constraints over a set $V$ of variables. Each constraint $c$ consists of a vector $\mathbf{a}_c \in \mathbb{Q}^V$ and a rational $b_c \in \mathbb{Q}$. The *feasibility problem* is to determine for such a program if there exists a vector $\mathbf{x} \in \mathbb{Q}^V$ such that $\mathbf{a}_c^T x \leq b_c$ for all $c \in C$. An instance of the *optimization problem* is obtained if we have, in addition, an objective function represented by a vector $\mathbf{f} \in \mathbb{Q}^V$. The aim is then to find an $\mathbf{x} \in \mathbb{Q}^V$ which satisfies all constraints in $C$ and maximizes the value of $\mathbf{f}^T\mathbf{x}$.

We want to consider instances of linear programming as relational structures that may act as interpretations for formulas of FPC. For this, we consider structures whose universe consists of three disjoint sets: $V$, $C$ and $B$. The last of these is equipped with a linear order and may be thought of as $\{0, \ldots, t\}$, i.e. an initial segment of the natural numbers where $t$ is large enough that all the rational numbers required to represent our instance can be written down using at most $t$ bits for both numerator and denominator. We can then encode the linear program through suitable relations for the numerators, denominators and signs of the values involved. For instance, we have a ternary relation $N$ so that for $c \in C$, $v \in V$ and $b \in B$, $N(c, v, b)$ holds if the $b$th bit in the numerator of $\mathbf{a}_c(v)$ is 1. It should be stressed that while the set $B$ is ordered, there is no order on the sets $V$ and $C$. If there were, then feasibility of such linear programming instances would be definable in FPC simply by the fact that all polynomial-time decidable properties of ordered structures are definable.

**Ellipsoid Method.** It has been shown[40] that the feasibility problem and, indeed, the optimization problem, for linear programming, is definable in FPC. The proof proceeds by means of expressing in the logic a version of Khachiyan's ellipsoid method[5]. In short, the ellipsoid method for determining the feasibility of a linear program proceeds by choosing a vector $\mathbf{x} \in \mathbb{Q}^V$ (one may as well begin from the vector of all zeroes) and calculating, based on the bit complexity of the instance, an ellipsoid around $\mathbf{x}$ which is guaranteed to include the polytope defined by the constraints $C$. Now, if $\mathbf{x}$ itself does not satisfy all the constraints in $C$, we can choose one $c \in C$ for which $\mathbf{a}_c^T\mathbf{x} > b_c$ and use it to calculate a new vector $\mathbf{x}'$ and an ellipsoid centred on $\mathbf{x}'$ which still includes the polytope defined by $C$. The construction guarantees that the volume of the new ellipsoid is at most half that of the original one. This means that in a number of steps that is bounded by a polynomial in the size of the instance, the process converges to a centre $\mathbf{x}$ that satisfies all the constraints in $C$ *or* the volume of the ellipsoid is small enough that we know for certain that the polytope is empty.

It is not difficult to show that all the calculations involved in computing the ellipsoid and centres can be expressed in FPC (see Holm[41] for details on how linear algebra using matrices without an order on the rows and columns can be expressed). The one step in the algorithm outlined above that causes difficulty is the *choice* of a violated constraint, which is used to define a new centre. There is no way, in any logical formalism that respects automorphisms of the structure on which it is interpreted (as any reasonable logic must) to choose an arbitrary element. However, the ellipsoid method is quite robust and it still works as long as, at each stage, we can construct some hyperplane that separates our current centre $\mathbf{x}$ from all points in the polytope defined by $C$. And, it is known[40] that such a construction can be done canonically by taking the set of all constraints in $C$ that are violated by $\mathbf{x}$ and taking their sum as our separating hyperplane. We thus get that feasibility of linear programs (and also the linear programming optimization problem) are definable in FPC.

So far, we have considered linear programs represented *explicitly*. That is, all the constraints are written out as part of the input structure. Relying on the robustness of the ellipsoid method, we can take this method further. In many applications of linear programming, we are not given an explicit constraint matrix, but some succinct description from which explicit constraints can be derived. In particular, the number of constraints may be exponential in the size of the input. The ellipsoid method can be used in such cases as long as we have a means of determining, for any vector $\mathbf{x}$, whether it is in the polytope $P$ described by the constraint matrix and, if it is not, a hyperplane that separates $\mathbf{x}$ from the polytope. This is known as a *separation oracle* for $P$. It can be shown[40] that, as long as a separation oracle for a polytope $P$ is itself definable in FPC, then the corresponding linear programming optimization problem can also be defined in FPC.

This reduction of optimization to separation reveals an interesting relationship of linear programming with the equivalence relations $\equiv^k$. We are given a polytope $P \subseteq \mathbb{Q}^V$ in the form of some relational structure $\mathbb{A}$ and an FPC interpretation $\varphi$ which acts as a separation oracle. That is, given an $\mathbf{x} \in \mathbb{Q}^V$ (and we assume that $V$ is part of the universe of $\mathbb{A}$, though the constraint set $C$ is not and may, in general, be exponentially larger than $\mathbb{A}$), $\varphi$ interpreted in a suitable expansion of $\mathbb{A}$ with $\mathbf{x}$ either determines that $\mathbf{x}$ is in the polytope $P$ or defines a hyperplane separating $\mathbf{x}$ from $P$. Then, there is some $k$ (the exact value will depend on $\varphi$) so that we can take the quotient $V' = V/_{\equiv^k}$ (where $\equiv^k$ is defined with respect to the structure $\mathbb{A}$) and project $P$ to a polytope $P' \subseteq \mathbb{Q}^{V'}$ in such a way that feasibility and optimization with respect to $P$ can be reduced to similar questions about $P'$.

**Graph Matching.** An important application[40] of the FPC definability of linear programming is to the *maximum matching problem*. Recall that a matching in a graph $G$ is a set of edges $M$ so that each vertex in $G$ is incident on at most one edge in $M$. The maximum matching problem is then to find a matching in $G$ of maximum size. Note that it would not be possible, in FPC or any other logic, to give a formula that would actually define the set $M$ that is a maximum matching. This is because $M$ would not, in general, be invariant under automorphisms of

$G$. For instance, if $G$ is $K_n$—the $n$-vertex complete graph—then it contains an exponential (in $n$) number of distinct maximum matchings all of which map to each other under automorphisms of $G$. However, it is shown[40] that the *size* of a maximum matching in $G$ can be defined in FPC. It is known, thanks to Edmonds[42] that we can associate with a graph $G$ a *matching polytope* with an exponential number of constraints (and Rothvoß[43] has shown that this is essentially optimal) so that optimizing over this polytope yields a maximum matching. Anderson et al.[40] show that a separation oracle for the matching polytope can be defined in FPC interpreted on the graph $G$ and hence deduce that the size of the maximum matching is definable. This settled a fifteen-year-old question posed by Blass et al.[44] who conjectured that the existence of a perfect matching in $G$ was not definable even in the stronger formalism of Choiceless Polynomial Time with Counting.

## 7. Finite Valued CSPs

In Section 5 above, we looked at constraint satisfaction problems and their definability. A paradigmatic such problem is 3-SAT, the problem of deciding satisfiability of a collection of Boolean clauses, each with 3 literals. Similarly, a much-studied problem in the context of constraint *optimization* is MAX-3-SAT, the problem of determining *how many* clauses in a given 3-CNF formula are *simultaneously* satisfiable. Similar generalizations of other constraint satisfaction problems lead us to the general definition of finite-valued CSP[45].

A *finite-valued constraint satisfaction problem* (VCSP) is given by a finite set $D$ and a collection $\Gamma$ of functions $f : D^k \to \mathbb{Q}_+$. An instance of the problem is a set $V$ of variables along with a set $C$ of constraints, each of which is a triple $c = (\mathbf{x}, f, w)$ with $f \in \Gamma$, $\mathbf{x} \in V^k$ where $k$ is the arity of $f$ and $w \in \mathbb{Q}_+$. The algorithmic problem is to find an assignment $h : V \to D$ of values to the variables which minimizes $\sum_{c \in C} w_c f_c(h\mathbf{x}_c)$. We should think of these as weighted constraints. In the original CSP, a constraint is a condition that must be satisfied. Now, instead, constraints are conditions that can be violated, but there is a cost to the violation. This is given by the function $f$. The instance also gives each constraint a weight $w$ and we aim to minimize the total weighted cost.

**Example 7.1.** Let MaxCut be the problem of determining the value of the maximum cut in a graph $G = (V, E)$ with edge weights $w : E \to \mathbb{Q}_+$. If we fix $D = \{0, 1\}$, and $\Gamma = \{f\}$ where $f(x, y) = 0$ if $x = y$, and $f(x, y) = 1$ if $x \neq y$, then an instance of MaxCut can be interpreted as an instance $I$ of VCSP$(D, \Gamma)$.

**Example 7.2.** Consider a finite $\tau$-structure $\mathcal{D}$ and the problem MaxCSP$(\mathcal{D})$ of determining, given an instance $\mathbb{A}$ of CSP$(\mathcal{D})$, the maximum number of constraints that can be simultaneously satisfied in $\mathbb{A}$. If we let $\Gamma$ contain for each $m$-ary relation $R \in \tau$ a function $f_R : D^m \to \{0, 1\}$ defined as $f(t) = 1$ if $t \in R$ and $f(t) = 0$ if $t \notin R$. Then VCSP$(D, \Gamma)$ is the problem MaxCSP$(\mathcal{D})$.

As usual, we can obtain a decision problem from the optimization problem by including with an instance an explicit threshold $t$. Thus, we think of VCSP$(D, \Gamma)$ as the decision problem of determining, given $(V, C, t)$ whether there is an assignment $h : V \to D$ such that $\sum_{c \in C} w_c f_c(h\mathbf{x}_c) \leq t$.

We wish to think of this as a class of finite structures and we do this as follows. We define the vocabulary $\tau_\Gamma$ as $\tau_\Gamma = \{(R_f)_{f \in \Gamma}, W, T, <\}$. An instance $I = (V, C, t)$ is then a structure $\mathbb{A}$ with a three-sorted universe: A sort for variables $V$; a sort of constraints $C$; and a bit sort $\{1, \ldots, B\}$ for some sufficiently large $B$. The relation $R_f^{\mathbb{A}} \subseteq V^k \times C$ (where $k$ is the arity of $f$) then contains a tuple $(\mathbf{s}, c)$ if $c \in C$ is a constraint of the form $(\mathbf{s}, f, w)$. Similarly, the relation $W^{\mathbb{A}} \subseteq C \times \{1, \ldots, B\}$ encodes the weight of each constraint $c = (\mathbf{s}, f, w)$, i.e. $W^I(c, \cdot) = \{k \in \{1, \ldots, B\} \mid \mathrm{Bit}(w, k) = 1\}$, and $T$ similarly codes the value of the threshold. Here, Bit$(a, k)$ is shorthand for the $k$th bit of $a$. Finally, $<$ is just interpreted as the standard linear order on $\{1, \ldots, B\}$.

This formulation allows us to speak of the definability, and also the counting width, of finite-valued CSPs. It leads to a dichotomy[7] along the lines of Corollary 5.2.

**Theorem 7.1.** *For every $D$ and $\Gamma$, either $\nu_{\mathrm{VCSP}(D,\Gamma)}(n) = O(1)$ or $\nu_{\mathrm{VCSP}(D,\Gamma)}(n) = \Omega(n)$.*

Moreover, whenever VCSP$(D, \Gamma)$ has bounded width, it is definable in FPC[46]. This is established by means of a remarkable link between this width dichotomy and linear programming relaxations. Fix $D$ and $\Gamma$ and consder an instance $(V, C)$ of VCSP$(D, \Gamma)$. We associate with this instance a linear program that we call the *basic linear programming relaxation* (BLP) of $(V, C)$. It is given as follows:

$$\begin{aligned}
&\max \sum_{c \in C} \sum_{\mathbf{x} \in D^a} \lambda_{c,\mathbf{x}} \cdot w \cdot f(\mathbf{x}) && \text{where } c = (\mathbf{s}, f, w) \text{and } a = |\mathbf{s}|, \text{ s.t.} \\
&\sum_{\mathbf{x} \in D^a; x_i = d} \lambda_{c,\mathbf{x}} = \mu_{s_i,d} && \forall c \in C, d \in D, i \in [a] \\
&\sum_{d \in D} \mu_{v,d} = 1 && \forall v \in V \\
&0 \leq \lambda_{c,\mathbf{x}} \leq 1 && \forall c \in C, \mathbf{x} \in D^a \\
&0 \leq \mu_{v,d} \leq 1 && \forall v \in V, d \in D.
\end{aligned}$$

A 0-1 valued solution to this (that is, a solution where all the variables $\lambda_{c,\mathbf{x}}$ and $\mu_{v,d}$ are either 0 or 1) corrsponds to an assignment $h : V \to D$ given by $h(v) = d$

if, and only if, $\mu_{v,d} = 1$. Thus, solving this for 0-1 solutions amounts to finding the optimum value for the VCSP instance $(V, C)$. In general, however, the BLP gives us an overestimate of the optimum value. Thapper and Živný[47] show that the BLP gives the exact optimum value exactly when VCSP$(D, \Gamma)$ is of bounded width. Moreover, in all other cases, the problem VCSP$(D, \Gamma)$ is NP-complete.

Picking up on this, we showed[46] that the BLP of an instance $(V, C)$, understood as a relational structure in the sense of Section 6 can be obtained by an FPC interpretation from $(V, C)$. Using the methods of that section for expressing the ellipsoid method on FPC, we obtain that whenever $\nu_{\mathrm{VCSP}(D,\Gamma)}(n) = O(1)$, then VCSP$(D, \Gamma)$ is, in fact, in FPC. For the cases where VCSP$(D, \Gamma)$ is of unbounded width, we know not only that it is provably not in FPC, but this gives us interesting lower bounds on approximation algorithms for the problem. This is the subject of the next section.

## 8. Lower Bounds on Semidefinite Relaxations

In Section 4, we mentioned the Sherali-Adams hierarchy of lifts of a linear program. This is only one of a number of lift-and-project hierarchies investigated in the context of combinatorial optimization as ways of strengthening a linear programming relaxation of an integer program. Among the most powerful considered are hierarchies of *semidefinite* programs. We begin with a quick review of semidefinite programming.

In general, semidefinite programming refers to a framework of constraint optimization problems where the search space is over the set of *positive semidefinite matrices*. More specifically, in a typical semidefinite program we are interested in the entries of a symmetric matrix $X \in \mathbb{Q}^{V \times V}$ that maximizes the value of an objective function $\langle C, X \rangle$ (i.e. the inner product of $C$ and $X$) subject to a set of constraints of the form $\langle A_i, X \rangle \leq b_i$ with the additional constraint that $X$ is a positive semidefinite matrix.

**Definition 8.1.** Let $V, M$ be sets, and let $V$ be non-empty. A *semidefinite program (SDP)* is given by an objective matrix $C \in \mathbb{Q}^{V \times V}$, a $\mathbb{Q}^{V \times V}$-valued vector $\mathcal{A} \in \mathbb{Q}^{M \times (V \times V)}$, and a vector $\mathbf{b} \in \mathbb{Q}^M$.

We call $\mathcal{F}_{\mathcal{A},b} := \{X \in \mathbb{Q}^{V \times V} \mid X \succeq 0, \langle A_i, X \rangle \leq b_i, A_i = \mathcal{A}(i), i \in M\}$ the set of *feasible solutions.*

Note that by the definition of inner product, the objective function $\langle C, X \rangle$ is a linear function over the variables $x_{u,v}$, $u, v \in V$. Likewise, constraints of the form $\langle A_i, X \rangle \leq b_i$ are also linear inequalities over the entries $x_{u,v}$. Hence, we can view semidefinite programs as a generalization of *linear programs*, with the additional constraint that the solution must define a positive semidefinite matrix. In fact, the semidefinite constraint $X \geq 0$ essentially imposes an *infinite* set of additional linear constraints, namely $\mathbf{a}^T X \mathbf{a} \geq 0$ for all $\mathbf{a} \in \mathbb{Q}^V$. The ellipsoid method for solving linear programs can equally well be used to optimize over semidefinite programs, subject to certain conditions. In particular we can show[7] the following:

**Lemma 8.1.** *There is an* FPC *formula that defines the optimum value for any semidefinite program given in explicit form, which is full dimensional (that is to say that there is a radius r such that a ball of radius r is contained inside the feasible region), and bounded (i.e. the feasible region is entirely contained in a ball of radius R).*

**Lasserre Hierarchy.** One of the most common applications of semidefinite programming is to give approximation algorithms for hard combinatorial problems. As we saw earlier, when the problem (such as a VCSP) is expressed as a 0-1 integer program, a generic way to find approximations is to drop the integrality condition so that a rational solution to the basic linear program can be efficiently computed. The value of the optimal rational solution serves as an upper bound to the optimal value of the integer problem and can be used as an approximation. The concept of relaxation hierarchies, such as the Sherali-Adams hierarchy we saw earlier, extends this idea further. Instead of solving the basic relaxation, these hierarchies define a sequence of linear or semidefinite programs that provide increasingly finer approximations to the original integer solution. Among the most powerful such hierarchies that has been studied is the Lasserre hierarchy (also sometimes called the sums-of-squares hierarchy) of semidefinite programs.

In the following, we write $\wp(V)$ to denote the power set of $V$ and $\wp_t(V)$ to mean the collection of subsets of $V$ with at most $t$ elements each.

**Definition 8.2.** Let $V$, $M$ be sets and $\mathcal{K} := \{\mathbf{x} \in \mathbb{Q}^V \mid A\mathbf{x} \geq \mathbf{b}\}$ a polytope given by $A \in \mathbb{Q}^{U \times V}, \mathbf{b} \in \mathbb{Q}^U$.

For a vector $\mathbf{y} \in \mathbb{Q}^{\wp(V)}$, and an integer $t$ with $1 \leq t \leq |V|$, we define the *t-th moment matrix* of $\mathbf{y}$, $M_t(\mathbf{y})$ as the $\wp_t(V) \times \wp_t(V)$-matrix with entries

$$M_t(\mathbf{y})_{I,J} := \mathbf{y}_{I \cup J}, \text{ for } |I|, |J| \leq t.$$

Similarly, the *t-th moment matrix of slacks* of $\mathbf{y}, A, \mathbf{b}$, and some $u \in U$ is given by

$$S_t^u(\mathbf{y})_{I,J} := \sum_{v \in V} A_{u,v} y_{I \cup J \cup \{v\}} - b_u y_{I \cup J}, \text{ for } |I|, |J| \leq t.$$

Finally, the *t-th level of the Lasserre hierarchy* of $\mathcal{K}$, $\mathrm{Las}_t(\mathcal{K})$ is the positive semidefinite set defined by

$$\mathrm{Las}_t(\mathcal{K}) := \{\mathbf{y} \in \mathbb{Q}^{\wp_{2t+1}(V)} \mid y_\emptyset = 1, M_t(\mathbf{y}) \succeq 0, S_t^u(\mathbf{y}) \succeq 0 \text{ for all } u \in U\}.$$

We write $\mathrm{Las}_t^\pi(\mathcal{K}) := \{y_{\{v\}}, v \in V \mid \mathbf{y} \in \mathrm{Las}_t(\mathcal{K})\}$ for the projection of $\mathrm{Las}_t(\mathcal{K})$ onto the original variables.

The general usage of the Lasserre hierarchy is as follows. Assume we have a 0–1 program where the feasible region is defined as $\mathcal{K} \cap \{0,1\}^V$. Now, instead of optimizing over the integer region, we can define $\mathrm{Las}_t(\mathcal{K})$ for some level $t$, and solve the corresponding SDP. For a fixed constant $t$, this SDP has $n^{O(t)}$ new variables, and the optimum can be obtained in polynomial time. This optimum, when projected down onto the original variables, serves as an approximation to the optimum in $\mathcal{K} \cap \{0,1\}^V$.

The following basic properties[48] of the Lasserre hierarchy establish that $\mathrm{Las}_t(\mathcal{K})$ is indeed a relaxation of $\mathcal{K} \cap \{0,1\}^V$. We write $\mathcal{K}^*$ for the polytope that is defined by the convex hull of the integer points in $\mathcal{K}$, i.e. $\mathcal{K}^* := \mathrm{conv}(\mathcal{K} \cap \{0,1\}^V)$.

**Lemma 8.2.** *Let* $\mathcal{K} = \{\mathbf{x} \in \mathbb{Q}^V \mid A\mathbf{x} \geq \mathbf{b}\}$. *Then,*

*(1)* $\mathcal{K}^* \subseteq \mathrm{Las}_t^\pi(\mathcal{K})$.
*(2)* $\mathrm{Las}_0(\mathcal{K}) \supseteq \mathrm{Las}_1(\mathcal{K}) \supseteq \ldots \supseteq \mathrm{Las}_{|V|}(\mathcal{K})$.
*(3)* $\mathrm{Las}_0^\pi(\mathcal{K}) \subseteq \mathcal{K}$, *and* $\mathcal{K}^* = \mathrm{Las}_{|V|}^\pi(\mathcal{K})$.

**Definition 8.3.** Let $I = (\mathbf{f}, \mathcal{K})$ be a 0-1 linear program with objective vector $\mathbf{f} \in \mathbb{Q}^V$ optimizing over a feasible region $\mathcal{K} \cap \{0,1\}^V$. We say that $I$ is *captured* at the $t^{\text{th}}$ level of the Lasserre hierarchy if $\mathrm{Las}_t^\pi(\mathcal{K}) = \mathcal{K}^*$. We write $l(I)$ for the minimum $t$, such that $I$ is captured at the $t$th level.

We see that at a sufficiently high level, namely at most at level $t = |V|$, any 0-1 program is captured by the $t$th level Lasserre relaxation. In those cases, the optimum of the Lasserre set yields not only an approximate optimum, but is the exact optimal value of the original 0-1 problem.

The connection with definability in FPC comes from showing that there is, for any $t$, an FPC interpretation which takes any linear program (given explicitly as a finite structure) defining the polytope $\mathcal{K}$ and maps it to a semidefinite program defining $\mathrm{Las}_t(\mathcal{K})$. Moreover, in the special case that the linear program we start with is the basic linear programming relaxation of a finite-valued CSP $(V, C)$, we can guarantee that this semidefinite program is bounded and full-dimensional. Combining the FPC interpretations gives us the key lemma[7].

**Lemma 8.3.** *For any* $(D, \Gamma)$ *and any* $t$, *there is an* FPC *formula* $\theta_t$ *which defines, in any instance* $(V, C)$ *of* VCSP$(D, \Gamma)$ *the optimal value of* $\mathrm{Las}_t(\mathcal{K})$ *where* $\mathcal{K}$ *is the* *polytope defined by* BLP$(V, C)$.

Moreover, it can be shown, by considering the width of the interpretations involved that the formula $\theta_t$ has $O(t)$ variables. Indeed, this is because the formula $\theta_{t(n)}$ can be easily adapted to give a formula with $O(t)$ variables that gives the exact value of an optimum solution to any instance of VCSP$(D, \Gamma)$. This allows us to conclude the following theorem.

**Theorem 8.1.** *If for some* $(D, \Gamma)$, $t : \mathbb{N} \to \mathbb{N}$ *is a function such that any instance* $(V, C)$ *of* VCSP$(D, \Gamma)$ *is solved exacly by considering the* $t(n)$*th Lasserre lift of the basic linear programming relaxation of* $(V, C)$, *then* $t(n) = \Omega(\nu_{\mathrm{VCSP}(D,\Gamma)}(n))$.

Combining this with Theorem 7.1, we obtain the following remarkable dichotomy result.

**Corollary 8.1.** *If for some* $(D, \Gamma)$, $t(n)$ *is the minumum value such that any instance* $(V, C)$ *of* VCSP$(D, \Gamma)$ *is solved exacly by considering the* $t(n)$*th Lasserre lift of the basic linear programming relaxation of* $(V, C)$, *then either* $t(n) = O(1)$ *or* $t(n) = \Omega(n)$.

In other words, either a constant lift suffices (indeed, in this case the basic linear programming relaxation suffices) or no sublinear level of lifts will suffice. Since linear Lasserre lifts are necessarily exponential in size, this tells us that, as far as solving hard VCSP is concerned, the Lasserre lift method cannot give us sub-exponential algorithms for solving the problem exactly. This is a strong lower bound on one of the most powerful algorithmic techniques we have. It should be noted that a similar and more general dichotomy, covering not only finite-valued CSP but more general valued CSP was obtained by Thapper and Živný[49] by very different methods.

## 9. Conclusion

The logic FPC emerged from the field of descriptive complexity and, as a logical formalism, it seems awkward and unwieldy. However, its usefulness as a tool for understanding algorithmic problems has been amply demonstrated and recent work, such as that linking it to symmetric circuits, has shown that the class of problems, i.e. properties of finite structures that are definable in FPC can be naturally understood as those that are decidable by algorithms that are simultaneously *efficient* and *symmetric*. This view prompts us to look at measures that quantify the degree of symmetry in a decision problem, much as we quantify the computational complexity of a problem. We are led to the measure that we call the *counting width* of a problem. This tells us how hard the problem is from the point of view of symmetric algorithms. There are computationally easy problems which have high counting width (such as 3XOR-SAT) and there are problems of arbitrarily high computationaly complexity of bounded counting width[21]. However, it turns out that in the context of constraint satisfaction problems, symmetry is a good explanation of tractability. While there are some decision problems like 3XOR-SAT that admit efficient algorithms that are not symmetric, when it comes to *optimization* problems, it appears that the only ones that admit efficient exact algorithms are the ones of bounded counting width.

Another way to view the results is that the classical method of Gaussian elimination, which can be used to solve systems of linear equations (and an instance of 3XOR-SAT can be understood as just such a system over the two-element field) is an algorithm that cannot be expressed symmetrically. Indeed, the unbounded counting width of 3XOR-SAT tells us this. On the other hand, we have seen that the ellipsoid method *can* be implemented in a symmetry preserving way. The gap between these two observations is very productive as it is the basis of showing the fundamental limitations of powerful approximation methods such as semidefinite programming with the Lasserre hierarchy. This is the symmetry gap referred to in the title of the present paper.

As an ending note, I would like to mention that recently Atserias and Ochremiak[50] have extended the FPC definability obtained in Lemma 8.1 by showing that the exact feasibility problem for semidefinite convex sets has bounded counting width, even when it is not necessarily solvable in polynomial time. From this, they derive the fact that semidefinite lifts of the graph isomorphism polytope presented in Section 4 cannot improve by more than a constant factor over the Sherali-Adams lifts.

## References

1. N. Immerman, Relational queries computable in polynomial time, *Information and Control* **68**, 86 (1986).
2. J.-Y. Cai, M. Fürer and N. Immerman, An optimal lower bound on the number of variables for graph identification, *Combinatorica* **12**, 389 (1992).
3. A. Dawar, The nature and power of fixed-point logic with counting, *ACM SIGLOG News* **2**, 8 (2015).
4. J. Edmonds, Paths, trees, and flowers, *Canadian Journal of Mathematics* **17**, 449 (1965).
5. L. G. Khachiyan, Polynomial algorithms in linear programming, *USSR Computational Mathematics and Mathematical Physics* **20**, 53 (1980).
6. N. Robertson and P. D. Seymour, Graph minors. XX. Wagner's conjecture, *J. Comb. Theory, Ser. B* **92**, 325 (2004).

7. A. Dawar and P. Wang, Definability of semidefinite programming and Lasserre lower bounds for CSPs, in *32nd Annual ACM/IEEE Symposium on Logic in Computer Science, LICS*, 2017.
8. A. Dawar and Y. Gurevich, Fixed-point logics, *Bulletin of Symbolic Logic* **8**, 65 (2002).
9. H.-D. Ebbinghaus and J. Flum, *Finite Model Theory*, 2nd edn. (Springer, 1999).
10. M. Y. Vardi, The complexity of relational query languages, in *Proc. of the 14th ACM Symp. on the Theory of Computing*, 1982.
11. M. Otto, *Bounded Variable Logics and Counting — A Study in Finite Models*, Lecture Notes in Logic, Vol. 9 (Springer-Verlag, 1997).
12. C. H. Papadimitriou, *Computational Complexity* (Addison-Wesley, 1994).
13. N. Immerman and E. S. Lander, Describing graphs: A first-order approach to graph canonization, in *Complexity Theory Retrospective*, ed. A. Selman (Springer-Verlag, 1990)
14. L. Hella, Logical hierarchies in PTIME, *Information and Computation* **129**, 1 (1996).
15. M. Grohe, Fixed-point logics on planar graphs, in *Proc. 13th IEEE Annual Symp. Logic in Computer Science*, 1998.
16. M. Grohe and J. Mariño, Definability and descriptive complexity on databases of bounded tree-width, in *Proc. 7th International Conference on Database Theory*, LNCS Vol. 1540 (Springer, 1999).
17. M. Grohe, *Descriptive Complexity, Canonisation, and Definable Graph Structure Theory*Lecture Notes in Logic, Lecture Notes in Logic (Cambridge University Press, 2017).
18. E. Dahlhaus, Reduction to NP–complete problems by interpretation, in *LNCS 171*, (Springer-Verlag, 1984) pp. 357–365.
19. L. Lovász and P. Gács, Some remarks on generalized spectra, *Zeitschrift für Mathematische Logik und Grundlagen der Mathematik* **23**, 27 (1977).
20. A. Dawar and E. Grädel, Properties of almost all graphs and generalized quantifiers, *Fundam. Inform.* **98**, 351 (2010).
21. A. Dawar, A restricted second order logic for finite structures, *Information and Computation* **143**, 154 (1998).
22. A. Atserias, A. Bulatov and A. Dawar, Affine systems of equations and counting infinitary logic, *Theoretical Computer Science* **410**, 1666 (2009).
23. L. Denenberg, Y. Gurevich and S. Shelah, Definability by constant-depth polynomial-size circuits, *Information and Control* **70**, 216 (1986).
24. M. Otto, The logic of explicitly presentation-invariant circuits, in *Computer Science Logic, 10th International Workshop, CSL '96, Annual Conference of the EACSL*, 1996.
25. M. Anderson and A. Dawar, On symmetric circuits and fixed-point logics, *Theory Comput. Syst.* **60**, 521 (2017).
26. A. A. Razborov, Lower bounds on the monotone complexity of some Boolean functions, *Dokl. Akad. Nauk. SSSR* **281**, 798 (1985).
27. A. Dawar and G. Wilsenach, Symmetric circuits for rank gates, (2018).
28. L. Babai, P. Erdős and S. M. Selkow, Random graph isomorphism, *SIAM Journal on Computing* **9**, 628 (1980).
29. B. Weisfeiler, On construction and identification of graphs, in *Lecture Notes in Mathematics*, 1976.

30. S. Abramsky, A. Dawar and P. Wang, The pebbling comonad in finite model theory, in *32nd Annual ACM/IEEE Symposium on Logic in Computer Science, LICS*, 2017.
31. M. V. Ramana, E. R. Scheinerman and D. Ullman, Fractional isomorphism of graphs, *Discrete Mathematics* **132**, 247 (1994).
32. H. Sherali and W. Adams, A hierarchy of relaxations between the continuous and convex hull representations for zero-one programming problems, *SIAM Journal on Discrete Mathematics* **3**, 411 (1990).
33. A. Atserias and E. N. Maneva, Sherali-Adams relaxations and indistinguishability in counting logics, *SIAM J. Comput.* **42**, 112 (2013).
34. P. N. Malkin, Sherali-Adams relaxations of graph isomorphism polytopes, *Discrete Optimization* **12**, 73 (2014).
35. M. Grohe and M. Otto, Pebble games and linear equations, *J. Symb. Log.* **80**, 797 (2015).
36. T. Feder and M. Vardi, Computational structure of monotone monadic SNP and constraint satisfaction: A study through Datalog and group theory, *SIAM Journal of Computing* **28**, 57 (1998).
37. L. Barto and M. Kozik, Constraint satisfaction problems solvable by local consistency methods, *J. ACM* **61**, 3:1 (2014).
38. A. A. Bulatov, A dichotomy theorem for nonuniform csps, in *58th IEEE Annual Symposium on Foundations of Computer Science, FOCS*, 2017.
39. D. Zhuk, A proof of CSP dichotomy conjecture, in *58th IEEE Annual Symposium on Foundations of Computer Science, FOCS*, 2017.
40. M. Anderson, A. Dawar and B. Holm, Solving linear programs without breaking abstractions, *J. ACM* **62** (2015).
41. B. Holm, Descriptive complexity of linear algebra, PhD thesis, University of Cambridge 2010.
42. J. Edmonds, Maximum matching and a polyhedron with 0, 1 vertices, *J. Research National Bureau of Standards* **69 B**, 125 (1965).
43. T. Rothvoß, The matching polytope has exponential extension complexity, in *Symp. Theory of Computing, STOC 2014*, 2014.
44. A. Blass, Y. Gurevich and S. Shelah, Choiceless polynomial time, *Annals of Pure and Applied Logic* **100**, 141 (1999).
45. D. A. Cohen, M. C. Cooper, P. Jeavons and A. A. Krokhin, The complexity of soft constraint satisfaction, *Artif. Intell.* **170**, 983 (2006).
46. A. Dawar and P. Wang, A definability dichotomy for finite valued CSPs, in *24th EACSL Annual Conference on Computer Science Logic, CSL 2015*, 2015.
47. J. Thapper and S. Živný, The complexity of finite-valued CSPs, in *Proceedings of the 45th ACM Symposium on the Theory of Computing (STOC)*, 2013.
48. T. Rothvoß, The Lasserre hierarchy in approximation algorithms, in *Lecture Notes for the MAPSP 2013 Tutorial*, 2013.
49. J. Thapper and S. Živny, The limits of SDP relaxations for general-valued CSPs, in *32nd Annual ACM/IEEE Symposium on Logic in Computer Science, LICS*, 2017.
50. A. Atserias and J. Ochremiak, Definable ellipsoid method, sums-of-squares proofs, and the isomorphism problem, in *33rd Annual ACM/IEEE Symposium on Logic in Computer Science, LICS*, 2018.

# Strategizing: A Meeting of Methods

Sujata Ghosh

*Indian Statistical Institute, Chennai, India*
*E-mail: sujata@isichennai.res.in*

This paper discusses a bridging technique of the different perspectives on modeling strategic reasoning, namely, experiments, logics and computational cognitive models. Empirical studies describe human strategic behavior. Logical studies on one hand facilitate the study of properties of such reasoning processes, on the other hand pave the way for implementation through formal languages. Computational cognitive models explore the essence of cognitive functionalities in the realm of strategic reasoning. Bridging these methodologies bring out a fresh perspective in terms of integrating different aspects of strategizing under one common standpoint.

*Keywords*: Strategic reasoning; Experiments; Logic; Computational cognitive models.

## 1. Introduction

Strategies are everywhere, not just in 'real' games like chess and bridge, but also in many social interactions in daily life: How can you create a win-win solution in *negotiations* in contexts as local as the sale of your house, or as global as an international treaty aimed at fighting global warming? How do you decide as a new company entering the market whether to give your articles a high price or a low one, reckoning with the strategy of the existing *competing* company? How do you decide to which place in New York to go and try to meet your friend, when only the meeting time and city have been *coordinated* in advance, and you cannot communicate? Unlike when playing a chess game, we usually do not know explicitly all the relevant details, leading to strategizing under partial information. Choosing between possible meeting points where a friend might show up, pricing of commodities without knowing the rival prices, or divulging fallback positions in negotiations without knowing its effects are natural examples.

### 1.1. *Modeling strategic reasoning in various fields*

Because strategies play out in so many different areas of life, the study of strategies has become an integral part of many areas of science: game theory itself, which is usually viewed as part of economics; ethics and social philosophy; the study of multi-agent systems in computer science; evolutionary game theory in biology; strategic reasoning in cognitive science; and the study of meaning in linguistics. There are various signs of interdisciplinary cooperation between these fields based on plausible viewpoints on the basic similarities between the perspectives on strategies. Let us first briefly describe some of the different perspectives on strategies.

*Game theory:* In this area of economics, strategies and their dependence on information form the main topics of study. One of the main focus of study is on strategies bringing about equilibrium play in conflicting as well as cooperating situations. Many such important concepts, such as Nash equilibrium, sub-game perfect equilibrium, sequential equilibrium and extensions and modifications thereof have been developed over the years by the game theorists while modeling various situations as games. In these games the players are considered with varied information content, e.g. perfect, imperfect, incomplete and their strategies are modeled accordingly. Such concepts have been developed by game theorists [1] and adopted by the other areas. More recently, work on psychological game theory [2] has gained momentum which considers the effect of human emotions on their strategies. This line of work often provides justifiable solutions to problems that come up with the orthodox assumptions of rationality on the players as cited by experimental studies on games. [3,4]

*Mathematics:* In mathematics, and in particular in set theory, game theory has received a warm reception. Set theorists have investigated infinite games, focusing mainly on the question of *determinacy*: does a particular game have a winning strategy for one of the players? This question has important repercussions for descriptive set theory. [5] The *axiom of determinacy* (AD) which states that 'a certain type of two player perfect information infinite games is determined, that is, one of the two players will always have a winning strategy' is inconsistent with the *axiom of choice* (AC), but certain related set of axioms which are actually consistent with AC have led to interesting results in terms of provability inside ZFC (Zermelo-Fraenkel set of axioms with AC) and more recently, in terms of the relationship to large cardinal axioms. [6]

*Philosophy:* David Hume (1740) was probably the first person to mention the role of *mutual knowledge* in *coordination* in his account of convention in *A treatise of Human Nature*. He argued that without the required amount mutual knowledge, beneficial social conventions would disappear. Subsequently, there have been a few researchers who noted the importance of the role of common knowledge in reasoning about one another in certain situations. However, Morris Friedell [7] and David Lewis, [8] in the late 1960s, were among the firsts who studied the concept of *coordination* using game-theoretic methods. For this, they formally introduced the concept of *common knowledge*, which came to be profitably used in economics. In the 1980s, Michael Bratman started the philosophical analysis of *intentions, plans, and practical reason.*[9] Flowing out of this line of work, the notions of *action* and *agency* form integral parts of the study on strategies.[10]

*Multi-agent systems:* In the begin of the 1990s, the field of multi-agent systems, investigating teamwork and other social interactions among software agents, started to flower. Part of the investigations concentrated on agents' planning and intentions, inspired by Bratman's philosophical work. The study of strategic reasoning forms another crucial ingredient of multi-agent systems. Agents, on the basis of some information, reason to devise strategies for ensuring maximal gain. Using the languages of logic and game theory, the models of strategic reasoning in multi-agent systems have led to new insights into the dynamics of observation, updating of knowledge and belief, preference change, and dialogues.[11] Researchers have also developed decidable/tractable formulations of strategies from the viewpoint of both strategy synthesis and strategy verification.[12]

*Logic:* Modeling social interaction has brought to the fore various logical systems to model agents' knowledge, beliefs, preferences, goals, intentions, common knowledge and belief.[13] When interactions are modeled as games, reasoning involves analysis of agents' long-term powers for influencing outcomes. While researching intelligent interaction, logicians have been interested in the questions of how an agent selects a particular strategy, what structure strategies may have, and how strategies are related to information. Thus, logicians have devised logical models in which strategies are 'first class citizens', rather than unspecified means to ensure outcomes.[14,15]

*Cognitive science:* In addition to idealized game-theoretic and logical studies on strategies in interactive systems, there have also been experimental studies on players' strategies and cognitive modeling of their reasoning processes. The classical game-theoretic perspective assumes that people are rational agents, maximizing their own utility by applying strategic reasoning. However, many experiments[3,4,16,17] have shown that people are not completely rational in this sense. Players may be altruists, giving more weight to the opponent's payoff; or they may try to salvage a good cooperative relation in case they meet the other agent again in future. Also, due to cognitive constraints such as working memory capacity,[18,19] people may be unable to perform optimal strategic reasoning, even if in principle they are willing to do so.Various cognitive models have been put forward to model such boundedly rational reasoning capabilities.[18]

*Linguistic semantics and pragmatics:* The concepts of strategies also play a role in language use and interpretation. For example, pragmatics can be explained in terms of a sender and receiver strategizing to understand and be understood, on the basis of concise and efficient messages.[20] Evolutionary game theory has been used to explain the evolution of language; for example, it has been shown that in signaling games, evolutionarily stable states occur when the sender's strategy is a one-one map from events to signals, and the receiver's strategy is the inverse map.[21]

### 1.2. *Logical approaches to modeling strategic reasoning*

The questions of how an agent selects a particular strategy, what structure strategies may have, and how strategies are constituted, are of utmost importance in the context of interaction situations, and as we see above, these questions arise in different forms in different subject areas. For a joint perspective on *strategizing* in related areas towards forming more realistic models of social and intelligent interaction, the need of the hour is to widen our scope of understanding to seemingly orthogonal viewpoints on strategizing, in other words, strategic reasoning. As a case in point, we describe below a meeting of game theory and logic for providing models of strategic interaction.

The study of strategic reasoning forms a crucial ingredient of the research area of intelligent interactive systems. Agents devise their strategies on how to interact so as to ensure maximal gain in the interaction process modeled as games. Their strategic reasoning is influenced by their knowledge, beliefs and intentions as well. Various logical studies of games and strategic reasoning have led to new insights into the dynamics of information,[22,23] updating of knowledge and belief,[24–28] preference change,[29–31] and processes of strategic interaction.[15,32–38] Studying *structures* in strategies, namely union, intersection, sequential composition and *response strategies* has led to describing top-down strategizing in large games played by perfectly-informed players with full as well as limited resources.[15,34,38] All these have now merged into broader studies of formal models of society, where computer science meets decision theory, game theory, and social choice theory.

### 1.3. *Cognitive science approaches to modeling strategic reasoning*

In cognitive science, the term 'strategy' is used much more broadly than in game theory. A well-known example is formed by George Polya's problem solving strategies (understanding the problem, developing a plan for a solution, carrying out the plan, and looking back to see what can be learned).[39] Nowadays, cognitive scientists construct fine-grained theories about human reasoning strategies,[40,41] based on which they construct computational cognitive models. These models can be validated by comparing the model's predicted outcomes to results from experiments with human subjects.[42] Cognitive models developed within this framework aim to explain certain aspects of cognition by assuming only general cognitive principles. Cognitive models of simple games exist in which it is important to know the opponent's behavior,[43–45] however, they do not take complex strategic reasoning into account.

The usual game-theoretic perspective assumes that people are rational agents, optimizing their gain by applying strategic reasoning. However, many experiments have shown that people are not completely rational in this sense. For example, McKelvey and Palfrey[4] have shown that in a traditional centipede game (cf. Figure 1) participants do not behave according to the Nash equilibrium reached by backward induction. In this version of the game, the payoffs are distributed in such a way that the optimal strategy is to always end the game at the first move. However, in McKelvey and Palfrey's experiment, participants stayed in the game for some rounds before ending the game: in fact, only 37 out of 662 games ended with the backward induction solution. One interpretation of this result is that the game-theoretic perspective fails to take into account the reasoning abilities of the participants. That is, perhaps, due to cognitive constraints such as working memory capacity, participants are unable to perform optimal strategic reasoning, even if in principle they are willing to do so. Thus, building up computational cognitive models for strategic reasoning, as suggested in our work,[46] provides a way to incorporate people's beliefs and constraints within the scope of their reasoning processes and also enhances our understanding of choices of strategies found in the empirical studies done by other researchers[4] as well as ourselves.[16,17]

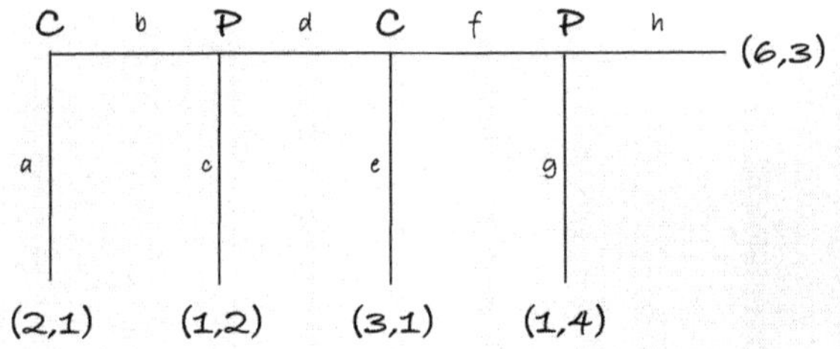


Figure 1. An example of a centipede game. For a detailed description, see [47].

In [48], a comparative study of various cognitive architectures has been provided. The quality of a cognitive model depends on its simplicity, its fit to the experimental data and predictions. These models sometimes have numerical parameters which can be tweaked, giving different outcomes. One can also specify the initial knowledge and actions of the model in a varied manner, resulting in different ways of modeling a single task. An advantage of having cognitive models, besides having statistical models, is that cognitive models can be broken down into mechanisms. Another advantage of a cognitive model is that one can compare the model's output with human data, and acquire a better understanding of individual differences. Strategic reasoning in complex interactive situations consists of multiple serial and concurrent cognitive functions, and thus it may be prone to great individual differences. Such differences become explicit in the modeling because of the reduction mechanism of the cognitive architecture of ACT-R.[42] It provides an excellent architecture to model learning and the application of cognitive skills. The theory has been validated by behavioral and neuro-scientific research.[49,50] From this computational cognitive modeling perspective, decision strategies like backward induction for turn-taking games have been used to capture second-order social reasoning.[45,51] For modeling strategic reasoning in such games, both declarative memory (to retrieve successive steps) and working memory (to store temporarily) have been used.

### 1.4. *Building bridges between empirical, logical and cognitive modeling*

In recent years, many researchers have questioned the idealization that a formal model undergoes while representing social reasoning methods (e.g. see [52,53]). Do formal methods represent human reasoning satisfactorily or should we concentrate on empirical studies and models based on those empirical data? A tension exists between the normative aspect of logic and the descriptive aspect of cognitive science.[54] A methodology for resolving this tension has been provided in our work in [46,55] which deals with human strategic reasoning in games of perfect information. Rather than thinking about formal and cognitive modeling as separate, one can consider them to be complementary and investigate how they can aid each other to bring about a more meaningful model of real-life scenarios. Game experiments will lead us to the behavioral strategies of humans having varying amounts of information. Such strategies will be modeled as logical formulas constituting a descriptive logic of strategies which will help in the construction of cognitive models. The cognitive models will predict human strategies which can be tested against the human data available from the experiments. The following is a schematic diagram of the process.

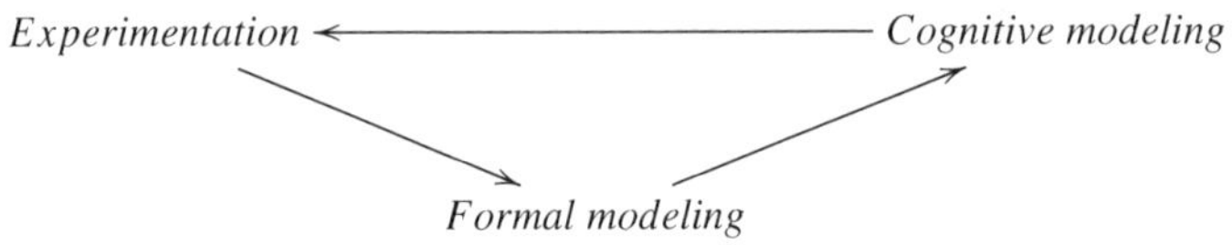


Figure 2. A schematic diagram of the bridging technique between different methodologies.

In [55], we provided an attempt to bridge the gap between experimental studies, cognitive modeling, and logical studies of strategic reasoning. In particular, a first study of a cognitive model of strategic reasoning that is constructed with the aid of a formal framework is discussed there. In [46], we extended the language that we introduced in [55] to represent strategies by a new belief component, so that we can describe reasoning about the opponent at a more fine-grained level. A new architecture PRIMs,[56] based on ACT-R, was used as the basis of the computational cognitive models. Actual implementations were made with respect to some strategy formulas and predictions were made based on the simulations about the data of the experiment reported in [16], closing the circle depicted in Figure 2.

To have a sustainable bridging technique between these methodologies, it is not enough to construct computational cognitive models corresponding to certain strategy formulas, but to come up with a translation system which, starting from a strategy represented in formal logic, automatically generates a computational model in the PRIMs cognitive architecture, which can then be run to generate decisions made in certain games, e.g. perfect informations games in our case. Such a translation system has been developed in [57,58] based on centipede-like games, a particular kind of perfect informations games. Since we wanted to make predictions on experiments like those reported in [16,17] which are based on centipede-like games, we concentrated on these games. These games are like centipede games (cf. Figure 1) in almost all respects, the only difference being that unlike centipede games, the sum of the points of players may not increase in the subsequent moves.[47]

In the remainder of the paper, based on the work reported in [46,55,57,58], we provide a systematic study of the endeavors underlying each of these *schematic arrows* depicted in Figure 2 providing pointers towards the complementary contributions of these methods in modeling human strategic behavior. To build up our study, we start with describing a particular kind of game experiment,[16,17] a strategy logic[46] and a computational cognitive model.[56]

## 2. Experiment, logic and computational cognitive model

In this section we briefly describe an experiment done in two phases, a language for describing strategic reasoning and a cognitive architecture, which are all essential ingredients for our study on the meeting of different methodologies involved in modeling strategic behavior of the humans.

### 2.1. *Experiments*

The experiments that we describe here are experiments on perfect information games reported in [16,17] to analyze human strategic behavior. Experimental studies in behavioral economics have shown that the backward induction outcome is often not reached in large centipede games (cf. Figure 1). Instead of immediately taking the 'down' option, people often show partial cooperation, moving right for several moves before eventually choosing 'down'.[4,59,60] Accordingly, in the experiments reported in [16,17], our main interest was to examine participants' behavior in centipede-like games following a deviation from backward induction behavior by their opponent right at the beginning of the game. We primarily asked the following questions:

(1) Are people inclined to use forward induction when they play such games?
(2) If not, what are they actually doing? What roles are played by risk attitudes and cooperativeness versus competitiveness?
(3) Do people take the perspective of their opponents and make use of theory of mind?
(4) Can they be reasonably divided into types of players?

The experiments were conducted at the Institute of Artificial Intelligence (ALICE) at the University of Groningen, The Netherlands. A group of 50 Bachelor's and Master's students from different disciplines at the university took part in each phase of the experiment. The participants had little or no knowledge of game theory, so as to ensure that neither backward induction nor forward induction reasoning was already known to them. The participants played the finite perfect-information games in a graphical interface on the computer screen (cf. Figure 3). In each case, the opponent was the computer, which had been programmed to play according to plans that were best responses to some plan of the participant. The participants were instructed accordingly. In each game, a marble was about to drop, and both the participant and the computer determined its path by controlling the orange and the blue trapdoors: The participant controlled the orange trapdoors, and the computer controlled the blue trapdoors. The participant's goal was that the marble should drop into the bin with as many orange marbles as possible. The computer's goal was that the marble should drop into the bin with as many blue marbles as possible.

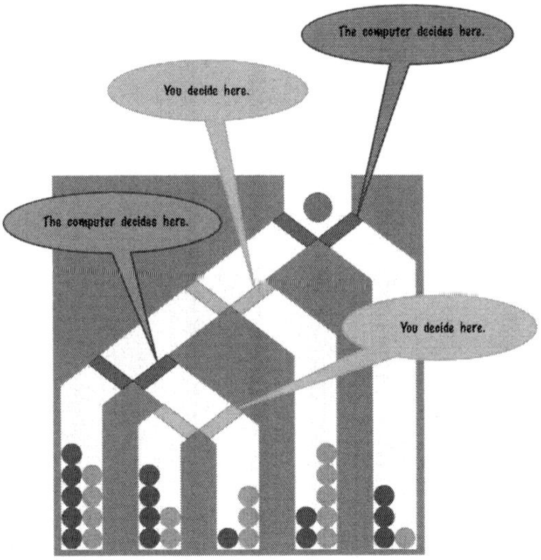


Figure 3. Graphical interface for the participants. The computer controls the blue trapdoors and acquires pay-offs in the form of blue marbles (represented as dark grey in a black and white print), while the participant controls the orange trapdoors and acquires pay-offs in the form of orange marbles (light grey in a black and white print).

In the experiment reported in [16], we investigated whether people are inclined to use forward induction in centipede-like games, rather than backward induction. We found that in the aggregate it appeared that participants showed forward induction behavior in response to deviation from backward induction behavior by their opponent, the computer, right at the beginning of the game. However, there exist alternative explanations for the choices of most participants; for example, choices could have been based on the extent of risk aversion that participants attributed to the computer in the remainder of the game, rather than to the sunk outside option that the computer has already foregone at the beginning of the game. Cardinal effects seemed to play a role as well: a number of participants might have been trying to maximize expected utility. For these reasons, the results of the experiment did not provide conclusive evidence for forward induction reasoning on the part of the participants.

So, we followed up with a similar experiment, reported in [17], where we designed centipede-like games with new payoff structures in order to make such cardinal effects less likely. We asked a number of questions to gauge the participants' reasoning about their own and the opponent's strategy at all decision nodes of a sample game. Even though in the aggregate, participants in the new experiment still tend to slightly favor the forward induction choice at their first decision node, their verbalized strategies most often depend on their own attitudes towards risk and those they assign to the computer opponent, sometimes in addition to considerations about cooperativeness and competitiveness.

The tasks that the participants had to perform in these experiments are mentioned in the table below in the order given there.

| | |
|---|---|
| Step 1 | Introduction and instructions. |
| Step 2 | Practice Phase: 14 games. |
| Step 3 | - Experimental Phase: 48 game items, divided into 8 rounds of 6 different games each, in terms of isomorphism class of pay-off structures.<br>- Each of the 6 games occurs once in each round; in [16], these games occur in the same order in each round; in [17], these games occur in a random order in each round.<br>- Question on computer's behavior in several rounds: For [16] Group A in rounds 3, 4, 7, 8; Group B in rounds 7, 8. For [17] Group A in the middle of the game at certain rounds; Group B at the end of the game in certain rounds. |
| Step 4 | Final Question(s): In [16], the question was about possible future moves of the computer; in [17], the questions were regarding decisions at all nodes of a sample game. |

## 2.2. *Logic*

With regard to the logical framework, we describe the one proposed in [46]. We restrict to the strategy specification language discussed there and present the same as that is the main ingredient of the formal framework involved in our bridge-building method. We start with describing extensive form games.

### 2.2.1. *Extensive form games*

Extensive form games are a natural model for representing *finite games* in an explicit manner. In this model, the game is represented as a finite tree where the nodes of the tree correspond to the game positions and edges correspond to moves of players. For this logical study, we will focus on game forms, and not on the games themselves, which come equipped with players' payoffs at the leaf nodes of the games. We present the formal definition below.

Let $N$ denote the set of players; we use $i$ to range over this set. For the time being, we restrict our attention to two player games, and we take $N = \{C, P\}$. We often use the notation $i$ and $\bar{\imath}$ to denote the players, where $\overline{C} = P$ and $\overline{P} = C$. Let $\Sigma$ be a finite set of action symbols representing moves of players; we let $a, b$ range over $\Sigma$.

### 2.2.2. *Game trees*

Let $\mathbb{T} = (S, \Rightarrow, s_0)$ be a tree rooted at $s_0$ on the set of vertices $S$ and let $\Rightarrow : (S \times \Sigma) \to S$ be a *partial* function specifying the edges of the tree. The tree $\mathbb{T}$ is said to be finite if $S$ is a finite set. For a node $s \in S$, let $\vec{s} = \{s' \in S \mid s \overset{a}{\Rightarrow} s'$ for some $a \in \Sigma\}$. A node $s$ is called a leaf node (or terminal node) if $\vec{s} = \emptyset$.

An *extensive form game tree* is a pair $T = (\mathbb{T}, \widehat{\lambda})$ where $\mathbb{T} = (S, \Rightarrow, s_0)$ is a tree. The set $S$ denotes the set of game positions with $s_0$ being the initial game position. The edge function $\Rightarrow$ specifies the moves enabled at a game position and the turn function $\widehat{\lambda} : S \to N$ associates each game position with a player. Technically, we need player labelling only at the non-leaf nodes. However, for the sake of uniform presentation, we do not distinguish between leaf nodes and non-leaf nodes as far as player labelling is concerned. An extensive form game tree $T = (\mathbb{T}, \widehat{\lambda})$ is said to be finite if $\mathbb{T}$ is finite. For $i \in N$, let $S^i = \{s \mid \widehat{\lambda}(s) = i\}$ and let *frontier*$(\mathbb{T})$ denote the set of all leaf nodes of $T$.

### 2.2.3. *Strategies*

A *strategy* for player $i$ is a function $\mu^i$ which specifies a move at every game position of the player, i.e. $\mu^i : S^i \to \Sigma$. A strategy $\mu^i$ can also be viewed as a subtree of $T$ where for each node belonging to player $i$, there is a unique outgoing edge and for nodes belonging to player $\bar{\imath}$, every enabled move is included.

A *partial strategy* for player $i$ is a partial function $\sigma^i$ which specifies a move at some (but not necessarily all) game positions of the player, i.e. $\sigma^i : S^i \rightharpoonup \Sigma$. As above, a partial strategy $\sigma^i$ can also be viewed as a subtree of $T$ where for some nodes belonging to player $i$, there is a unique outgoing edge and for other nodes belonging to player $i$ as well as nodes belonging to player $\bar{\imath}$, every enabled move is included.

### 2.2.4. *Syntax for extensive form game trees*

We now build a syntax for game trees. We use this syntax to parametrize the belief operators given below so as to distinguish between belief operators for players at each node of a finite extensive form game. Let *Nodes* be a finite set. The syntax for specifying finite extensive form game trees is given by:

$$\mathbb{G}(Nodes) ::= (i, x) \mid \Sigma_{a_m \in J}((i, x), a_m, t_{a_m})$$

where $i \in N$, $x \in Nodes$, $J$(finite) $\subseteq \Sigma$, and $t_{a_m} \in \mathbb{G}(Nodes)$.

Given $h \in \mathbb{G}(Nodes)$, we define the tree $T_h$ generated by $h$ inductively as follows (see Figure 4 for an example):

- $h = (i, x)$: $T_h = (S_h, \Rightarrow_h, \widehat{\lambda}_h, s_x)$ where $S_h = \{s_x\}$, $\widehat{\lambda}_h(s_x) = i$.
- $h = ((i, x), a_1, t_{a_1}) + \cdots + ((i, x), a_k, t_{a_k})$: Inductively we have trees $T_1, \ldots T_k$ where for $j : 1 \leq j \leq k$, $T_j = (S_j, \Rightarrow_j, \widehat{\lambda}_j, s_{j,0})$.

  Define $T_h = (S_h, \Rightarrow_h, \widehat{\lambda}_h, s_x)$ where

  - $S_h = \{s_x\} \cup S_{T_1} \cup \ldots \cup S_{T_k}$;
  - $\widehat{\lambda}_h(s_x) = i$ and for all $j$, for all $s \in S_{T_j}$, $\widehat{\lambda}_h(s) = \widehat{\lambda}_j(s)$;
  - $\Rightarrow_h = \bigcup_{j:1 \leq j \leq k}(\{(s_x, a_j, s_{j,0})\} \cup \Rightarrow_j)$.

Given $h \in \mathbb{G}(Nodes)$, let $Nodes(h)$ denote the set of distinct pairs $(i, x)$ that occur in the expression of $h$.

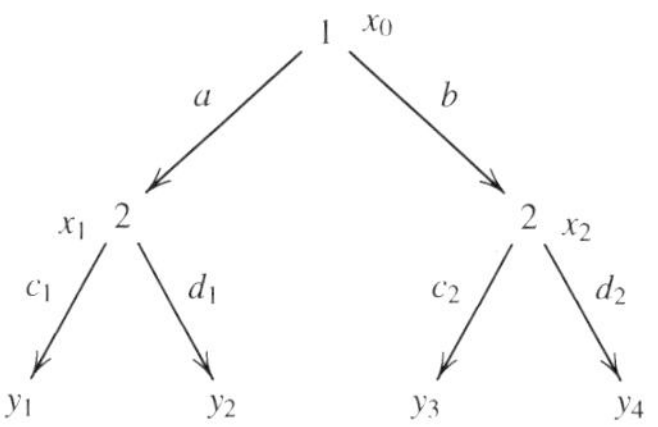


Figure 4. Extensive form game tree. The nodes are labelled with turns of players and the edges with the actions. The syntactic representation of this tree can be given by:
$h = ((1, x_0), a, t_1) + ((1, x_0), b, t_2)$, where
$t_1 = ((2, x_1), c_1, (2, y_1)) + ((2, x_1), d_1, (2, y_2))$;
$t_2 = ((2, x_2), c_2, (2, y_3)) + ((2, x_2), d_2, (2, y_4))$.

### 2.2.5. *Strategy specifications*

We are now ready to describe the strategy specification language. The main case specifies, for a player, which conditions she tests before making a move. In what follows, the pre-condition for a move depends on observables that hold at the current game position, some belief conditions, as well as some simple finite past-time conditions and some finite look-ahead that each player can perform in terms of the structure of the game tree. Both the past-time and future conditions may involve some strategies that were or could be enforced by the players. These pre-conditions are given by the syntax defined below.

For any countable set $X$, let $BPF(X)$ (the boolean, past and future combinations of the members of $X$) be sets of formulas given by the following syntax:

$$BPF(X) ::= x \in X \mid \neg\psi \mid \psi_1 \vee \psi_2 \mid \langle a^+ \rangle\psi \mid \langle a^- \rangle\psi,$$

where $a \in \Sigma$, a countable set of actions.

Formulas in $BPF(X)$ can be read as usual in a dynamic logic framework and are interpreted at game positions. The formula $\langle a^+ \rangle\psi$ (respectively, $\langle a^- \rangle\psi$) refers to one step in the future (respectively, past). It asserts the existence of an $a$ edge after (respectively, before) which $\psi$ holds. Note that future (past) time assertions up to any bounded depth can be coded by iteration of the corresponding constructs. The 'time free' fragment of $BPF(X)$ is formed by the boolean formulas over $X$. We denote this fragment by $Bool(X)$.

For each $h \in \mathbb{G}(Nodes)$ and $(i, x) \in Nodes(h)$, we now add a new operator $\mathbb{B}_h^{(i,x)}$ to the syntax of $BPF(X)$ to form the set of formulas $BPF_b(X)$. The formula $\mathbb{B}_h^{(i,x)}\psi$ can be read as 'in the game tree $T_h$, player $i$ *believes* at node $x$ that $\psi$ holds'. We reiterate our disclaimer from [46]. One might feel that it is not elegant that the belief operator is parametrized by the nodes of the tree. However, our main aim is not to propose a logic for the sake of its nice properties, but to have a logical language that can be used suitably for constructing computational cognitive models corresponding to participants' strategic reasoning.

Let $P^i = \{p_0^i, p_1^i, \ldots\}$ be a countable set of observables for $i \in N$ and $P = \bigcup_{i \in N} P^i$. To this set of observables we add two kinds of propositional variables $(u_i = q_i)$ to denote 'player $i$'s utility (or payoff) is $q_i$' and $(r \leq q)$ to denote that 'the rational number $r$ is less than or equal to the rational number $q$'. The syntax of strategy specifications is given by:

$$Strat^i(P^i) ::= [\psi \mapsto a]^i \mid \eta_1 + \eta_2 \mid \eta_1 \cdot \eta_2,$$

where $\psi \in BPF_b(P^i)$. The basic idea is to use the above constructs to specify properties of strategies as well as to combine them to describe a play of the game. For instance, the interpretation of a player $i$'s specification $[p \mapsto a]^i$ where $p \in P^i$, is to choose move $a$ at every game position belonging to player $i$ where $p$ holds.

At positions where $p$ does not hold, the strategy is allowed to choose any enabled move. The strategy specification $\eta_1 + \eta_2$ says that the strategy of player $i$ conforms to the specification $\eta_1$ or $\eta_2$. The construct $\eta_1 \cdot \eta_2$ says that the strategy conforms to specifications $\eta_1$ and $\eta_2$.

## 2.3. *Computational cognitive model*

We now provide a brief description of the cognitive architectures at the basis of our computational cognitive models. We first provide a description of ACT-R based on which the architecture of PRIMs has been developed, followed by a description of PRIMs, especially pointing out the differences from ACT-R. The description of ACT-R is based on what we provided in [55], and that of PRIMs in [46].

### 2.3.1. *ACT-R*

ACT-R, *Adaptive Control of Thought - Rational*, is an integrated theory of cognition as well as a cognitive architecture that many cognitive scientists use.[42] It consists of modules that link with cognitive functions, for example, vision, motor processing, and declarative processing. Each module maps onto a specific brain region. Furthermore, each module is associated with a buffer and the modules communicate via these buffers. Importantly, cognitive resources are bounded in ACT-R models: Each buffer can store just one piece of information at a time. Consequently, if a model has to keep track of more than one piece of information, it has to move the pieces of information back and forth between two important modules: *declarative memory* and the *problem state*. Moving information back and forth comes with a time cost, in some cases causing a cognitive bottleneck.[18]

The *declarative memory* module represents long-term memory and stores information encoded in so-called *chunks*, representing knowledge structures. For example, a chunk can be represented as a formal expression with a defined meaning. Each chunk in declarative memory has an activation value that determines the speed and success of its retrieval. Whenever a chunk is used, the activation value of that chunk increases. As the activation value increases, the probability of retrieval increases and the latency (time delay) of retrieval decreases. Therefore, a chunk representing a comparison between two payoffs will have a higher probability of retrieval, and will be retrieved faster, if the comparison has been made *recently*, or *frequently* in the past.[61] Anderson[42] provided a formalization of the mechanism that produces the relationship between the probability and speed of retrieval. If the activation value drops below a certain minimal value (the retrieval threshold), the related information is no longer accessible. In that case, the system will report a retrieval failure after a constant time factor. If the activation value is above the retrieval threshold, the information is accessible. Then, the higher the activation value, the faster the retrieval will be. As soon as a chunk is retrieved from declarative memory, it is placed into the declarative module's buffer. As mentioned earlier, each ACT-R module has a buffer that may contain one chunk at a time. On a functional level of description, the chunks that are stored in the various buffers are the knowledge structures of which the cognitive architecture is aware.

The *problem state* module also contains a buffer that can hold one chunk in which information can be temporarily stored. Typically, the problem state stores a sub-solution to the problem at hand. In the case of a social reasoning task, this may be the outcome of a reasoning step that will be relevant in subsequent reasoning. Storing information in the problem state buffer is associated with a time cost (typically 200 ms).

A central procedural system recognizes patterns in the information stored in the buffers, and responds by sending requests to the modules, for example, 'retrieve a fact from declarative memory'. This condition-action mechanism is implemented in production rules. Production rules have so-called utility values. The model receives reward or punishment depending on the correctness of its response. Both reward and punishment propagate back to previously fired production rules, and the utility values of these production rules are increased in case of reward and decreased in case of punishment by a process called *utility learning*.[42] If two or more production rules match a particular game state, the production rule with the highest utility is selected.

### 2.3.2. *PRIMs*

PRIM, the *primitive elements theory*, is a recent cognitive theory developed by Taatgen, who implemented it in the computational cognitive architecture PRIMs.[56] It builds on ACT-R, using ACT-R modules, buffers and mechanisms such as production compilation. However, in contrast to ACT-R, PRIMs is suited for modeling general reasoning strategies that are not included in the basic cognitive architecture shared by all humans, but that are at the same time more general than *ad hoc* task-specific reasoning rules. Thereby, PRIMs is especially suitable for modeling the nature and transfer of cognitive skills. Because of our need to model participants' beliefs about the opponent's beliefs, we decided to use PRIMs rather than ACT-R as cognitive architecture in [46] to model more sophisticated reasoning strategies.

More specifically, PRIM breaks down the complex production rules typically used in ACT-R models into the smallest possible elements (PRIMs) that move, compare

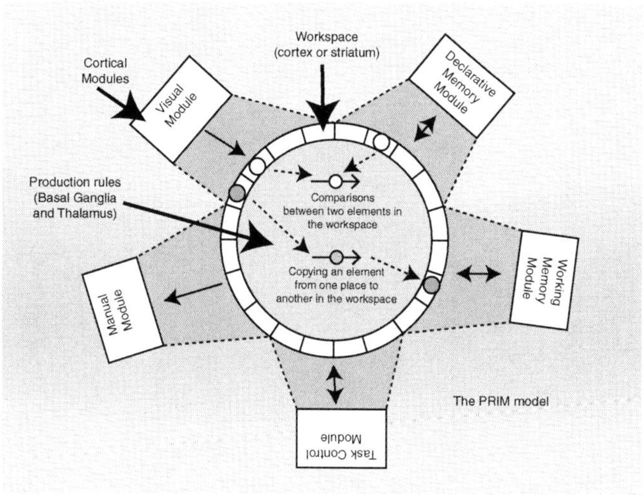


Figure 5. The PRIMs cognitive architecture, from [56]. Reprinted with permission from the author. For this work, PRIMs models of strategic reasoning are used as 'virtual participants' that play centipede-like games.

or copy information between modules (cf. Figure 5). There is a fixed number of PRIMs in the architecture. When PRIMs are used often over time, production compilation combines them to form more complex production rules. While those PRIMs may have some task-specific elements, PRIMs also have task-general elements that can be used by other tasks. Taatgen[56,62] showed the predictive power of PRIMs by modeling a variety of transfer experiments such as text editing, arithmetic, and cognitive control. The architecture has been used to model children's development of theory of mind,[63] transfer between the 'take the best' heuristic and the balance beam task,[64] and children's mistakes in arithmetic.[65] PRIMs models can also be run to predict the estimated time to complete certain tasks. Like ACT-R, PRIMs models cognitive resources as being bounded.

## 3. Details of the bridging techniques

To describe the essentials of the bridging between experiments, logic and computational cognitive models we will mainly focus on the centipede-like games presented in Figure 6, which are the Games 1, 4 and 1′ used in the experiment reported in [17].

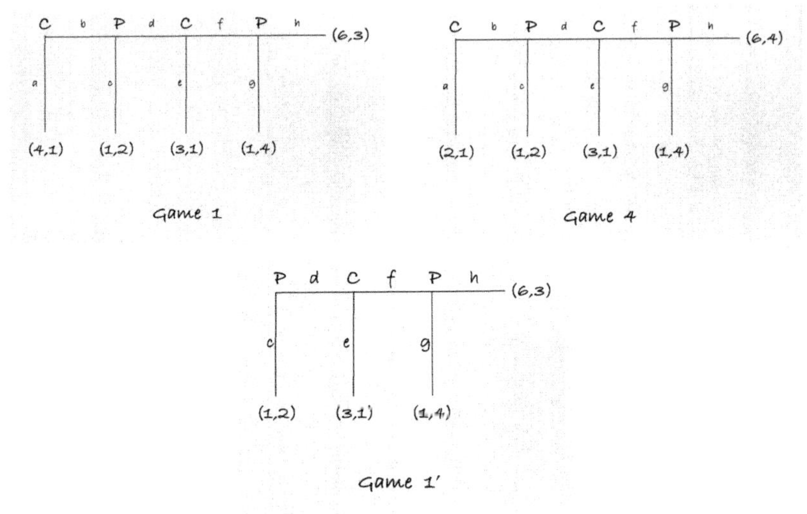


Figure 6. Games 1, 4 and 1′ from [17].

### 3.1. *Experimentation* ⟶ *Formal modeling*

Empirical studies on games can describe human strategic behavior. Logical studies on one hand facilitate the study of properties of such reasoning processes in games, on the other hand pave the way for implementation through formal languages. It is this second feature that we utililize for our bridging methodology. We will use a suitably defined formal language to express human strategic behavior as found in the empirical studies. As a case in point, we consider some of the strategies used in playing games 1, 1′ and 4 as reported in [17], and represent them in the language described in Section 2.2.

We start with fixing the preliminary notions. Let us assume that actions are part of the observables, that is, $\Sigma \subseteq P$. The semantics for the actions can be defined appropriately. Let $n_1, \dots, n_4$ denote the four decision nodes of games 1 and 4 of Figure 6, with $C$ playing at $n_1$ and $n_3$, and $P$ playing at the remaining two nodes $n_2$ and $n_4$. Since game 1′ is a subgame of game 1, the three decision nodes are denoted by $n_2$, $n_3$ and $n_4$. We have four belief operators for the games 1 and 4, namely two per player. We abbreviate some formulas that describe the payoff structure of the games:

$\alpha := \langle d\rangle\langle f\rangle\langle h\rangle((u_C = p_C) \wedge (u_P = p_P))$
(from the current node, a $d$ move followed by an $f$ move followed by an $h$ move lead to the payoff $(p_C, p_P)$ )

$\beta := \langle d\rangle\langle f\rangle\langle g\rangle((u_C = q_C) \wedge (u_P = q_P))$
(from the current node, a $d$ move followed by an $f$ move followed by a $g$ move lead to the payoff $(q_C, q_P)$ )

$\gamma := \langle d\rangle\langle e\rangle((u_C = r_C) \wedge (u_P = r_P))$
(from the current node, a $d$ move followed by an $e$ move lead to the payoff $(r_C, r_P)$ )

$\delta := \langle c\rangle((u_C = s_C) \wedge (u_P = s_P))$
(from the current node, a $c$ move leads to the payoff $(s_C, s_P)$ )

$\chi := \langle b^-\rangle\langle a\rangle((u_C = t_C) \wedge (u_P = t_P))$
(the current node can be accessed from another node by a $b$ move from where an $a$ move leads to the payoff $(t_C, t_P)$ )

Now we can define the conjunction of these descriptions to describe the payoff structures of the games in Figure 6:

$$\varphi_1 := \alpha_1 \wedge \beta_1 \wedge \gamma_1 \wedge \delta_1 \wedge \chi_1 \qquad \varphi_4 := \alpha_4 \wedge \beta_4 \wedge \gamma_4 \wedge \delta_4 \wedge \chi_4$$

$$\varphi_{1'} := \alpha_{1'} \wedge \beta_{1'} \wedge \gamma_{1'} \wedge \delta_{1'}.$$

Let $\psi_i^j$ denote the conjunction of all the order relations of the rational payoffs for player $i$ ($\in \{P, C\}$) given in Game j ($\in \{1, 4, 1'\}$) of Figure 6.

Strategy specifications describing forward induction (extensive form rationalizable (EFR))[66] reasoning of player $P$ at the node $n_2$ in games 1 and 4 are as follows:

$$\eta_P^1 : [(\varphi_1 \wedge \psi_P^1 \wedge \psi_C^1 \wedge \langle b^-\rangle\mathbf{root} \wedge \mathbb{B}_{g1}^{n_2,P}\langle d\rangle\neg e \wedge \mathbb{B}_{g1}^{n_2,P}\langle d\rangle\langle f\rangle g) \mapsto d]^P$$

$$\eta_P^4 : [(\varphi_4 \wedge \psi_P^4 \wedge \psi_C^4 \wedge \langle b^-\rangle\mathbf{root} \wedge \mathbb{B}_{g4}^{n_2,P}\langle d\rangle e \wedge \mathbb{B}_{g4}^{n_2,P}\langle d\rangle\langle f\rangle g) \mapsto c]^P.$$

Backward induction reasoning at the same node $n_2$ for these games can be formulated as follows:

$$\zeta_P^1 : [(\varphi_1 \wedge \psi_P^1 \wedge \psi_C^1 \wedge \langle b^-\rangle\mathbf{root} \wedge \mathbb{B}_{g1}^{n_2,P}\langle d\rangle e \wedge \mathbb{B}_{g1}^{n_2,P}\langle d\rangle\langle f\rangle g) \mapsto c]^P$$

$$\zeta_P^4 : [(\varphi_4 \wedge \psi_P^4 \wedge \psi_C^4 \wedge \langle b^-\rangle\mathbf{root} \wedge \mathbb{B}_{g4}^{n_2,P}\langle d\rangle f \wedge \mathbb{B}_{g4}^{n_2,P}\langle d\rangle\langle f\rangle h) \mapsto d]^P.$$

We note here that in game 1, $P$ has a unique extensive form rationalizable strategy and a unique backward induction strategy and they are not identical, whereas in game 4 all strategies can be considered as both backward induction and extensive form rationalizable strategies. Thus the formulas for game 4 provide some example cases.

We now describe some other simple strategies of players with different kinds of restrained reasoning capabilities. Note that such limited reasoning is ubiquitous in our daily life (see e.g. [45]). A *myopic* (or near-sighted) player can be considered as one who only considers her current node and the next one to compare her payoffs and act rationally depending on those payoffs without being able to look further into the game (cf. [67]). Such a player-strategy can be described for game 1′ as follows:

$$\kappa_P^{1'} : [(\delta_{1'} \wedge \gamma_{1'} \wedge (1 \leq 2) \wedge \mathbf{root}) \mapsto c]^P.$$

One can also consider players who are only capable or interested to look at their own payoffs and do not consider the opponent's payoffs at all and move wherever they get more payoff (cf. [68]). Their strategy in game 1′ can be described as follows:

$$\chi_P^{1'} : [(\alpha_{1'} \wedge \beta_{1'} \wedge \delta_{1'} \wedge \gamma_{1'} \wedge (1 \leq 2) \wedge (2 \leq 4) \wedge (2 \leq 3) \wedge \mathbf{root}) \mapsto d]^P.$$

Note that in the above set of formulas, we only consider the relevant pay-offs, e.g. $\delta$ and $\gamma$ in case of the $\kappa$ formula, and $\alpha$, $\beta$, $\delta$, and $\gamma$ for the $\chi$ formula. In fact, one could ignore the payoffs for $C$ for the $\chi$ formula. We will come back to these strategies in Section 3.3 when we validate the model predictions with the experimental results. The experimental findings [17] showed that $c$ was played about 35% in game 1, about 28% in game 4, and about 41% in game 1′ at the first decision node for the participant $P$.

### 3.2. *Formal modeling ⟶ Cognitive modeling*

Computational cognitive models provide ways to explore the essence of cognitive functionalities in the realm of strategic reasoning. The current method of constructing computational cognitive models is basically ad hoc - created by hand. For example, decision strategies like backward induction for turn-taking games have been used to capture second-order social reasoning. [45,51] Using the language described in Section 2.2, Top [57] has devised a method that automatically generates cognitive models from strategy formulas, without human intervention. In this section, we describe the main points of this approach - for details, see [58]. The models are constructed based on the PRIMs architecture.

In what follows, we provide for each component in the logical language the corresponding behavior of a PRIMs model generated using that component.

- $\langle a^+ \rangle$ and $\langle a^- \rangle$: A model translated from a formula containing these operators uses focus actions[63] to move its visual attention to the specified location. Focus actions take time to complete similar to human gazing, causing these operators to increase the model's reaction time.
- **root**: When a strategy formula contains the proposition **root**, the PRIMs model will visually inspect the specified node to determine whether it is the root of the tree.
- $\mathbf{turn}_i$: When a strategy formula contains a proposition $\mathbf{turn}_i$, where $i$ is C or P, the PRIMs model will read the player name from the specified node in the game tree, and compare it to $i$.
- $(u_i = q_i)$: The proposition $(u_i = q_i)$ states that player $i$'s payoff is equal to $q_i$ at a certain location. The PRIMs model will compare $q_i$ to a value in its visual input. Because this value may be required for future comparisons, it is also stored in an empty slot of working memory.
- $(r \leqslant q)$: A PRIMs model cannot instantly access each value in a visual display: it has to remember them by placing them in working or declarative memory before it can compare them. A proposition $(u_i = q_i)$ causes such a value to be stored in working memory. A proposition $(r \leqslant q)$ then sends two of these values from working memory to declarative memory, to try and remember which one is bigger. When a model is created, its declarative memory is filled with facts about single-digit comparisons, such as $(0 \leqslant 3)$ and $(2 \leqslant 2)$.
- $\mathbb{B}^{(i,x)}_{\mathbb{h}}$ and $a$: To describe the PRIMs model behavior, let us consider an example of a belief formula:

  $$\mathbb{B}^{(C,n1)}_{g1} \langle b^+ \rangle c.$$

  This formula can be read as 'In Game 1, at node 1, player C believes that after playing $b$, $c$ will be played'. To verify such a belief, a model employs a plan of action similar to the ones used by models in [69]. When a model is created, it contains several strategies in its declarative memory. When a model verifies a belief, it sends a *partial* sequence of actions to declarative memory, corresponding to the assumptions of the belief, in an attempt to retrieve a *full* sequence of actions, which conforms to a strategy. Using the formula above as an example, the assumptions of the belief are that $b$ is played. Therefore the model sends the sequence $b$ to declarative memory. All sequences $b$-$c$, $b$-$d$-$e$, $b$-$d$-$f$-$g$ and $b$-$d$-$f$-$h$ could be retrieved, depending on the strategies present in declarative memory. However, only $b$-$c$ verifies $B^{(C,n1)}_{g1} \langle b^+ \rangle c$. All the other sequences falsify it.

As we have mentioned in Section 3.1, strategies such as BI and EFR can have multiple solutions in one game, when there are payoff ties (cf. game 4 in Figure 6). In this case one has to exhaustively list all solutions for the specified strategy. Because of this, the translation system as reported in [58] allows for a strategy to consist of a list of multiple strategy formulas. The PRIMs model generated from this list tries to verify each formula in it, using the behavior described above, until it finds one it can verify, and play the action prescribed by the formula it verified. There is no need to specify what the model has to do when it cannot verify any of the formulas in the list - the list is exhaustive, and at least one of the formulas holds.

A pertinent question here could be as follows: How to make this exhaustive list of strategy formulas? Doing it with hand is just moving the burden from the ad hoc construction of computational cognitive models to the ad hoc listing of strategy formulas. This could be taken care of quite easily by implementing a software for generating this exhaustive list of formulas, given the alphabet of the language, formation rules and the finite game(s) under consideration.

While describing the bridging methodologies, (i) experiment to logic, and (ii) logic to cognitive modeling, we have only used the syntax of the logic as presented in Section 2.2. We neither needed the semantics of the logic, nor any properties within the purview of logic. So we could have easily restricted to some suitable formal language doing away with the logic aspect altogether. The reason that we do not do so, and logic is an integral part of this bridge-building technique is that, presenting the whole logical framework, that is, the language plus semantics provides a sanity check on, for example, the expressivity of the language. Moreover, without the semantics it would not have been possible to attach the respective roles of the different language components in the PRIMs models. Because of space restrictions we did not present the semantics of the language described in Section 2.2, it is available in [46].

### 3.3. *Cognitive modeling ⟶ Experimentation*

Towards completing the full circle in our schematic diagram of Figure 2, based on the work done in [58], we now provide a comparison of how the computational cognitive models built by hand as well as by the translation system described in Section 3.2 performed with respect to the human participants in the experiment reported in [46].

We consider four models: a handmade myopic and own-payoff model, and an automatically generated myopic and own-payoff model (cf. Section 3.1). These automatically generated myopic and own-payoff models are generated from the myopic and own-payoff strategy formulas for game 1′. These models play Game 1′ (see Figure 6) only. They play against computer opponents who play pre-specified moves. The models play as player P, whereas the computer opponent plays as player C. Each model was run 50 times, where it plays 50 games, to simulate 50 virtual participants who play 50 games each. Reaction times and decisions were recorded.

The reaction times for the four models, as well as the human participants in [46], can be found in Figure 7. It indicates that the myopic models are faster than the own-payoff models, and the generated models are faster than the handmade models. Human players tend to be faster than these models, but they may not use the myopic and own-payoff strategies, which is why there is no point in making direct comparisons for now.

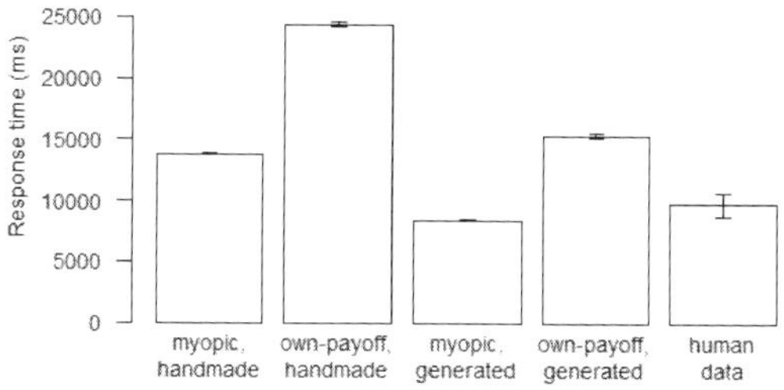


Figure 7. Reaction times for handmade and automatically generated myopic and own-payoff models, as well as the human participants in [46], when making their first decision in Game 1′.

The abilities of the translation system were also investigated by looking at novel automatically generated models playing games 1 and 4 (see Figure 6). For both games, two models were generated: one that uses backward induction (BI), and one that uses extensive-form rationalizability (EFR). These strategy formulas not only contain payoffs and comparisons, like the myopic and own-payoff strategy formulas, but also contain beliefs. Because both BI and EFR strategies have multiple solutions in game 4, exhaustive strategy formulas are required to describe these strategies.

The reaction times for the BI and EFR models can be found in Figure 8. These exhaustive models are a lot slower than the myopic and own-payoff models. Furthermore, reaction times in Game 1 are faster than reaction times in Game 4. It seems that reaction times are a function of the number of formulas required to create the exhaustive strategy formula: for both BI and EFR, in Game 1, only one formula is needed. In Game 4, BI requires two formulas, and EFR requires four formulas. To test this, a simple linear regression using number of formulas was performed to predict reaction times. A significant regression equation is found ($F(1, 189) = 432.6, p < 2.2 \cdot 10^{-16}$), with an $R^2$ of 0.696. Predicted reaction time in milliseconds is equal to 10401 + 50453 · (number of formulas).

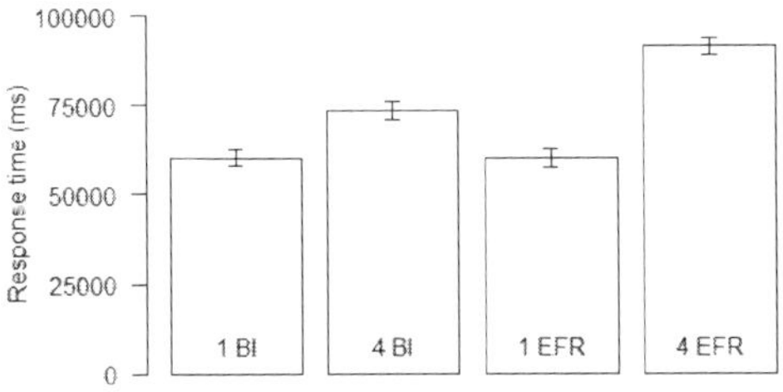


Figure 8. Reaction times for automatically generated BI and EFR models, when making their first decision. Here, '1 BI' denote the reaction time for the BI model in Game 1.

## 4. Conclusion and future work

In this paper we have presented a broad overview of a bridging methodology between the so-called orthogonal approaches towards modeling human reasoning - descriptive reasoning methods vis-à-vis normative or idealized reasoning methods. In [55], these ideas were presented for the first time, and we followed it up with certain extensions in [46], providing the first ever description of how the cycle depicted in Figure 2 can be completed in terms of connecting the different methods. Based on subsequent work [17,57,58] we provide here more detailed explorations of the underlying ideas for each of the arrows in Figure 2. They provide further proofs of concept of the methodology under consideration.

It is evident that an in-depth study of these connections between different methods also provide novel insight into the individual methods. The experimental findings on strategic reasoning of humans suggest for incorporating newer operators in the logical languages to deal with such kind of reasoning. For example, risk-taking and risk-averseness form a major part of the considerations of the participants in the experiments reported in [16,17]. This suggests that a graded belief operator might model the participants' behavior in a better way in comparison to the belief operators that we use in the current work. From the computational cognitive modeling perspective, it is useful to note that the performance of automated models is better than the corresponding handmade models (cf. Figure 7). This could speed up research to a great extent in this area. Also, it is found that the reaction times of the models are too slow compared to human reaction times (cf. Figure 7). This suggests the need of further development of the focus actions in the PRIMs model.[58] Finally, from the experimentation viewpoint, given the exhaustive list of strategy formulas and the corresponding cognitive models, it is useful to have a comparative study of these strategies empirically and then to have a comparison with the predictions of the cognitive models.

Marr[70] has influentially argued that any task computed by a cognitive system must be analyzed at the following three levels of explanation (in order of decreasing abstraction):

**the computational level:** identification of the goal and of the information-processing task as an input - output function;
**the algorithmic and representational level:** specification of an algorithm which computes the function;
**the implementation level:** physical or neural implementation of the algorithm.

According to Isaac et al.,[71] logic can be of use at each of Marr's three levels, but in the history of cognitive science, logic has been especially useful at the computational level. Baggio and colleagues[72] provide some fruitful examples in which computational level theories based on appropriate logics predict and explain behavioral data and even EEG data in the cognitive neuroscience of reasoning and language. As to computational cognitive modeling, Cooper and Peebles[73] argue that computational cognitive architectures such as ACT-R through their theoretical commitments constrain declarative and procedural learning, thereby constraining both the functions that can be computed (the computational level) and the way that they can be computed (the algorithmic level). Through our bridging methodology we show that this study based on logic, experiment and computational cognitive model can play a fruitful role at all these levels and at the interfaces between them.

A natural continuation for this line of work is to model partially-informed agents' strategies which provide interesting challenges - incorporating information structures and memory restrictions leading to evolution of strategies in formal frameworks, testing and introducing such information states in experimental subjects, and modeling such states as cognitive modules. As before, one should start with developing logical frameworks, possibly using automata-theoretic techniques. Then one can produce computational cognitive models based on these formal frameworks, and finally validate those models based on game experiments that have been performed earlier to investigate human strategic reasoning under partial information.

The focus would be to overcome interdisciplinary modeling challenges and construct computationally useful and efficient models of human society.

## References


1. M. Osborne and A. Rubinstein, *A Course in Game Theory* (MIT Press, Cambridge, MA, 1994).
2. G. Attanasi and R. Nagel, A survey of psychological games: Theoretical findings and experimental evidence, in *Games, Rationality and Behaviour: Essays on Behavioural Game Theory and Experiments*, eds. A. Innocenti and P. Sbriglia (Palgrave McMillan, 2008) pp. 204–232.
3. C. Bicchieri, Common knowledge and backward induction: A solution to the paradox, in *TARK '88: Proceedings of the 2nd Conference on Theoretical Aspects of Reasoning About Knowledge*, ed. M. Vardi (Morgan Kaufmann Publishers Inc., 1988).
4. R. McKelvey and T. Palfrey, An experimental study of the centipede game, *Econometrica* **60(4)**, 803 (1992).
5. A. Kanamori, *The Higher Infinite*, second edn. (Springer, Hiedelberg, 2008).
6. P. Koellner and W. H. Woodin, Large cardinals from determinacy, in *Handbook of Set Theory*, eds. M. Foreman and A. Kanamori (Springer-Verlag, 2010) pp. 2–202.
7. M. Friedell, On the structure of shared awareness, *Behavioral Science* **14**, 28 (1969).
8. D. Lewis, *Convention: A Philosophical Study* (Harvard University Press, 1969).
9. M. Bratman, *Intention, Plans, and Practical Reason* (CSLI Publications, 1999).
10. J. F. Horty and N. Belnap, The deliberative stit: A study of action, omission, ability and obligation, *Journal of Philosophical Logic* **24**, 583 (1995).
11. M. Wooldridge, *An Introduction to Multi-agent Systems*, second edn. (John Wiley & Sons, 2009).
12. J. Pilecki, M. Bednarczyk and W. Jamroga, Synthesis and verification of uniform strategies for multi-agent systems, in *Computational Logic in Multi-Agent Systems. CLIMA 2014 Proceedings*, ed. N. B. et. al., LNCS, Vol. 8624 (Springer, 2014).
13. J. van Benthem, *Logic in Games* (MIT Press, 2016).
14. K. Chatterjee, T. Henzinger and N. Piterman, Strategy logic, in *Proceedings of the 18th International Conference on Concurrency Theory (CONCUR 07)*, eds. L. Caires and V. Vasconcelos, LNCS, Vol. 4703 (Springer Verlag, Berlin, 2007).
15. R. Ramanujam and S. Simon, A logical structure for strategies, in *Logic and the Foundations of Game and Decision Theory (LOFT 7)*, eds. G. Bonanno, W. van der Hoek and M. Wooldridge, Texts in Logic and Games, Vol. 3 (Amsterdam University Press, Amsterdam, 2008) pp. 183–208.
16. S. Ghosh, A. Heifetz and R. Verbrugge, Do players reason by forward induction in dynamic perfect information games?, in *Proceedings of the Fifteenth Conference on Theoretical Aspects of Rationality and Knowledge, TARK*, ed. R. Ramanujam, EPTCS, Vol. 2152015.
17. S. Ghosh, A. Heifetz, R. Verbrugge and H. de Weerd, What drives people's choices in turn-taking games, if not game-theoretic rationality?, in *Proceedings of the Sixteenth Conference on Theoretical Aspects of Rationality and Knowledge, TARK*, ed. J. Lang, EPTCS, Vol. 2512017.

18. J. Borst, N. Taatgen and H. van Rijn, The problem state: A cognitive bottleneck in multitasking, *Journal of Experimental Psychology: Learning, Memory, & Cognition* **36**, 363 (2010).
19. G. Bergwerff, B. Meijering, J. Szymanik, R. Verbrugge and S. Wierda, Computational and algorithmic models of strategies in turn-based games, in *Proceedings of the 36th Annual Conference of the Cognitive Science Society*, 2014.
20. A. Benz, G. Jäger and R. van Rooij (eds.), *Game Theory for Pragmatics* (Palgrave Macmillan, London, 2006).
21. G. Jäger, Applications of game theory to linguistics, *Language and Linguistics Compass* **2**, 406 (2008).
22. J. Gerbrandy, Bisimulation on planet Kripke, PhD thesis, University of Amsterdam (1999).
23. J. van Benthem, J. van Eijck and B. Kooi, Logics of communication and change, *Information and Computation* **204**, 1620 (2006).
24. J. van Benthem, Games in dynamic epistemic logic, *Bulletin of Economic Research* **53**, 219 (2001).
25. G. Bonanno, Belief revision in a temporal framework, in *New Perspectives on Games and Interaction*, eds. K. Apt and R. van Rooij, Text in Logic and Games, Vol. 4 (Amsterdam University Press, Amsterdam, 2009).
26. J. van Benthem, Dynamic logic for belief revision, *Journal of Applied Non-Classical Logic* **17**, 129 (2007).
27. T. Ågotnes and N. Alechina, The dynamics of syntactic knowledge, *Journal of Logic and Computation* **17**, 83 (2007).
28. W. Jamroga and W. van der Hoek, Agents that know how to play, *Fundamenta Informaticae* **63**, 185 (2004).
29. F. Liu, Changing for the better: Preference dynamics and agent diversity, PhD thesis, University of Amsterdam (2008).
30. O. Roy, Thinking before acting, PhD thesis, University of Amsterdam (2008).
31. P. Girard, Modal logic for belief and preference change, PhD thesis, Department of Philosophy (Stanford University, Stanford, CA, USA, 2008).
32. R. Parikh, The logic of games and its applications, *Annals of Discrete Mathematics* **24**, 111 (1985).
33. W. van der Hoek, A. Lomuscio and M. Wooldridge, On the complexity of practical ATL model checking, in *Proceedings of the Fifth International Joint Conference on Autonomous Agents and Multiagent Systems (AAMAS 06)*, eds. P. Stone and G. Weiss (ACM Inc., New York, 2006).
34. S. Ghosh, Strategies made explicit in dynamic game logic, in *Proceedings of the Workshop on Logic and Intelligent Interaction, ESSLLI*, eds. J. van Benthem and E. Pacuit (2008).
35. J. van Benthem, In praise of strategies, in *Games, Actions and Social Software*, eds. J. van Eijck and R. Verbrugge, Texts in Logic and Games, FoLLI subseries of LNCS (Springer Verlag, Berlin, 2012) pp. 96–116.
36. D. Walther, W. van der Hoek and M. Wooldridge, Alternating-time temporal logic with explicit strategies, in *Proceedings of the XIth Conference on Theoretical Aspects of Rationality and Knowledge (TARK)*, ed. D. Samet (ACM Inc., New York, 2007).
37. S. Ghosh, R. Ramanujam and S. Simon, Playing extensive form games in parallel, in

*Proceedings of the 11th International Workshop on Computational Logic in Multi-Agent Systems (CLIMA XI)*, eds. J. Dix and others, LNCS, Vol. 6245 (Springer, Berlin, 2010).
38. R. Ramanujam and S. Simon, Dynamic logic on games with structured strategies, in *Proceedings of the 11th International Conference on Principles of Knowledge Representation and Reasoning (KR-08)*, eds. G. Brewka and J. Lang (AAAI Press, Menlo Park, CA, 2008).
39. G. Polya, *How to Solve It. A New Aspect of Mathematical Method* (Princeton University Press, Princeton, N.J., 1945).
40. M. Lovett, A strategy-based interpretation of Stroop, *Cognitive Science* **29**, 493 (2005).
41. I. Juvina and N. Taatgen, Modeling control strategies in the n-back task, in *Proceedings of the 8th International Conference on Cognitive Modeling*, eds. R. Lewis, T. Polk and J. Laird (Psychology Press, 2007).
42. J. Anderson, *How Can the Human Mind Occur in the Physical Universe?* (Oxford University Press, New York (NY), 2007).
43. C. Lebiere, D. Wallach and R. West, A memory-based account of the prisoner's dilemma and other 2x2 games, in *Proceedings of Third International Conference on Cognitive Modeling*, eds. N. Taatgen and J. Aasman (Universal Press, 2000).
44. R. West, C. Lebiere and D. Bothell, Cognitive architectures, game playing, and human evolution, in *Cognition and Multi-Agent Interaction: From Cognitive Modeling to Social Simulation*, ed. R. Sun (Cambridge University Press, New York (NY), 2006) pp. 103–123.
45. B. Meijering, N. A. Taatgen, H. van Rijn and R. Verbrugge, Modeling inference of mental states: As simple as possible, as complex as necessary, *Interaction Studies* **15**, 455 (2014).
46. S. Ghosh and R. Verbrugge, Studying strategies and types of players: Experiments, logics and cognitive models, *Synthese*, 1 (2017).
47. R. Rosenthal, Games of perfect information, predatory pricing and the chain-store paradox, *Journal of Economic theory* **25**, 92 (1981).
48. N. Taatgen and J. Anderson, The past, present, and future of cognitive architectures, *Topics in Cognitive Science* **2**, 693 (2010).
49. J. Anderson, D. Bothell, M. Byrne, S. Douglass, C. Lebiere and Y. Qin, An integrated theory of the mind, *Psychological Review* **111 (4)**, 1036 (2004).
50. J. Anderson and C. Lebiere, The Newell test for a theory of cognition, *Behavioral and Brain Science* **26**, 1 (2003).
51. L. van Maanen and R. Verbrugge, A computational model of second-order social reasoning, in *Proceedings of the 10th International Conference on Cognitive Modeling*, eds. D. Salvucci and G. Gunzelmann (2010).
52. B. Edmonds, Social intelligence and multi-agent systems, in *Invited talk, The Multi-Agent Logics, Languages, and Organisations Federated Workshops*, 2010.
53. R. Verbrugge, Logic and social cognition: The facts matter, and so do computational models, *Journal of Philosophical Logic* **38**, 649 (2009).
54. J. van Benthem, Logic and reasoning: Do the facts matter?, *Studia Logica* **88**, 67 (2008), Special issue on logic and the new psychologism, edited by H. Leitgeb.
55. S. Ghosh, B. Meijering and R. Verbrugge, Strategic reasoning: Building cognitive models from logical formulas, *Journal of Logic, Language and Information* **23**, 1 (2014).

56. N. A. Taatgen, The nature and transfer of cognitive skills, *Psychological Review* **120**, 439 (2013).
57. J. Top, Bridging the gap between logic and cognition: A translation method for centipede games, Master's thesis, University of Groningen (2017).
58. J. Top, R. Verbrugge and S. Ghosh, An automated method for building cognitive models for turn-based games from a strategy logic, *Games* **9** (2018), 44.
59. R. Nagel and F. F. Tang, Experimental results on the centipede game in normal form: An investigation on learning, *Journal of Mathematical Psychology* **42**, 356 (1998).
60. C. Camerer, *Behavioral Game Theory* (Princeton University Press, 2003).
61. J. R. Anderson and L. J. Schooler, Reflections of the environment in memory, *Psychological Science* **2**, 396 (1991).
62. N. A. Taatgen, Between architecture and model: Strategies for cognitive control, *Biologically Inspired Cognitive Architectures* **8**, 132 (2014).
63. B. Arslan, S. Wierda, N. Taatgen and R. Verbrugge, The role of simple and complex working memory strategies in the development of first-order false belief reasoning: A computational model of transfer of skills, in *Proceedings of the 13th International Conference on Cognitive Modeling*, 2015.
64. L. Gittelson and N. Taatgen, Transferring primitive elements of skill within and between tasks, in *Proceedings of the 36th Annual Meeting of the Cognitive Science Society*, (Cognitive Science Society, 2014).
65. T. Buwalda, J. Borst, H. v. d. Maas and N. Taatgen, Explaining mistakes in single digit multiplication: A cognitive model, in *Proceedings of the 14th International Conference on Cognitive Modeling*, 2016.
66. D. Pearce, Rationalizable strategic behaviour and the problem of perfection, *Econometrica* **52**, 1029 (1984).
67. T. Hedden and J. Zhang, What do you think I think you think? Strategic reasoning in matrix games, *Cognition* **85**, 1 (2002).
68. M. Raijmakers, D. Mandell, S. van Es and M. Counihan, Children's strategy use when playing strategic games, *Synthese* **191**, 355 (2014).
69. C. Stevens, J. Daamen, E. Gaudrain, T. Renkema, J. Top, F. Cnossen and N. Taatgen, Using cognitive agents to train negotiation skills, *Frontiers in Psychology* **9** (2018).
70. D. Marr, *Vision* (Freeman and Company, New York, 1982).
71. A. Isaac, J. Szymanik and R. Verbrugge, Logic and complexity in cognitive science, in *Johan van Benthem on Logic and Information Dynamics*, (Springer, 2014) pp. 787–824.
72. G. Baggio, M. v. Lambalgen and P. Hagoort, Logic as Marr's computational level: Four case studies, *Topics in Cognitive Science* **7**, 287 (2015).
73. R. Cooper and D. Peebles, Beyond single-level accounts: The role of cognitive architectures in cognitive scientific explanation, *Topics in Cognitive Science* **7**, 243 (2015).

# Muchnik Degrees and Medvedev Degrees of Randomness Notions

Kenshi Miyabe

*Math Department, Meiji University,*
*Kawasaki, 214-8571, Japan*
*E-mail: research@kenshi.miyabe.name*
*http://kenshi.miyabe.name/wordpress/*

The main theme of this paper is computational power when a machine is allowed to access random sets. The computability depends on the randomness notions and we compare them by Muchnik and Medvedev degrees. The central question is whether, given a random oracle, one can compute a more random set. The main result is that, for each Turing functional, there exists a Schnorr random set whose output is not computably random.

*Keywords*: Muchnik degree, Medvedev degree, Schnorr randomness, computable randomness.

## 1. Introduction

In mathematical logic or theoretical computer science, a function $f :\subseteq \omega \to \omega$ is *computable* if the function $f$ is computable by a Turing machine. Computability on natural numbers is formalized in several ways that have been proved to be equivalent. By the Church-Turing thesis, we believe that this is the correct definition of computability. The definition naturally extends to a function $f :\subseteq 2^\omega \to 2^\omega$.

When one is writing a code in a programming language, one may use a random generator and may say that a function is computable even if the function uses a random generator. We formalize computational power of functions $f$ on Cantor space when the functions are allowed to access random sets. In computational complexity theory, there are some such classes such as BPP and RP. In this paper, we focus on computability theory.

One classical approach is the following. Suppose a set $X$ is Turing computable with an oracle in a class of random sets. The set of random sets (for any reasonable randomness notion) has measure 1. Thus, $\{Y \in 2^\omega \ : \ Y \geq_T X\}$ has measure 1. A classical theorem due to Sacks says that if a set $Z \in 2^\omega$ is not computable, then $\{Y \in 2^\omega \ : \ Y \geq_T Z\}$ has Lebesgue measure 0 (Ref. 1). Hence, the set $X$ should be a computable set, which means that it is already computable without random oracles. This is the case where we are trying to compute an individual sequences via random oracles.

In the case that we are using random oracles to compute a set in a given collection of sequences, the results are very different. The computability with random access can be formalized as *mass problems*. Here we think of a mass problem as a collection of sets. We provide a solution by identifying some sequence in the collection. Consider two problems $P, Q \subseteq 2^\omega$. If each solution $f \in Q$ can compute a solution $g \in P$, then we can say that constructing a solution of $P$ is not more difficult than constructing a solution of $Q$. The induced degree of $P$ will represent computational power when allowed to access an element of $P$. When the computation is uniform, we say that $P$ is *Medvedev (or strongly) reducible to* $Q$ and it is denoted by $P \leq_s Q$. When the computation can be nonuniform, $P$ is *Muchnik (or weakly) reducible to* $Q$ and denoted by $P \leq_w Q$. For a formal definition, see Definition 2.1.

Many randomness notions have been studied in the theory of algorithmic randomness. Each randomness notion is a subset of Cantor space, so we can directly study its degree. If a randomness notion $P$ is weaker than $Q$ ($P \supseteq Q$), then it straightforwardly implies $P \leq_s Q$ and $P \leq_w Q$. In other words, for any $P, Q \subseteq 2^\omega$, we have

$$P \supseteq Q \Rightarrow P \leq_s Q \Rightarrow P \leq_w Q.$$

So, the real problem is whether each reduction is strict. In other words, we ask whether one can compute a more random set from a given random set.

The following is (a part of) well-known hierarchy of randomness notions:

$$\text{WR} \supsetneq \text{SR} \supsetneq \text{CR} \supsetneq \text{MLR} \supsetneq \text{DiffR} \begin{array}{c} \nearrow \text{W2R} \searrow \\ \\ \searrow \text{DemR} \nearrow \end{array} \text{2R}$$

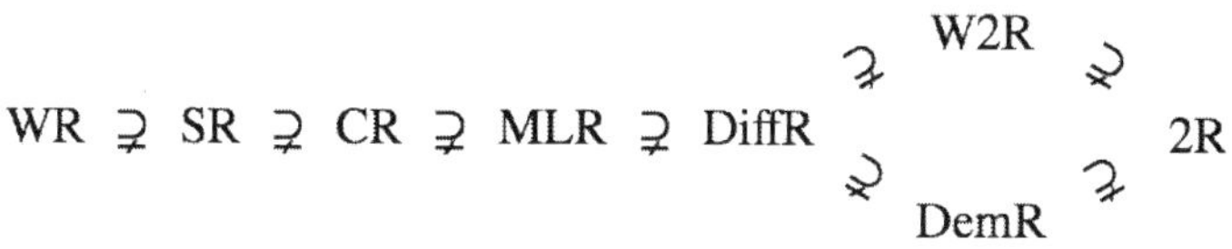

Figure 1. The hierarchy of the randomness notions.

Here, W2R and DemR are incomparable. The abbreviations of randomness notions above will be explained in the next section. The following is the structure of the corresponding Muchnik and Medvedev degrees, which we establish in this paper: The main result is SR $<_s$ CR and will be proved in Theorem 3.9.

Related work can be found in Simpson[2], which mainly studied the Muchnik degrees of $\Pi^0_1$-class and the relation with the Muchnik degree of the class of all ML-random sets. Similar questions can be asked in the context of Weihrauch degrees[3].

The rest of the paper is devoted to the proofs of these results.

$$WR <_w SR \equiv_w CR <_w MLR \equiv_w DiffR \quad \begin{matrix} \swarrow_w & W2R & \searrow_w \\ & & \\ \searrow_w & & \swarrow_w \\ & Demr & \end{matrix} \quad 2R$$

$$SR <_s CR, \quad MLR <_s DiffR$$

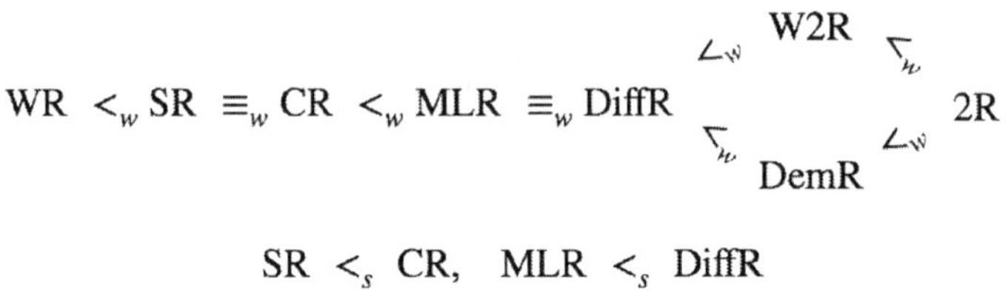

Figure 2. Muchnik and Medvedev degrees of the randomness notions.

## 2. Preliminaries

### 2.1. Computability theory

We fix some notation. We denote the set of finite binary strings by $2^{<\omega}$, and the set of infinite binary sequences by $2^\omega$. We often identify an infinite sequence $X \in 2^\omega$ with a set of natural numbers $Y \subseteq \omega$ by $X(n) = 1 \iff n \in Y$. The *jump* of a sequence $A$ is denoted by $A'$ and is defined by $A' = \{n \in \omega \ : \ \Phi_n^A(n) \downarrow\}$, where $\Phi_n$ is the $n$-th Turing machine.

For more background on computability theory, see textbooks such as Refs. 4,5.

We define Muchnik and Medvedev reducibility on $2^\omega$ (we can define these notions in $\omega^\omega$, but since we are interested in randomness notions on $2^\omega$, we will restrict our attention to $2^\omega$).

**Definition 2.1 (Muchnik[6], Medvedev[7]).** *Let $P, Q \subseteq 2^\omega$.*

*(1) We say that $P$ is* Muchnik reducible (or weakly reducible) *to $Q$, denoted by $P \leq_w Q$, if, for every $f \in Q$, there is an element $g \leq_T f$ in $P$.*

*(2) We say that $P$ is* Medvedev reducible (or strongly reducible) *to $Q$, denoted by $P \leq_s Q$, if there is a Turing functional $\Phi$ such that $\Phi^f \in P$ for every $f \in Q$.*

A rough idea is that $P \leq_w Q$ if each solution in $Q$ can Turing compute a solution in $P$, but the reduction need not be uniform. In contrast $P \leq_s Q$ if one Turing reduction $\Phi$ can compute a solution in $P$ from a solution in $Q$ uniformly. The reductions $\leq_w$ and $\leq_s$ are pre-orders. We write $P \equiv_w Q$ to mean $P \leq_w Q$ and $Q \leq_w P$, and $P \equiv_s Q$ similarly. Muchnik degrees and Medvedev degrees are equivalence classes derived from the equivalence relations $\equiv_w$ and $\equiv_s$, respectively.

### 2.2. The theory of algorithmic randomness

We review some definitions from the theory of algorithmic randomness. For more details, see monographs such as Refs. 8,9.

Cantor space is the space $2^\omega$ equipped with the topology generated by the cylinder sets $[\sigma] = \{X \in 2^\omega \ : \ \sigma \prec X\}$ for $\sigma \in 2^{<\omega}$, where $\prec$ is the prefix relation.

We define $X \oplus Y = \{2n : n \in X\} \cup \{2n+1 : n \in Y\}$. We usually consider the fair-coin measure $\lambda$ on $2^\omega$.

An open set $V$ is *c.e.* if $V = \bigcup_{\sigma \in S}[\sigma]$ for a computable set $S$. A Martin-Löf test (ML-test) is a sequence $\{U_n\}_{n\in\omega}$ of uniformly c.e. open sets such that $\lambda(U_n) \leq 2^{-n}$ for all $n$. A sequence $X \in 2^\omega$ is *ML-random* if $X \notin \bigcap_n U_n$ for each ML-test. When one is allowed to use an oracle $A$ in the computation of the open sets in the definition of ML-tests, the sequence is called ML-random relative to $A$ and we just call it $A$-ML-random. Van Lambalgen's theorem[10] (see Corollary 6.9.3 in Ref. 9) says that $A \oplus B$ is ML-random if and only if $A$ is ML-random and $B$ are $A$-ML-random. The class of ML-random sequences is denoted by MLR. Note that MLR is a subset of $2^\omega$. We compare this class MLR with other randomness notions in Muchnik degrees and Medvedev degrees.

A *martingale* is a function $d : 2^{<\omega} \to \mathbb{R}^+$ such that $2d(\sigma) = d(\sigma 0) + d(\sigma 1)$ for every $\sigma \in 2^{<\omega}$. Here, $\mathbb{R}^+$ denotes the set of non-negative reals. A set $X$ is not *computably random* if there exists a computable martingale $d$ such that $\sup_n d(X \upharpoonright n) = \infty$. A set $X$ is not *Schnorr random* if and only if there is a computable martingale $d$ and a computable function $f$ such that $d(A \upharpoonright f(n)) \geq n$ for infinitely many $n$ (Ref. 11).

We denote the classes of Kurtz random sets, Schnorr random sets, computably random sets, difference random sets, weakly 2-random sets, Demuth random sets, and 2-random sets by WR, SR, CR, DiffR, W2R, DemR, and 2R, respectively. We will give the definition of the remaining randomness notions as they come up in the next section. In the proofs that follows we need only their characterizations or some of their basic properties.

The randomness notions defined here depends on the measure $\lambda$. A measure $\nu$ on the Cantor space is *computable* if the function $\sigma \mapsto \nu([\sigma])$ is computable uniformly in $\sigma \in 2^{<\omega}$. If one replaces $\lambda$ with a computable measure $\nu$ in the definition of ML-randomness, then we can define ML-randomness with respect to $\nu$, or $\nu$-ML-randomness. For a computable measure $\nu$, a *$\nu$-martingale* is a function $d : 2^\omega \to \mathbb{R}^+$ such that $\nu(\sigma)d(\sigma) = \nu(\sigma 0)d(\sigma 0) + \nu(\sigma 1)d(\sigma 1)$ for every $\sigma \in 2^{<\omega}$. Schnorr randomness and computable randomness with respect to a computable measure $\nu$ can be defined via $\nu$-martingales.

## 3. Proofs

### 3.1. *Separations in Muchnik degrees*

Here, we show equivalences and separations of randomness notions in the Muchnik degrees. In the proofs we make use of some notions in Turing degrees.

We start with the proof of CR $<_w$ MLR. The proof is probably simplest and it is appropriate to explain the intuition from this. We look for a Turing degree **a** that contains a computably random set but no Turing degree below **a** contains a ML-random set. We will choose this degree to be a high minimal degree. We use the notion of highness to prove that the degree contains a computably random set, and the one of minimality to prove that the degrees do not contain any ML-random set.

A Turing degree $\mathbf{a} > \mathbf{0}$ is *minimal* if no degree **b** satisfies $\mathbf{0} < \mathbf{b} < \mathbf{a}$. We say that a Turing degree **a** is ML-random if it contains a ML-random set. Note that no ML-random degree is minimal, since van Lambalgen's theorem guarantees that if $Z = X \oplus Y$ is ML-random, then $X$ is ML-random and strictly below $Z$ in the Turing degrees.

A sequence $X \in 2^{\omega}$ is called *high* if $\emptyset'' \leq_T X'$. If $X \equiv_T Y$ and $X$ is high, then $Y$ is also high because $X' \equiv_T Y'$. A degree **a** is called high if **a** contains a high sequence, or equivalently if all sequences in **a** are high. Highness is an important property in the study of computable randomness. In particular, Nies, Stephan and Terwijn[12] showed that every high degree contains a computably random set.

The last piece is finding a high minimal degree, whose existence is known in computability theory (Ref. 13).

**Theorem 3.1.**

$$\mathrm{CR} <_w \mathrm{MLR}.$$

**Proof.** We begin by showing that $\mathrm{CR} \leq_w \mathrm{MLR}$. Every ML-random set is computably random, so every ML-random set computes a computably random set. In fact $\mathrm{CR} \leq_s \mathrm{MLR}$. Let Id be the Turing functional such that $\mathrm{Id}(X) = X$ for all $X \in 2^{\omega}$. Then, $\Phi^X$ is computably random for every ML-random sequence $X$. Thus, $\mathrm{CR} \leq_s \mathrm{MLR}$.

Suppose, for a contradiction, that $\mathrm{MLR} \leq_w \mathrm{CR}$. Let **a** be a high minimal Turing degree, whose existence is shown in Ref. 13. Then, there exists a computably random set $X \in \mathbf{a}$. Thus, there should be $Y \in \mathrm{MLR}$ such that $Y \leq_T X$. Since the Turing degree of $X$ is minimal and $Y$ can not be computable, we have $X \equiv_T Y$ and the Turing degree of $Y$ is minimal, which contradicts the fact $Y \in \mathrm{MLR}$. □

Notice that $\mathrm{MLR} <_w \mathrm{CR}$ implies that MLR and CR are distinct, because, otherwise, they would have the same Muchnik and Medvedev degrees. In contrast, different sets can have the same Muchnik or Medvedev degree as we will see $\mathrm{MLR} \equiv_w \mathrm{DiffR}$ later. Thus, the fact that MLR and CR are different is not enough to show $\mathrm{MLR} <_w \mathrm{CR}$. Some proof ideas are already implicit in proofs of the strict inclusions of the randomness notions. One can see this paper as a list of proofs of the strict inclusions only by degree properties, not by construction of a particular set.

Next, we show WR $<_w$ SR. A sequence is *Kurtz random* (or weakly random) if it is contained in every c.e. open set with measure 1. Here we use the fact that every non-high Schnorr random set is ML-random, which is shown in Nies, Stephan and Terwijn [12]. Then we can make a similar argument by minimal degrees.

Another notion we need is hyperimmunity. A Turing degree **a** is *hyperimmune* if **a** computes a function $f : \omega \to \omega$ that is not dominated by a computable function. Every hyperimmune contains a Kurtz random set that is not Schnorr random (Corollary 8.11.10 in Ref. 9).

Now it suffices to show the existence of a degree that is minimal, non-high, and hyperimmune.

**Theorem 3.2.**

$$\text{WR} <_w \text{SR}.$$

**Proof.** Since WR $\supseteq$ SR, we have WR $\leq_w$ SR. Assume, for a contradiction, that SR $\leq_w$ WR.

Let **a** be a minimal (Turing) degree below $\mathbf{0}'$, whose existence was shown in Sacks [14]. Note that **a** is not high because no minimal degree below $\mathbf{0}'$ can be high (Ref. 13). Furthermore, since every nonzero degree $\mathbf{0}'$ is hyperimmune (Ref. 15), **a** is hyperimmune. Let $X \in \mathbf{a}$ such that $X \in$ WR $\setminus$ SR. This is possible because every hyperimmune degree contains a Kurtz random set that is not Schnorr random.

Since SR $\leq_w$ WR, there exists a Schnorr random set $Y \leq_T X \in$ WR. Since **a** is minimal and $Y$ can not be computable, $Y \in \mathbf{a}$ and $Y$ is not high. Since $Y$ is non-high Schnorr random, $Y$ is already ML-random, and with minimal degree. This is a contradiction. □

Next we show DiffR $<_w$ W2R. Both notions have characterizations by a degree property in ML-randomness, which is suitable for our purpose. A set $X$ is *difference random* if and only if $X$ is ML-random and $X$ is Turing incomplete [16]. A set $X$ is called *weakly 2-random* if it is contained in every $\Sigma^0_2$ class (computable unions of co-c.e. closed sets) with measure 1. Equivalently, a set $X$ is weakly 2-random if and only if $X$ is ML-random and the degree of $X$ forms a minimal pair with $\mathbf{0}'$ (Ref. 17). We say that two incomputable sets $A, B$ form a minimal pair if every set $Z \leq_T A, B$ is computable. We again make use of van Lambalgen's theorem for ML-randomness.

**Theorem 3.3.**

$$\mathrm{DiffR} <_w \mathrm{W2R}.$$

**Proof.** Since every weakly 2-random set is difference random, we have DiffR $\leq_w$ W2R.

For a proof of the strictness, let $\Omega$ be a halting probability. Recall that $\Omega$ is ML-random and $\Omega \equiv_T \emptyset'$. Let $\Omega_0 \oplus \Omega_1 = \Omega$.

We claim that $\Omega_0$ is difference random Turing below $\mathbf{0}'$. First of all, by van Lambalgen's theorem, $\Omega_0$ is ML-random. Notice that $\Omega_0$ is incomplete, otherwise $\Omega_1 \leq_T \Omega \equiv_T \emptyset' \leq_T \Omega_0$ and this contradicts the fact that $\Omega_1$ is ML-random relative to $\Omega_0$ by van Lambalgen's theorem. Hence, $\Omega_0$ is difference random. Furthermore, $\Omega_0 \leq_T \Omega \equiv_T \emptyset'$.

Now, suppose that W2R $\leq_w$ DiffR. Then, there exists $X \in$ W2R such that $X \leq_T \Omega_0 \leq_T \emptyset'$, but $X$ cannot be a minimal pair with $\mathbf{0}'$, because $X \leq_T X$ and $X \leq_T \emptyset'$ is not computable and $X$ itself is a counterexample. □

Next, we show W2R $<_w$ 2R. A set is *2-random* if it is ML-random relative to $\emptyset'$. If **a** is not hyperimmune, the degree **a** is called *hyperimmune-free*. Clearly, if $A \leq_T B$ and $B$ has a hyperimmune-free degree, then $A$ has a hyperimmune-free degree.

**Theorem 3.4.**

$$\mathrm{W2R} <_w 2\mathrm{R}.$$

**Proof.** Since every weakly 2-random set is 2-random, we have W2R $\leq_w$ 2R.

For a proof of the strictness, we use the following facts:

- There exists a ML-random set of hyperimmune-free degree (Theorem 8.1.3 in Ref. 9).
- If $A$ has hyperimmune-free degree, then $A$ is Kurtz random if and only if $A$ is weakly 2-random (Yu, see Theorem 8.11.12 in Ref. 9).
- Every 2-random degree is hyperimmune (Kautz[18], see Theorem 8.21.2 in Ref. 9).

The above two facts imply that there exists a weakly 2-random set of hyperimmnue-free degree, which can not compute a hyperimmune set, nor a 2-random set. □

Finally, we see the relation with Demuth randomness. A *Demuth test* is a sequence $\{W_{g(n)}\}$ of c.e. open sets such that $\lambda(W_{g(n)}) \leq 2^{-n}$ for all $n$ and $g$ is a $\omega$-c.e.

function. Here, $\{W_e\}$ is a fixed enumeration of all c.e. open sets. A function $g$ is called *ω-c.e.* if there are a computable function $f$ and a computable approximation $g(n) = \lim_s g(n,s)$ such that $|\{s : g(n,s+1) \neq g(n,s)\}| < f(n)$ for all $n$. A sequence $X \in 2^\omega$ is *Demuth random* if $X \notin W_{g(n)}$ for almost all $n$ for every Demuth test $\{W_{g(n)}\}$.

**Theorem 3.5.**

$$\mathrm{DiffR} <_w \mathrm{DemR} <_w 2\mathrm{R}.$$

*Furthermore,* W2R *and* DemR *are incomparable in Muchnik degrees.*

**Proof.** Since every difference random set is Demuth random and every Demuth random set is 2-random, we have DiffR $\leq_w$ DemR $\leq_w$ 2R.

There exists a weakly 2-random set $X$ of hyperimmune-free degree. Any set $Y \leq_T X$ is of hyperimmune-free degree, while no Demuth random set is of hyperimmune-free degree (p.316 in Ref. 9). Thus, DemR $\not\leq_w$ W2R.

There exists a Demuth random set $Z$ in $\Delta^0_2$ (Theorem 7.6.3 in Ref. 9). Any set $W \leq_T Z$ cannot be a minimal pair with $\mathbf{0}'$. Thus, W2R $\not\leq_w$ DemR.

Thus, W2R and DemR are incomparable in Muchnik degrees.

If DemR $\leq_w$ DiffR, then this would imply DemR $\leq_w$ DiffR $<_w$ W2R by Theorem 3.3. This contradicts the fact DemR $\not\leq_w$ W2R. Hence, DiffR $<_w$ DemR.

If 2R $\leq_w$ DemR, then this would imply W2R $<_w$ 2R $\leq_w$ DemR by Theorem 3.4. This contradicts the fact W2R $\not\leq_w$ DemR. Hence, DemR $<_w$ 2R. □

### 3.2. *ML-randomness and difference randomness*

We have seen the separations of the adjacent randomness notions in Muchnik degrees. From now on we show the equivalence between MLR and DiffR in Muchnik degrees. The equivalence MLR $\equiv_w$ DiffR means that one can compute a difference random set from every ML-random set. The proof is given by constructing a reduction.

**Theorem 3.6.**

$$\mathrm{MLR} \equiv_w \mathrm{DiffR}.$$

**Proof.** Since every difference random set is ML-random, we have MLR $\leq_w$ DiffR.

For a proof of the converse, let $X = Y \oplus Z$ be a ML-random set. We claim that at least one of $Y$ or $Z$ should be difference random.

Suppose that $Y$ is not difference random. By van Lambalgen's theorem, $Y$ is ML-random, so $Y$ is Turing above $\mathbf{0}'$. Again, by van Lambalgen's theorem, $Z$ is

ML-random relative to $Y$. Thus, $Z$ is $\mathbf{0}'$-ML-random, thus 2-random and difference random.

Note that $Y, Z \leq_T X$. Then $X$ can compute a difference random set, which is either $Y$ or $Z$. □

In this proof, we do not know which of $Y$ or $Z$ is difference random, so the reduction is not uniform. In fact, we cannot do this uniformly.

**Theorem 3.7.**

$$\mathrm{MLR} <_s \mathrm{DiffR}.$$

Since every difference random set is ML-random, we have $\mathrm{MLR} \leq_s \mathrm{DiffR}$. It remains to prove the strictness.

The task is as follows. For every Turing functional $\Phi$, we need to construct a ML-random set $X$ such that $\Phi(X)$ is not difference random.

The key notion in the proof is the push-forward measure. Let $(X_1, A_1)$ and $(X_2, A_2)$ be two measurable spaces where $A_1$ and $A_2$ are $\sigma$-algebras on $X_1$ and $X_2$, respectively. Given a measurable map $f : X_1 \to X_2$ and a measure $\nu$ on $X_1$, the *push-forward m* measure of $\nu$ by $f$ is defined to be

$$m(B) = \nu(f^{-1}(B)) \text{ for } B \in A_2.$$

If two measurable functions are equal almost everywhere, then they induce the same push-forward measure.

In the following proof, we consider the push-forward measure $\mu$ of the uniform measure $\lambda$ by the Turing functional $\Phi$. Roughly speaking, if the measure $\mu$ is very different from $\lambda$, then we can find such $X$ because randomness with respect to $\lambda$ is not preserved by $\Phi$. If the measure $\mu$ is close to $\lambda$ in some sense, then we can construct such $X$ because some properties are preserved by $\Phi$. This strategy is also used in the proof of $\mathrm{SR} <_s \mathrm{CR}$.

In this case, more concretely, we consider two cases depending on whether the push-forward measure $\mu$ has an atom. Given a measure $\nu$ on a space $X$, an element $x \in X$ is an *atom* if $\nu(\{x\}) > 0$. In this case we say that $\nu$ has an atom. If the measure does not have an atom, then the measure is called *continuous*. If $\mu$ has an atom $Y$, then $Y$ should be computable and we can find a ML-random set $X$ such that $Y = \Phi(X)$. If $\mu$ is continuous, then the classes of the random sets with respect to $\mu$ and $\lambda$ are not so different in the sense of Turing degree, so we can find a ML-random set $Y \geq_T \emptyset'$ such that $Y = \Phi(X)$ for some ML-random set $X$.

It is known that every atom for a computable measure is computable (Lemma 6.12.7 in Ref. 9). Any computable point can not be random (in any sense) with respect to the uniform measure $\lambda$. Moreover, the inverse image of the atom should have a random point because its measure is positive.

For two continuous computable measures, their random points have 1-1 correspondence in some sense. In particular, Levin-Kautz theorem (Refs. 18,19 and Theorem 6.12.9(iii) in Ref. 9) says that, if $\nu$ is a continuous computable measure and $\mathbf{a} > 0$ then $\mathbf{a}$ contains a $\lambda$-ML-random set iff $\mathbf{a}$ contains a $\nu$-ML-random set. Here, $\lambda$ is the uniform measure.

The last theorem we use in the proof is a no-randomness-from-nothing result, which is a converse of the randomness conservation theorem.

The randomness conservation theorem is the following. Let $\nu$ be a computable computable measure on $2^\omega$. Let $\Phi :\subseteq 2^\omega \to 2^\omega$ be a Turing functional defined $\nu$-almost-everywhere. if $X \in 2^\omega$ is $\nu$-ML-random, then $\Phi(X)$ is ML-random with respect to the push-forward measure of $\nu$ by $\Phi$ The push-forward measure is computable in this case. For a proof, see Theorem 3.2 in Ref. 20 etc.

No-randomness-from-nothing result is the following. Let $\nu$ be a computable computable measure on $2^\omega$. Let $\Phi :\subseteq 2^\omega \to 2^\omega$ be a Turing functional defined $\nu$-almost-everywhere. Let $\mu$ be the push-forward measure of $\nu$ by $\Phi$. If $Y \in 2^\omega$ is $\mu$-ML-random, then there exists a $\nu$-ML-random set $X$ such that $\Phi(X) = Y$. For a proof, see Theorem 3.5 in Ref. 20, according to which this result is due to Shen.

This result is really useful in our proof. We do not need to construct $X$; we only have to find a ML-random set $Y$ that is not difference random.

**Proof of Theorem 3.7.** Since MLR $\supseteq$ DiffR, we have MLR $\leq_s$ DiffR. Assume, for a contradiction, that DiffR $\leq_s$ MLR. Then, there exists a Turing functional $\Phi :\subseteq 2^\omega \to 2^\omega$ such that $\Phi(X)$ is difference random for every ML-random set $X$.

Let $\lambda$ be the fair-coin measure on $2^\omega$. Let $\mu$ be the push-forward measure of $\lambda$ by $\Phi$, that is, $\mu(B) = \lambda(\Phi^{-1}(B))$ for every Borel set $B$ on $2^\omega$. The measure $\mu$ is computable because the measures $\mu([\sigma])$ are uniformly left-c.e. and $\sum_{\sigma \in 2^n} \mu([\sigma]) = 1$ for every $n$.

Suppose that $\mu$ has an atom, that is, $\mu(\{Y\}) > 0$ for some $Y \in 2^\omega$. Then, $\lambda(\Phi^{-1}(\{Y\})) > 0$. Hence, there exists a ML-random set $X \in \Phi^{-1}(\{Y\})$. Then, $Y = \Phi(X)$ and $Y$ is difference random by the assumption. This contradicts the fact that every atom for a computable measure is computable.

Now suppose that $\mu$ is continuous (no atom). Notice that the degree $\mathbf{0}' > \mathbf{0}$ contains a ML-random set such as Chaitin's constant $\Omega$. By Levin-Kautz theorem, $\mathbf{0}'$ contains a $\mu$-ML-random set $Y$. By no-randomness-from-nothing for ML-randomness, there exists a $\lambda$-ML-random set $X$ such that $Y = \Phi(X)$. However, $\Phi(X)$ is not difference random because $Y \in \mathbf{0}'$, which is a contradiction. $\square$

### 3.3. *Schnorr randomness and computable randomness*

Now we turn to the relation between Schnorr randomness and computable randomness. As the relation between ML-randomness and difference randomness, we will see $\mathrm{SR} \equiv_w \mathrm{CR}$ and $\mathrm{SR} <_s \mathrm{CR}$. The proof of the former is not difficult, but the one of the latter needs a little work.

Recall that the difference between Schnorr randomness and computable randomness is the rate of divergence for computable martingales. In the definition of Schnorr randomness, we can replace $d(A \upharpoonright f(n)) \geq n$ with $d(A \upharpoonright f(n)) \geq 2^n$ by making $f$ grow faster. Futhermore, we can assume $d$ has the *savings property*: for each $\sigma, \tau$ we have

$$d(\sigma\tau) \geq d(\sigma) - 2.$$

First we see their equivalence in Muchnik degrees.

**Theorem 3.8.**

$$\mathrm{SR} \equiv_w \mathrm{CR}.$$

**Proof.** Since every Schnorr random set is computably random, we have $\mathrm{SR} \leq_w \mathrm{CR}$.

For a proof of the converse, let $X \in \mathrm{SR}$. If $X$ is high, then there exists a computably random set $Y \equiv_T X$, becausė every high degree contains a computably random set[12]. If $X$ is not high, $X$ is already ML-random (again by Ref. 12), thus computably random. □

Note that the proof is not uniform again. The reduction can not be uniform as the following theorem indicates. This is the main theorem of this paper.

**Theorem 3.9.**

$$\mathrm{SR} <_s \mathrm{CR}.$$

The goal is to show the following: for each Turing functional $\Phi :\subseteq 2^\omega \to 2^\omega$, there exists a Schnorr random set $X$ such that $\Phi(X)$ is (undefined or) not computably random. If $\Phi = \mathrm{id}$ (the identity map), this is equivalent to saying that there exists a Schnorr random set that is not computably random. In fact, we extend the method of the construction of $X \in \mathrm{SR} \setminus \mathrm{CR}$ in Ref. 12.

To give a proof idea, let us recall the "martingale strategy". The terminology of martingales in theory of algorithmic randomness and probability theory comes from this strategy in a coin flipping game. In this game, the player predicts whether the next toss of the coin comes up a head or a tail. The player bets some money, and the capital increases the same amount of money if the player is correct and the capital decreases the amount if the player is incorrect. The martingale strategy is the following strategy. The player first bets 1 dollar (say) to a head (say again). If incorrect, the player bets the money doubled from the previous bet, again and again. The player will be correct at some turn, say at $n$-th turn and increase his capital by

$$2^n - (1 + 2 + 2^2 + \cdots + 2^{n-1}) = 1.$$

We call this strategy *martingale strategy*. By repeating this martingale strategy, the player will increase the capital to infinity almost surely.

The strategy sounds good, but does not work in reality. This is because the player needs infinite amount of money in hand. If the initial capital is finite and the capital should be non-negative at any turn, the player can not continue the strategy from some point almost surely.

Suppose we know that a head will appear at least one turn in $\log(n)$ turns for some reason. Then, by making the first bet $\frac{1}{n}$, the amount of money lost is bounded by

$$\sum_{k=1}^{\log(n)} \frac{2^{k-1}}{n} = \frac{2^{\log(n)} - 1}{n} < 1.$$

The capital will increase $\frac{1}{n}$ in $\log(n)$ turns. By repeating this for $n = 1, 2, \ldots$, the player will increase the capital to infinity.

The construction of $X \in \mathrm{SR} \setminus \mathrm{CR}$ makes use of this strategy as follows. Construct a martingale $V$ that dominates all computable martingales. If $V$ is bounded along a set $X$, then $X$ should be computably random. If the rate of divergence of $V$ along a set $X$ is slower than any computable function, then $X$ should be Schnorr random. We construct a set $X$ so that $V$ does not increase along $X$ except the positions $\{a_n\}$ where we set $X(a_n) = 0$. The positions are too sparse so no computable martingale succeeds fast enough in Schnorr sense. The numbers of candidates of the positions are small enough for some computable martingale to succeed by iterating the martingale strategy.

In our case, for each almost-everywhere-defined functional $\Phi$, we need to construct a set $X \in \mathrm{SR}$ such that $\Phi(X) \notin \mathrm{CR}$. We force $\Phi(X)(a_n) = 0$ in some positions similarly. The difficulty here is that the measure of the class of such sets $X$ (loosely speaking) may be very small or even 0, and some computable martingale succeeds on $X$ fast enough.

We consider two cases. Let $\mu$ be the push-forward measure of $\lambda$ by $\Phi$. The first case is that $\mu$ is "far from" the fair-coin measure $\lambda$ and we can construct $X \in \mathrm{SR}$ such that $\Phi(X) \notin \mathrm{CR}$ by a separate argument. The second case is that $\mu$ is "close to" the fair-coin measure and we can apply the method of the case that $\Phi$ is the identity functional. The two cases are separated by the condition $\mathrm{CR}(\mu) \subseteq \mathrm{CR}(\lambda)$, which was (essentially) suggested by Laurent Bienvenu.

Let us consider two examples that illustrate these two cases. First, consider the map $\Phi : 2^\omega \to 2^\omega$, $X \mapsto 0X$. In this case, $\mathrm{CR}(\mu) \subsetneq \mathrm{CR}(\lambda)$. Look for a set $Y \in (\mathrm{SR} \setminus \mathrm{CR}) \cap [0]$. Then, the set $X$ such that $Y = 0X$ is in $\mathrm{SR} \setminus \mathrm{CR}$. Next, consider the map $\Phi : 2^\omega \to 2^\omega$, $X \mapsto 0^\omega$. In this case, $0^\omega \in \mathrm{CR}(\mu) \setminus \mathrm{CR}(\lambda)$ and $\mathrm{CR}(\mu) \nsubseteq \mathrm{CR}(\lambda)$. Then, there exists a Schnorr random set $X$ such that $\Phi(X) = 0^\omega$.

**Proof of Theorem 3.9.** Suppose for a contradiction that there exists a Turing reduction $\Phi :\subseteq 2^\omega \to 2^\omega$ such that $\Phi(X) \in \mathrm{CR}$ for every $X \in \mathrm{SR}$. Since $\Phi$ is defined almost everywhere, the push-forward measure $\mu$ is computable.

*Case 1:* $\mathrm{CR}(\mu) \nsubseteq \mathrm{CR}(\lambda)$

By $\mathrm{CR}(\mu)$ and $\mathrm{CR}(\lambda)$, we denote the class of sets that are computably random relative to $\mu$ and $\lambda$ respectively. Then, there exists a set $Y \in \mathrm{CR}(\mu) \setminus \mathrm{CR}(\lambda)$. By the no-randomness-from-nothing result for computable randomness by Rute[21], there exists a set $X \in \mathrm{CR}(\lambda)$ such that $\Phi(X) = Y$. Thus, $X \in \mathrm{SR}$ but $\Phi(X) \notin \mathrm{CR}$. □

For a proof of the other case, we use some analytical lemmas. Let $\nu, \mu$ be measures on a measurable space. We say that $\mu$ is *absolutely continuous* with respect to $\nu$ (denoted by $\mu \ll \nu$) if $\mu(A) = 0$ for every set such that $\nu(A) = 0$. If $\mu \ll \nu$ and $\nu \ll \mu$, then we say that $\mu$ and $\nu$ are *equivalent*. The Radon-Nikodym theorem says that, if $\mu \ll \nu$, then there is a measurable function $f$ on the space to $\mathbb{R}$ such that

$$\mu(A) = \int_A f \, d\nu$$

for any measurable set $A$. The function $f$ is called the *Radon-Nikodym derivative* and denoted by $\frac{d\mu}{d\nu}$.

Bienvenu and Merkle[22] observed that, for computable measures $\mu$ and $\nu$ on $2^\omega$, we have

$$\mathrm{CR}(\mu) = \mathrm{CR}(\nu) \;\Rightarrow\; \mathrm{MLR}(\mu) = \mathrm{MLR}(\nu) \;\Rightarrow\; \mu \text{ and } \nu \text{ are equivalent} \qquad (1)$$

where the former implication follows from Theorem 9.7 in Muchnik, Semenor, and Uspensky[23]. The theorem by Muchnik, Semenor, and Uspensky actually shows that, for computable measures $\mu$ and $\nu$, we have

$$\mathrm{MLR}(\nu) \cap \mathrm{CR}(\mu) \subseteq \mathrm{MLR}(\mu).$$

Thus, if $\mathrm{CR}(\nu) \subseteq \mathrm{CR}(\mu)$, we have

$$\mathrm{MLR}(\nu) = \mathrm{MLR}(\nu) \cap \mathrm{CR}(\nu) \subseteq \mathrm{MLR}(\nu) \cap \mathrm{CR}(\mu) \subseteq \mathrm{MLR}(\mu).$$

The proof of the latter implication of (1) actually showed that, if there exists $\mathcal{X}$ such that $\mu(\mathcal{X}) = 0$ and $\nu(\mathcal{X}) > 0$ for two computable measures $\mu, \nu$, then $\mathrm{MLR}(\nu) \not\subseteq \mathrm{MLR}(\mu)$. Hence, we have the following implications.

**Lemma 3.1.** *Let $\mu, \nu$ be computable measures. Then, we have*

$$\mathrm{CR}(\nu) \subseteq \mathrm{CR}(\mu) \quad \Rightarrow \quad \mathrm{MLR}(\nu) \subseteq \mathrm{MLR}(\mu) \quad \Rightarrow \quad \nu \ll \mu.$$

The following lemma is the key observation. If $\mathrm{CR}(\mu) \subseteq \mathrm{CR}(\lambda)$, then $\mu \ll \lambda$. By the Radon-Nikodym theorem, there exists a measurable function $f : 2^\omega \to \mathbb{R}^+$ such that $\mu(A) = \int_A f \, d\lambda$ for a Borel set $A$. By this existence of $f$, we can find positions $n$ such that the measure of $\{X \in 2^\omega : \Phi(X)(n) = 0\}$ is roughly half in measure. This process can be done repeatedly and we can find the positions on which the measure is close to the fair-coin measure as follows.

**Lemma 3.2.** *Let $\Phi :\subseteq 2^\omega \to 2^\omega$ be a Turing functional almost everywhere defined. Let $\mu$ be the push-forward measure of $\lambda$ by $\Phi$ where $\lambda$ is the fair-coin measure on $2^\omega$. Assume that $\mu \ll \lambda$. Then, for each $\sigma \in 2^{<\omega}$, we have*

$$\lim_{n\to\infty} \lambda\{X \in [\sigma] : \Phi(X)(n) = 0\} = \frac{1}{2}\lambda(\sigma).$$

**Proof.** Let $\mu_\sigma$ be the measure defined by

$$\mu_\sigma(\tau) = \lambda(\Phi^{-1}([\tau]) \cap [\sigma]) = \lambda\{X \in [\sigma] : \Phi(X) \in [\tau]\}.$$

Then,

$$\mu_\sigma \ll \mu \ll \lambda.$$

Let $g = \dfrac{d\mu_\sigma}{d\lambda}$ be a Radon-Nikodym derivative and let

$$g_n(X) = \frac{1}{2^{-n}} \int_{[X\restriction n]} g \, d\lambda.$$

By Lévy's zero-one law (see textbooks in probability theory such as Ref. 24), $\{g_n\}$ converges in the $L^1$-norm, that is,

$$\lim_n \int |g_n - g| \, d\lambda = 0.$$

For $n \in \omega$ and $i = 0, 1$, let

$$D(n, i) = \{Y \in 2^\omega : \Phi(Y)(n) = i\}.$$

Since $g_{n-1}$ looks only at $X \upharpoonright (n-1)$, we have $\int_{D(n,0)} g_{n-1} d\lambda = \int_{D(n,1)} g_{n-1} d\lambda$. Then,

$$\begin{aligned}
&\left| \int_{D(n,0)} g_n d\lambda - \int_{D(n,1)} g_n d\lambda \right| \\
&\leq \left| \int_{D(n,0)} g_n d\lambda - \int_{D(n,0)} g_{n-1} d\lambda \right| + \left| \int_{D(n,1)} g_{n-1} d\lambda - \int_{D(n,1)} g_n d\lambda \right| \\
&\leq \int_{D(n,0)} |g_n - g_{n-1}| d\lambda + \int_{D(n,1)} |g_n - g_{n-1}| d\lambda \\
&= \int |g_n - g_{n-1}| d\lambda.
\end{aligned}$$

Notice that

$$\int_{D(n,0)} g_n d\lambda + \int_{D(n,1)} g_n d\lambda = \mu_\sigma(2^\omega) = \lambda(\sigma).$$

Thus,

$$\lim_n \int_{D(n,0)} g_n d\lambda = \frac{1}{2}\lambda(\sigma).$$

Finally notice that

$$\int_{D(n,0)} g_n d\lambda = \int_{D(n,0)} g d\lambda = \mu_\sigma(D(n,0)) = \lambda\{X \in [\sigma] \ : \ \Phi(X)(n) = 0\}.$$

□

The lemma enables us to find, for each $\sigma$, a position $n$ such that the measure $\lambda\{X \in [\sigma] \ : \ \Phi(X)(n) = 0\}$ is roughly a half of $\lambda(\sigma)$. So when one looks only at these positions, $\mu$ looks like the uniform measure. We define such positions inductively. The positions are just the candidates for forcing. There are many candidates for $\sigma$, but still only finitely many. Thus, we consider all candidates for $\sigma$ and look for $n$ sufficiently large for all $\sigma$.

We now complete Case 2 of the proof.

**Proof of Theorem 3.9 (continued).**

*Case 2:* $\mathrm{CR}(\mu) \subseteq \mathrm{CR}(\lambda)$

First, we construct a computable function $g : \omega \to \omega$. The values of $g$ will be the possible candidates of the positions where we force $\Phi(X)(n) = 0$. The function $g$ depends only on $\Phi$. We will define $g(n)$ inductively on $n$. Let $G = \{g(n) \ : \ n \in \omega\}$.

Let $\epsilon$ be a positive rational number sufficiently small. Let $b_n$ be a computable strictly increasing sequence of rationals such that $0 < b_n < 1$ and $\prod_n b_n > 1-\epsilon$. The values $b_n$ are close to 1. They control the ratio between $\lambda\{X \in [\sigma] \ : \ \Phi(X)(n) = 0\}$

and $\lambda(\sigma)$. The values $b_n$ also control how many bits are needed to compute $\Phi(X)(n)$ as explained below.

As a warmup, we define $g(0)$ first. By Lemma 3.2, we can compute $n \in \omega$ such that

$$\lambda\{X \in 2^\omega \ : \ \Phi(X)(n) = i\} > \frac{1}{2}b_0$$

for $i = 0, 1$. We wish to define $n$ to be $g(0)$. However, there are many possibilities of $X$ such that $\Phi(X)(n) = 0$. In order to construct $X$ inductively, we would like to restrict possibilities to be finite. Notice that we can enumerate all strings $\tau$ such that $\Phi(\tau)(n)$ is defined. Since $\Phi$ is almost everywhere defined, the measure of all such strings is 1 for each $n$. In some fixed order search for sufficiently large $n$ and the least $s$ such that

$$\lambda(\bigcup\{[\tau] \ : \ \Phi(\tau)(n)[s] = i\}) > \frac{b_0}{2}$$

for both $i = 0, 1$. Then we define $g(0)$ to be this $n$. For this $s$, let $a(0)$ be the maximum of the lengths of the strings in the set

$$\{\tau \ : \ \Phi(\tau)(n)[s] \downarrow\}.$$

We also let

$$T(0) = \{\sigma \in 2^{a(0)} \ : \ \tau \preceq \sigma, \ \Phi(\tau)(n)[s] \downarrow\}.$$

We define $g(k)$, $a(k)$, $T(k)$ inductively on $k$. For each $\sigma \in T(k-1)$, search for $n > g(k-1)$ and $s$ such that

$$\lambda(\bigcup\{[\tau] \ : \ \sigma \prec \tau, \ \Phi(\sigma)(n)[s] = i\}) > \frac{b_k}{2}\lambda(\sigma)$$

for both $i = 0, 1$. For each $\sigma$, we can find such $s$ by taking sufficiently large $n$. Since the set $T(k-1)$ is finite, we can computably find sufficiently large $n$ and the least $s$ such that the above inequality holds for all $\sigma \in T(k-1)$. Then we define $g(k)$ to be this $n$. For this $s$, let $a(k)$ be the maximum of the lengths of the strings in the set

$$\bigcup_{\sigma \in T(k-1)} \{\tau \ : \ \sigma \prec \tau, \ \Phi(\tau)(n)[s] \downarrow\}.$$

We also let

$$T(k) = \{\rho \in 2^{a(k)} \ : \ \sigma \prec \tau \preceq \rho, \ \Phi(\tau)(n)[s] \downarrow\}.$$

Note that, for each $\sigma \in T(k-1)$,

$$|\{\rho \in T(k) \ : \ \sigma \preceq \rho\}| > b_k 2^{a(k)-a(k-1)}.$$

Similarly,

$$|\{\rho \in T(k) \ : \ \sigma \preceq \rho,\ \Phi(\rho)(n) = 0\}| > \frac{b_k}{2} 2^{a(k)-a(k-1)}.$$

Now we essentially follow the construction of $X \in \mathrm{SR} \setminus \mathrm{CR}$ in Ref. 12. We define $\psi$ as follows:

$$\psi(e, x) = \begin{cases} \langle e, x, s\rangle + 1 & \text{where } s \text{ is the smallest such that } \Phi_e(x)[s] \downarrow, \\ \uparrow & \text{if } \Phi_e(x) \uparrow . \end{cases}$$

Here, $\langle i, j\rangle$ is a number coding of the ordered pair and define $\langle i, j, k\rangle = \langle\langle i, j\rangle, k\rangle$. The function $\Phi_e$ is the $e$-th partial computable function in a fixed order. Notice that $\psi$ is one-to-one where defined, $\psi(e, x) > x$, and the relation $n \in \mathrm{rng}\psi$ is decidable. Furthermore, the numbers $e, x$ such that $\psi(e, x) = n$ is computable from $n$.

We define a computable function $p : \omega \to \omega$ as follows:

$$p(n) = \begin{cases} p(x) + 1 & \text{if } \exists x < n, \exists e < \log p(x) - 1, \psi(e, x) = n, \\ n + 4 & \text{otherwise.} \end{cases}$$

Loosely speaking, the number of candidates of forcing positions at the stage starting from $n$ is $\log p(n) - 1$. Notice that $\lim_n p(n) = \infty$.

We define a set $H_x$ as follows:

$$H_x = \{\psi(e, x) \ : \ \psi(e, x) \downarrow, e < \log p(x) - 1\}.$$

The set $H_x$ is the candidates of forcing positions at the stage starting from $x$. We assume $\Phi_0$ is total and $H_x$ is not empty for each $x \in \omega$. Notice that $\{H_x\}_{x\in\omega}$ is pairwise disjoint and $|H_x| \leq \log p(x) - 1$.

Let $h(x) = \max(H_x)$. We will force $\Phi(X)(g(h(x))) = 0$. The function $h$ may not be monotone, but $h$ grows faster than any computable order in the following sense. Let $f : \omega \to \omega$ be an increasing unbounded computable function. Let $e$ be such that $\Phi_e(x) = f(x)$. Then, $\psi(e, x) > f(x)$. Since $\lim_x p(x) = \infty$, we have $e < \log p(x) - 1$ for almost all $x$. For such $x$, we have $h(x) > f(x)$. Let $\widehat{h}$ be the reordering of $h$, that is, the strictly increasing function $\widehat{h}$ such that the range of $\widehat{h}$ is equal to the range of $h$. Note that $\widehat{h}$ also dominates all computable functions.

Let $\{M_n\}$ be a non-effective enumeration of $\mathbb{Q}_2$-valued computable martingales with initial capital 1. Let $V = \sum_n 2^{-n-1} M_n$. Then, $V$ is a martingale.

Now, we construct $X \in 2^\omega$. We define $\sigma_{-1} \prec \sigma_0 \prec \cdots \prec \sigma_n \prec X$. Let $\sigma_{-1}$ to be the empty string and $T(-1)$ to be the set containing only the empty string. Suppose we have chosen $\sigma_{n-1} \in T(n-1)$. The definition of $\sigma_n$ varies depending on whether there exists $k$ such that $\widehat{h}(k) = n$.

If no $k$ satisfies $\widehat{h}(k) = n$, we would like to define $\sigma_n$ so that $V$ does not increase rapidly along $\sigma_n$ after $\sigma_{n-1}$. This is possible because the measure of $\bigcup\{[\tau] : \sigma_{n-1} \prec \tau \in T(n)\}$ is close to the measure of $[\sigma_{n-1}]$. We claim that there exists $\tau \in T(n)$ such that $\sigma_{n-1} \preceq \tau$ and

$$V(\tau) \leq \frac{1}{b_n} V(\sigma_{n-1}).$$

Suppose not. Since the the number of possibilities of $\tau$ is more than $b_n 2^{a(n)-a(n-1)}$,

$$\sum\{V(\tau) : \tau \in T(n), \sigma_{n-1} \preceq \tau\} > \frac{1}{b_n} V(\sigma_{n-1}) \cdot b_n 2^{a(n)-a(n-1)}. \tag{2}$$

In contrast, the fairness condition in the definition of martingales imply

$$\begin{aligned} V(\sigma_{n-1}) =& 2^{-(a(n)-a(n-1))} \sum\{V(\tau) : \tau \in 2^{a_n}, \sigma_{n-1} \preceq \tau\} \\ \geq& 2^{-(a(n)-a(n-1))} \sum\{V(\tau) : \tau \in T(n),\ \sigma_{n-1} \preceq \tau\}, \end{aligned}$$

which contradicts the inequality (2) above. Then, define $\sigma_n$ to be such a string $\tau$.

If there exists $k$ such that $\widehat{h}(k) = n$, we need to define $\sigma_n$ so that $\Phi(\sigma_n)(g(n)) = 0$. We can show that there exists $\tau \in T(n)$ such that $\sigma_{n-1} \preceq \tau$, $\Phi(\tau)(g(n)) = 0$, and

$$V(\tau) \leq \frac{2}{b_n} V(\sigma_{n-1})$$

in the same way as above. Then, define $\sigma_n$ to be such a string $\tau$.

We claim that $X$ is Schnorr random. Suppose that there exists a computable martingale $M_t$ and a computable order $f$ such that $M_t(X \upharpoonright f(m)) > 2^m$ for infinitely many $m$. Let $c(m)$ be the smallest natural number such that $f(m) \leq a(c(m))$. Since $f$ and $a$ are increasing and computable, $c$ is also computable. By the savings property of $V$,

$$\begin{aligned} V(X \upharpoonright a(c(m))) \geq& V(X \upharpoonright f(m)) - 2 \geq 2^{-t-1} M_t(X \upharpoonright f(m)) - 2 \\ \geq& 2^{m-t-1} - 2. \end{aligned} \tag{3}$$

By contrast, by the construction of $X$, the martingale $V$ increases only a little along $X$ except at the positions $g(n)$ such that $\widehat{h}(k) = n$ for some $k$. The number of such numbers $k$ is bounded by $|\{k : \widehat{h}(k) \leq c(m)\}|$ and $V$ at most doubles at each position. Hence, we have

$$V(X \upharpoonright a(c(m))) \leq \prod_{n=1}^{c(m)} \frac{1}{b_n} \cdot 2^{|\{k : \widehat{h}(k) \leq c(m)\}|} < \frac{1}{1-\epsilon} 2^{|\{k : \widehat{h}(k) \leq c(m)\}|}. \tag{4}$$

By these inequality (3) and (4), we have

$$|\{k : \widehat{h}(k) \leq c(m)\}| > \log((1-\epsilon)(2^{m-t-1} - 2)).$$

Since $t$ and $\epsilon$ are fixed constants, the right-hand side is a computable function of $m$. Since $c$ is a computable function and $\widehat{h}$ dominates all computable functions, the left-hand side grows more slowly than any computable function, which contradicts the inequality.

Finally, we claim that $\Phi(X)$ is not computably random. We construct a computable martingale $M$ that succeeds on $X$. The initial capital of $M$ is 1. The martingale $M$ uses the martingale strategy $S$. We describe the strategy of $M$ for each $Z = Z(0)Z(1)Z(2)\cdots \in 2^\omega$. The strategy can be divided into countably many stages.

At 0-th stage, the strategy uses the martingale strategy with initial bet $\frac{1}{p(0)} = \frac{1}{4}$ only at positions in $g(H_0)$ until the prediction is correct. Notice that $g(H_0)$ is a finite set and the relation $x \in g(H_0)$ is computable. A more formal definition of $M$ at the 0-th stage is as follows. If $|\sigma| \notin g(H_0)$, then define $M(\sigma 0) = M(\sigma 1) = M(\sigma)$. If $|\sigma| \in g(H_0)$, then define $M(\sigma 0) = M(\sigma) + \frac{2^k}{p(0)}$ and $M(\sigma 1) = M(\sigma) - \frac{2^k}{p(0)}$ with an exception explained below. Here $k$ is the number of positions taking risks so far, that is, $k = |\{n \leq |\sigma| \ : \ n \in H_0\}|$. The exception is the case that $M(\sigma) - \frac{2^k}{p(0)} < 0$ for some $\sigma$. In this case, define $M(\tau) = M(\sigma)$ for all $\tau > \sigma$ and 0-th stage never ends. Clearly, $M$ keeps the capital non-negative. The 0-th stage ends at the position $g(n)$, that is, the smallest number such that $n \in H_0$ and $Z(g(n)) = 0$, or never ends if no such $n$ exists. Note that the position $g(n)$ need not be $g(h(0))$. Since $|H_0| \leq \log p(0)$, by the property of the martingale strategy, the capital $M(Z \upharpoonright g(n))$ at the end of the stage is $1 + \frac{1}{4}$. Since $n \in H_0$, $p(n) = p(0) + 1 = 5$. Next stage starts from the next position, say, $g(n) + 1$.

The strategy at $m$-th stage for $m \geq 1$ as follows. Suppose that $(m-1)$-th stage ends the position $g(n)$ and assume that $p(n) = m + 4$. Later we will prove that this assumption holds by induction by $m$. The strategy uses the martingale strategy with initial bet $\frac{1}{m+4}$ only at $g(H_n)$ until the prediction is correct. Note that each element in $g(H_n)$ is larger than $g(n)$. Since $|H_n| \leq \log p(n) = \log(m+4)$, the capital $M(Z \upharpoonright g(n))$ at the end of the $m$-th stage is $1 + \sum_{i=0}^{m} \frac{1}{i+4}$.

Suppose that the $m$-th stage ends at the position $g(n')$. We claim that $p(n') = m + 5$. This is true for $m = 0$. For $m \geq 1$, let $n$ be such that $n' \in H_n$. Notice that such an $n$ is unique because $\{H_x\}_{x \in \omega}$ is pairwise disjoint. Then, there exists $e < \log p(n) - 1$ such that $n' = \psi(e, n)$. By the definition of $p$, since $n < n'$, we have

$$p(n') = p(n) + 1 = (m - 1 + 5) + 1 = m + 5.$$

Finally, we prove that $M$ succeeds on the set $\Phi(X)$. By construction of $X$, we have $\Phi(X)(g(h(x))) = 0$ for all $x$. Thus, for each $m$, there exists $n \in H_m$ such that

$\Phi(X)(g(n)) = 0$. Hence, each stage ends at some point and the supremum of $M$ along $X$ is infinity. □

### 3.4. *KL-randomness and ML-randomness*

We give a comment on the relation between Kolmogorov-Loveland randomness (KL-randomness) and ML-randomness. We know that each ML-random set is KL-random but it is a long open question whether the inclusion is strict. Merkle, Miller, Nies, Reimann and Stephan showed that, if $X = Y \oplus Z$ is KL-random, then at least one of $Y$ and $Z$ is ML-random (Theorem 12 in Ref. 25). This fact immediately implies that KL $\equiv_w$ MLR where KL is the class of all KL-random sets. We do not know whether KL $\equiv_s$ MLR.

## Acknowledgement

This research initiated when Rupert Hölzl visited the author after NII Shonan meeting of "Algorithmic Randomness and Complexity" in September, 2014. A crucial idea of the proof of Theorem 3.9 was brought in the discussion with Laurent Bienvenu in the CIRM conference of "Computability, Randomness and Applications" in June 2016. The author was supported by JSPS KAKENHI Grant Number 26870143, and by the Research Institute for Mathematical Sciences, a Joint Usage/Research Center located in Kyoto University.

## References

1. K. de Leeuw, E. F. Moore, C. F. Shannon and N. Shapiro, Computability by probabilistic machines, *Automata Studies* **34**, 183 (1956).
2. S. G. Simpson, Mass Problems and Randomness, *The Bulletin of Symbolic Logic* **11**, 1 (2005).
3. V. Brattka, G. Gherardi and R. Hölzl, Probabilistic Computability and Choice, *Information and Computation* **242**, 249 (2015).
4. S. B. Cooper, *Computability Theory* (CRC Press, 2004).
5. R. I. Soare, *Turing Computability* Theory and Applications of Computability, Theory and Applications of Computability (Springer, 2016).
6. A. A. Muchnik, On strong and weak reducibilities of algorithmic problems, *Sibirskii Matematicheskii Zhurnal* **4**, 1328 (1963).
7. Y. T. Medvedev, Degrees of difficulty of the mass problem, *Doklady Akademii Nauk SSSR (N.S.)* **104**, 501 (1955).
8. A. Nies, *Computability and randomness*, Oxford Logic Guides, Vol. 51 (Oxford University Press, Oxford, 2009).
9. R. G. Downey and D. R. Hirschfeldt, *Algorithmic randomness and complexity* Theory and Applications of Computability, Theory and Applications of Computability (Springer, New York, 2010).

10. M. van Lambalgen, Random sequences, PhD thesis, University of Amsterdam (1987).
11. J. N. Y. Franklin and F. Stephan, Schnorr trivial sets and truth-table reducibility, *Journal of Symbolic Logic* **75**, 501 (2010).
12. A. Nies, F. Stephan and S. Terwijn, Randomness, relativization and Turing degrees, *Journal of Symbolic Logic* **70**, 515 (2005).
13. S. B. Cooper, Minimal degrees and the jump operator, *The Journal of Symbolic Logic* **38**, 249 (1973).
14. G. Sacks, On the degrees less than $\mathbf{0}'$, *Annals of Mathematics* **77**, 211 (1963).
15. W. Miller and D. A. Martin, The degrees of hyperimmune sets, *Zeitschrift für Mathematische Logic und Grundlagen der Mathematik* **14**, 159 (1968).
16. J. N. Y. Franklin and K. M. Ng, Difference randomness, *Proceedings of the American Mathematical Society* **139**, 345 (2011).
17. R. Downey, A. Nies, R. Weber and L. Yu, Lowness and $\Pi^0_2$ nullsets, *J. Symbolic Logic* **71**, 1044 (2006).
18. S. Kautz, Degrees of Random Sets, PhD thesis, Cornell University (1991).
19. A. K. Zvonkin and L. A. Levin, The complexity of finite objects and the development of the concepts of information and randomness by means of the theory of algorithms, *Russian Mathematical Surveys* **25**, 83 (1970).
20. L. Bienvenu and C. Porter, Strong reductions in effective randomness, *Theoretical Computer Science* **459**, 55 (2012).
21. J. Rute, When does randomness come from randomness?, *Theoretical Computer Science* **635**, 35 (2016).
22. L. Bienvenu and W. Merkle, Constructive equivalence relations on computable probability measures, *Annals of Pure and Applied Logic* **160**, 238 (2009).
23. A. A. Muchnik, A. L. Semenov and V. A. Uspensky, Mathematical metaphysics of randomness, *Theoretical Computer Science* **207**, 263 (1998).
24. R. Durrett, *Probability: Theory and Examples*, 4th edition edn. (Cambridge University Press, 2010).
25. W. Merkle, J. Miller, A. Nies, J. Reimann and F. Stephan, Kolmogorov-Loveland randomness and stochasticity, *Annals of Pure and Applied Logic* **138**, 183 (2006).

# Two Results on Cardinal Invariants at Uncountable Cardinals

Dilip Raghavan*

*Department of Mathematics, National University of Singapore,*
*Singapore 119076*
*E-mail: raghavan@math.nus.edu.sg*
*http://www.math.nus.edu.sg/~raghavan*

Saharon Shelah†

*Institute of Mathematics, The Hebrew University,*
*Jerusalem 9190401, Israel*
*E-mail: shelah@math.huji.ac.il*
*http://shelah.logic.at/*

We prove two ZFC theorems about cardinal invariants above the continuum which are in sharp contrast to well-known facts about these same invariants at the continuum. It is shown that for an uncountable regular cardinal $\kappa$, $\mathfrak{b}(\kappa) = \kappa^+$ implies $\mathfrak{a}(\kappa) = \kappa^+$. This improves an earlier result of Blass, Hyttinen, and Zhang[1]. It is also shown that if $\kappa \geq \beth_\omega$ is an uncountable regular cardinal, then $\mathfrak{d}(\kappa) \leq \mathfrak{r}(\kappa)$. This result partially dualizes an earlier theorem of the authors[2].

*Keywords*: Cardinal invariants, almost disjoint family, reaping number, revised GCH.

## 1. Introduction

The theory of cardinal invariants at uncountable regular cardinals remains less developed than the theory at $\omega$. One of the first papers to explore the situation above $\omega$ was by Cummings and Shelah[3]. In that paper, they considered the direct analogues of the bounding and dominating numbers. They also considered bounding and domination modulo the club filter, a notion which has no counterpart at $\omega$ but which becomes very natural at uncountable regular cardinals. Recall the following definitions.

**Definition 1.** Let $\kappa > \omega$ be a regular cardinal. Let $f, g \in \kappa^\kappa$. $f \leq^* g$ means that $|\{\alpha < \kappa : g(\alpha) < f(\alpha)\}| < \kappa$ and $f \leq_{\mathrm{cl}} g$ means that $\{\alpha < \kappa : g(\alpha) < f(\alpha)\}$ is

*First author was partially supported by the Singapore Ministry of Education's research grant number MOE2017-T2-2-125.

†Both authors were partially supported by European Research Council grant 338821. Publication 1135 on Shelah's list.

non-stationary. We say that $F \subset \kappa^\kappa$ is $*$-*unbounded* if $\neg\exists g \in \kappa^\kappa \forall f \in F\,[f \leq^* g]$ and we say that $F$ is cl-*unbounded* if $\neg\exists g \in \kappa^\kappa \forall f \in F\,[f \leq_{\text{cl}} g]$. Define

$$\mathfrak{b}(\kappa) = \min\{|F| : F \subset \kappa^\kappa \wedge F \text{ is } *\text{-unbounded}\},$$
$$\mathfrak{b}_{\text{cl}}(\kappa) = \min\{|F| : F \subset \kappa^\kappa \wedge F \text{ is cl-unbounded}\}.$$

We say that $F \subset \kappa^\kappa$ is $*$-*dominating* if $\forall g \in \kappa^\kappa \exists f \in F\,[g \leq^* f]$ and we say that $F$ is cl-*dominating* if $\forall g \in \kappa^\kappa \exists f \in F\,[g \leq_{\text{cl}} f]$. Define

$$\mathfrak{d}(\kappa) = \min\{|F| : F \subset \kappa^\kappa \text{ and } F \text{ is } *\text{-dominating}\}.$$
$$\mathfrak{d}_{\text{cl}}(\kappa) = \min\{|F| : F \subset \kappa^\kappa \text{ and } F \text{ is cl-dominating}\}.$$

Cummings and Shelah[3] proved that for any regular $\kappa$, $\kappa^+ \leq \text{cf}(\mathfrak{b}(\kappa)) = \mathfrak{b}(\kappa) \leq \text{cf}(\mathfrak{d}(\kappa)) \leq \mathfrak{d}(\kappa) \leq 2^\kappa$, and that these are the only relations between $\mathfrak{b}(\kappa)$ and $\mathfrak{d}(\kappa)$ that are provable in ZFC, thereby generalizing a classical result of Hechler from the case $\kappa = \omega$. Quite remarkably, they also showed that for every regular $\kappa > \omega$, $\mathfrak{b}(\kappa) = \mathfrak{b}_{\text{cl}}(\kappa)$, and that if $\kappa \geq \beth_\omega$ is regular, then $\mathfrak{d}(\kappa) = \mathfrak{d}_{\text{cl}}(\kappa)$. The question of whether $\mathfrak{d}_{\text{cl}}(\kappa) < \mathfrak{d}(\kappa)$ is consistent for any $\kappa$ was left open; as far as we are aware, it remains open.

Other early papers which studied the splitting number at uncountable cardinals revealed interesting differences with the situation at $\omega$. Recall the following definitions.

**Definition 2.** Let $\kappa > \omega$ be a regular cardinal. For $A, B \in \mathcal{P}(\kappa)$, $A \subset^* B$ means $|A \setminus B| < \kappa$. For a family $F \subset [\kappa]^\kappa$ and a set $B \in \mathcal{P}(\kappa)$, $B$ is said to *reap* $F$ if for every $A \in F$, $|A \cap B| = |A \cap (\kappa \setminus B)| = \kappa$. We say that $F \subset [\kappa]^\kappa$ is *unreaped* if there is no $B \in \mathcal{P}(\kappa)$ that reaps $F$.

$$\mathfrak{r}(\kappa) = \min\{|F| : F \subset [\kappa]^\kappa \text{ and } F \text{ is unreaped}\}.$$

A family $F \subset \mathcal{P}(\kappa)$ is called a *splitting family* if

$$\forall B \in [\kappa]^\kappa \exists A \in F\,[|B \cap A| = |B \cap (\kappa \setminus A)| = \kappa].$$

$$\mathfrak{s}(\kappa) = \min\{|F| : F \subset \mathcal{P}(\kappa) \text{ and } F \text{ is a splitting family}\}.$$

For instance, Suzuki[4] showed that for a regular cardinal $\kappa > \omega$, $\mathfrak{s}(\kappa) \geq \kappa$ iff $\kappa$ is strongly inaccessible and $\mathfrak{s}(\kappa) \geq \kappa^+$ iff $\kappa$ is weakly compact. Zapletal[5] additionally showed that the statement that there exists some regular uncountable cardinal $\kappa$ for which $\mathfrak{s}(\kappa) \geq \kappa^{++}$ has large consistency strength, significantly more than a measurable cardinal. More recently, the authors proved in[2] that $\mathfrak{s}(\kappa) \leq \mathfrak{b}(\kappa)$ for all regular $\kappa > \omega$. This is in marked contrast to the situation at $\omega$, where it is known that $\mathfrak{s}(\omega)$ and $\mathfrak{b}(\omega)$ are independent. More information about cardinal invariants at $\omega$ can be found in[6].

Blass, Hyttinen, and Zhang[1] is a work about the almost disjointness number at regular uncountable cardinals. Let us recall the definition of maximal almost disjoint families.

**Definition 3.** Let $\kappa > \omega$ be a regular cardinal. $A, B \in [\kappa]^{\kappa}$ are said to be *almost disjoint* or *a.d.* if $|A \cap B| < \kappa$. A family $\mathscr{A} \subset [\kappa]^{\kappa}$ is said to be *almost disjoint* or *a.d.* if the members of $\mathscr{A}$ are pairwise a.d. Finally $\mathscr{A} \subset [\kappa]^{\kappa}$ is called *maximal almost disjoint* or *m.a.d.* if $\mathscr{A}$ is an a.d. family, $|\mathscr{A}| \geq \kappa$, and $\mathscr{A}$ cannot be extended to a larger a.d. family in $[\kappa]^{\kappa}$.

$$\mathfrak{a}(\kappa) = \min\{|\mathscr{A}| : \mathscr{A} \subset [\kappa]^{\kappa} \text{ and } \mathscr{A} \text{ is m.a.d.}\}.$$

Blass, Hyttinen, and Zhang[1] proved that if $\kappa > \omega$ is regular, then $\mathfrak{d}(\kappa) = \kappa^+$ implies $\mathfrak{a}(\kappa) = \kappa^+$. This is potentially different from the situation at $\omega$: it remains an open problem whether $\mathfrak{d}(\omega) = \aleph_1$ implies $\mathfrak{a}(\omega) = \aleph_1$, while Shelah[7] showed the consistency of $\mathfrak{d}(\omega) = \aleph_2 < \aleph_3 = \mathfrak{a}(\omega)$ (see also Question 15).

There is also a well-developed theory of duality for cardinal invariants at $\omega$. Thus, for example, $\mathfrak{b}(\omega)$ and $\mathfrak{d}(\omega)$ are dual to each other, while $\mathfrak{s}(\omega)$ and $\mathfrak{r}(\omega)$ are duals. The ZFC inequality $\mathfrak{s}(\omega) \leq \mathfrak{d}(\omega)$ dualizes to the inequality $\mathfrak{b}(\omega) \leq \mathfrak{r}(\omega)$, and indeed even the proof of $\mathfrak{s}(\omega) \leq \mathfrak{d}(\omega)$ dualizes to the proof of $\mathfrak{b}(\omega) \leq \mathfrak{r}(\omega)$. It is possible to make this notion of duality precise using Galois-Tukey connections. We refer the reader to[6] for further details about duality of cardinal invariants at $\omega$. It is unclear at present if there can be a smooth theory of duality for cardinal invariants at uncountable cardinals too. For example, if we try to naïvely dualize Suzuki's result mentioned above that $\mathfrak{s}(\kappa)$ is small for most $\kappa$, then we would be trying to show that $\mathfrak{r}(\kappa)$ is large for most $\kappa$. In other words, we might expect to show that if $\kappa$ is not weakly compact, then $\mathfrak{r}(\kappa) = 2^{\kappa}$. However it is still an open problem whether the inequality $\mathfrak{r}(\aleph_1) < 2^{\aleph_1}$ is consistent (see Question 17). Nevertheless, it is of interest to ask whether for all regular $\kappa > \omega$ the result from[2] that $\mathfrak{s}(\kappa) \leq \mathfrak{b}(\kappa)$ can be dualized to the result that $\mathfrak{d}(\kappa) \leq \mathfrak{r}(\kappa)$.

We present two further ZFC theorems on cardinal invariants at uncountable regular cardinals in the paper. Our first result, Theorem 5, says that if $\kappa > \omega$ is regular, then $\mathfrak{b}(\kappa) = \kappa^+$ implies $\mathfrak{a}(\kappa) = \kappa^+$. This improves the above mentioned result of Blass, Hyttinen, and Zhang[1]. It also shows that $\omega$ is unique among regular cardinals in that it is the only such $\kappa$ where $\mathfrak{b}(\kappa) = \kappa^+ < \kappa^{++} = \mathfrak{a}(\kappa)$ is consistent. Our next result, Theorem 13, is a partial dual to our earlier result from[2]. It says that for all regular cardinals $\kappa \geq \beth_{\omega}$, $\mathfrak{d}(\kappa) \leq \mathfrak{r}(\kappa)$. Thus for sufficiently large $\kappa$, the invariants $\mathfrak{s}(\kappa), \mathfrak{b}(\kappa), \mathfrak{d}(\kappa)$, and $\mathfrak{r}(\kappa)$ are provably comparable and ordered as

$\mathfrak{s}(\kappa) \leq \mathfrak{b}(\kappa) \leq \mathfrak{d}(\kappa) \leq \mathfrak{r}(\kappa)$. The proof of our first theorem makes use of the equality $\mathfrak{b}(\kappa) = \mathfrak{b}_{\mathrm{cl}}(\kappa)$ of Cummings and Shelah[3] discussed before. Their theorem that $\mathfrak{d}(\kappa) = \mathfrak{d}_{\mathrm{cl}}(\kappa)$ for all regular $\kappa \geq \beth_\omega$ is not directly used. However the main idea of the proof of our Theorem 13 is similar to the main idea in the proof of $\mathfrak{d}(\kappa) = \mathfrak{d}_{\mathrm{cl}}(\kappa)$ – both results use the revised GCH of Shelah, which is a striking application of PCF theory exposed in[8].

Finally one word about our notation, which is standard. $X \subset Y$ means that $\forall x\,[x \in X \implies x \in Y]$. So the symbol "$\subset$" does not mean "proper subset". If $f$ is a function and $X \subset \mathrm{dom}(f)$, then $f''X$ is the image of $X$ under $f$, that is $f''X = \{f(x) : x \in X\}$.

## 2. The bounding and almost disjointness numbers: A ZFC result

We will quote the following well-known result of Cummings and Shelah[3].

**Theorem 4 (see Theorem 6 of[3]).** *For every regular cardinal $\kappa > \omega$, $\mathfrak{b}(\kappa) = \mathfrak{b}_{\mathrm{cl}}(\kappa)$.*

**Theorem 5.** *Let $\kappa > \omega$ be a regular cardinal. If $\mathfrak{b}(\kappa) = \kappa^+$, then $\mathfrak{a}(\kappa) = \kappa^+$.*

**Proof.** The hypothesis and Theorem 4 imply that there exists a sequence $\langle f_\delta : \delta < \kappa^+\rangle$ of functions in $\kappa^\kappa$ with the property that for any $g \in \kappa^\kappa$, there is a $\delta < \kappa^+$ such that $\{\alpha < \kappa : g(\alpha) < f_\delta(\alpha)\}$ is stationary in $\kappa$. For any $E \subset \kappa$, if $\mathrm{otp}(E) = \kappa$, then let $\langle \mu_{E,\xi} : \xi < \kappa\rangle$ be the increasing enumeration of $E$. For each $\delta < \kappa^+$, let $C_\delta = \{\alpha < \kappa : \alpha \text{ is closed under } f_\delta\}$. Recall that $C_\delta$ is a club in $\kappa$. Also, fix a sequence $\langle e_\delta : \kappa \leq \delta < \kappa^+\rangle$ of bijections $e_\delta : \kappa \to \delta$. We will construct a sequence $\langle\langle A_\delta, E_\delta\rangle : \delta < \kappa^+\rangle$ satisfying the following conditions for each $\delta < \kappa^+$:

(1) $A_\delta \in [\kappa]^\kappa$ and $E_\delta \subset C_\delta$ is a club in $\kappa$;
(2) $\forall \gamma < \delta \left[\left|A_\gamma \cap A_\delta\right| < \kappa\right]$;
(3) if $\kappa \leq \delta$, then $A_\delta = \bigcup_{\xi<\kappa} B_{\delta,\xi}$, where for each $\xi < \kappa$, $B_{\delta,\xi}$ is defined to be

$$\left\{\mu_{E_\delta,\xi} \leq \alpha < \mu_{E_\delta,\xi+1} : \forall \nu < \mu_{E_\delta,\xi}\left[\alpha \notin A_{e_\delta(\nu)}\right]\right\}.$$

Suppose for a moment that such a sequence can be constructed. Let $\mathscr{A} = \{A_\delta : \delta < \kappa^+\}$. By (1) and (2), $\mathscr{A}$ is an a.d. family in $[\kappa]^\kappa$ of size $\kappa^+$. We claim that it is maximal. To see this, fix $B \in [\kappa]^\kappa$. Define a function $g : \kappa \to \kappa$ by stipulating that for each $\mu \in \kappa$, $g(\mu) = \sup\left(\{\min(B \setminus (\mu+1))\} \cup \{f_\nu(\mu) : \nu \leq \mu\}\right)$. Find $\delta < \kappa^+$ such that $S = \{\mu \in \kappa : g(\mu) < f_\delta(\mu)\}$ is stationary in $\kappa$. Note that $\kappa \leq \delta$. Therefore the consequent of (3) applies to $\delta$. Let $I = \{\xi < \kappa : B_{\delta,\xi} \cap B \neq 0\}$. If $|I| = \kappa$, then $|A_\delta \cap B| = \kappa$, and we are done. So assume that $|I| < \kappa$. Then $\{\mu_{E_\delta,\xi} : \xi \in I\} \subset E_\delta \subset \kappa$ and $\left|\{\mu_{E_\delta,\xi} : \xi \in I\}\right| \leq |I| < \kappa$. Therefore $\sup\left(\{\mu_{E_\delta,\xi} : \xi \in I\}\right) = \nu_0 < \kappa$. Now $\{\mu \in$

$E_\delta : \mu > \nu_0\}$ is a club in $\kappa$ and $T = S \cap \{\mu \in E_\delta : \mu > \nu_0\}$ is stationary in $\kappa$. Consider any $\mu \in T$. There exists $\xi \in \kappa \setminus I$ with $\mu = \mu_{E_\delta,\xi}$. Note that $B_{\delta,\xi} \cap B = 0$ because $\xi \notin I$. On the other hand, $\mu_{E_\delta,\xi} = \mu < \min(B \setminus (\mu+1)) \le g(\mu) < f_\delta(\mu) < \mu_{E_\delta,\xi+1}$ because $\mu \in S$ and because $\mu_{E_\delta,\xi+1} \in C_\delta$. Since $\min(B \setminus (\mu+1)) \notin B_{\delta,\xi}$, it follows from the definition of $B_{\delta,\xi}$ that $\exists \nu < \mu \left[\min(B \setminus (\mu+1)) \in A_{e_\delta(\nu)}\right]$. Thus we have proved that for each $\mu \in T$, $\exists \nu < \mu \exists \beta \in B \left[\mu < \beta \wedge \beta \in A_{e_\delta(\nu)}\right]$. Since $T$ is stationary in $\kappa$, there exist $T^* \subset T$ and $\nu$ such that $T^*$ is stationary in $\kappa$ and for each $\mu \in T^*$, $\nu < \mu$ and $\exists \beta \in B \left[\mu < \beta \wedge \beta \in A_{e_\delta(\nu)}\right]$. It now easily follows that $\left|A_{e_\delta(\nu)} \cap B\right| = \kappa$. This proves the maximality of $\mathscr{A}$. Since $|\mathscr{A}| = \kappa^+$, we have $\mathfrak{a}(\kappa) \le \kappa^+$, while standard arguments (see Theorem 1.2 of[9]) show that $\kappa^+ \le \mathfrak{a}(\kappa)$. Hence we have $\mathfrak{a}(\kappa) = \kappa^+$.

Thus it suffices to construct a sequence satisfying (1)–(3) above. Let $\langle A_\gamma : \gamma \in \kappa \rangle$ be any partition of $\kappa$ into $\kappa$ many pairwise disjoint pieces of size $\kappa$. For each $\gamma < \kappa$, let $E_\gamma = C_\gamma$. It is clear that the sequence $\langle \langle A_\gamma, E_\gamma \rangle : \gamma < \kappa \rangle$ satisfies (1)–(3). Now fix $\kappa^+ > \delta \ge \kappa$ and assume that $\langle \langle A_\gamma, E_\gamma \rangle : \gamma < \delta \rangle$ satisfying (1)–(3) is given. We construct $A_\delta$ and $E_\delta$ as follows. Let $\theta$ be a sufficiently large regular cardinal. Let $x = \{\kappa, \langle f_\delta : \delta < \kappa^+ \rangle, \langle C_\delta : \delta < \kappa^+ \rangle, \langle e_\delta : \kappa \le \delta < \kappa^+ \rangle, \delta, \langle \langle A_\gamma, E_\gamma \rangle : \gamma < \delta \rangle\}$. Let $\langle N_\xi : \xi < \kappa \rangle$ be such that

$$\begin{aligned}
&(1)\ \forall \xi < \kappa \left[N_\xi \prec H(\theta) \wedge x \in N_\xi\right];\\
&(2)\ \forall \xi < \kappa \left[\left|N_\xi\right| < \kappa \wedge \mu_\xi = N_\xi \cap \kappa \in \kappa\right];\\
&(3)\ \forall \xi < \xi + 1 < \kappa \left[\langle N_\zeta : \zeta \le \xi \rangle \in N_{\xi+1}\right];\\
&(4)\ \forall \xi < \kappa \left[\xi \text{ is a limit ordinal} \implies N_\xi = \textstyle\bigcup_{\zeta<\xi} N_\zeta\right].
\end{aligned}$$

Observe that these conditions imply that $\forall \zeta < \xi < \kappa \left[N_\zeta \in N_\xi \wedge N_\zeta \subset N_\xi\right]$. Observe also that $E_\delta = \{\mu_\xi : \xi < \kappa\}$ is a club in $\kappa$ and that $\mu_{E_\delta,\xi} = \mu_\xi$, for all $\xi < \kappa$. Next for each $\xi < \kappa$, $C_\delta \in N_\xi$. It follows that $\mu_\xi \in C_\delta$ because $C_\delta$ is a club in $\kappa$. So $E_\delta \subset C_\delta$. Now define $A_\delta = \bigcup_{\xi<\kappa} B_{\delta,\xi}$, where for each $\xi < \kappa$, $B_{\delta,\xi}$ is

$$\left\{\mu_\xi \le \alpha < \mu_{\xi+1} : \forall \nu < \mu_\xi \left[\alpha \notin A_{e_\delta(\nu)}\right]\right\}.$$

It is clear that (3) is satisfied by definition and that $A_\delta \subset \kappa$. So to complete the proof, it suffices to check that $|A_\delta| = \kappa$ and that $\forall \gamma < \delta \left[\left|A_\gamma \cap A_\delta\right| < \kappa\right]$. To see the second statement, fix any $\gamma < \delta$. Since $e_\delta : \kappa \to \delta$ is a bijection, we can find $\nu \in \kappa$ with $e_\delta(\nu) = \gamma$. Find $\zeta < \kappa$ with $\nu < \mu_\zeta$. Consider any $\xi < \kappa$ so that $\zeta \le \xi$. Then $\nu < \mu_\zeta \le \mu_\xi$. It follows that $A_\gamma \cap B_{\delta,\xi} = A_{e_\delta(\nu)} \cap B_{\delta,\xi} = 0$. Therefore, $A_\gamma \cap A_\delta = \bigcup_{\xi<\kappa}\left(A_\gamma \cap B_{\delta,\xi}\right) = \bigcup_{\xi<\zeta}\left(A_\gamma \cap B_{\delta,\xi}\right) \subset \bigcup_{\xi<\zeta} B_{\delta,\xi}$. For each $\xi < \zeta$, $\left|B_{\delta,\xi}\right| < \kappa$. So $\bigcup_{\xi<\zeta} B_{\delta,\xi}$ is the union of $\le |\zeta| \le \zeta < \kappa$ many sets each of size $< \kappa$. Since $\kappa$ is regular, we conclude that $\left|\bigcup_{\xi<\zeta} B_{\delta,\xi}\right| < \kappa$. So $\left|A_\gamma \cap A_\delta\right| < \kappa$, as needed.

Finally we check that for each $\xi < \kappa$, $B_{\delta,\xi} \ne 0$. This will imply that $|A_\delta| = \kappa$.

Fix any $\xi < \kappa$. Note that for each $\nu < \mu_\xi$, $\left|A_{e_\delta(\mu_\xi)} \cap A_{e_\delta(\nu)}\right| < \kappa$. Therefore $R_\xi = \bigcup_{\nu<\mu_\xi}\left(A_{e_\delta(\mu_\xi)} \cap A_{e_\delta(\nu)}\right)$ is the union of at most $\left|\mu_\xi\right| \leq \mu_\xi < \kappa$ many sets each having size $< \kappa$. Since $\kappa$ is regular, it follows that $\left|R_\xi\right| < \kappa$. Hence there is an $\alpha \in A_{e_\delta(\mu_\xi)} \setminus R_\xi$ with $\mu_\xi \leq \alpha$ because $\left|A_{e_\delta(\mu_\xi)}\right| = \kappa$. Since $N_{\xi+1} \prec H(\theta)$ and since all the relevant parameters belong to $N_{\xi+1}$, we conclude that there exists $\alpha \in N_{\xi+1}$ such that $\alpha \in \kappa$, $\mu_\xi \leq \alpha$, and $\forall \nu \in \mu_\xi \left[\alpha \notin A_{e_\delta(\nu)}\right]$. Now we have that $\mu_\xi \leq \alpha < \mu_{\xi+1}$ and so $\alpha \in B_{\delta,\xi}$. This shows that $B_{\delta,\xi} \neq 0$ and concludes the proof. □

## 3. The reaping and dominating numbers: An application of PCF theory

We begin with a well-known fact, whose proof we include for completeness.

**Definition 6.** Let $\kappa > \omega$ be a regular cardinal. If $A \in [\kappa]^\kappa$, then we let $e_A : \kappa \to A$ be the order isomorphism from $\langle \kappa, \in \rangle$ to $\langle A, \in \rangle$. We also define a function $s_A : \kappa \to A$ by setting $s_A(\alpha) = \min(A \setminus (\alpha + 1))$, for each $\alpha \in \kappa$. We also write $\lim(\kappa) = \{\alpha < \kappa : \alpha \text{ is a limit ordinal}\}$ and $\mathrm{succ}(\kappa) = \{\alpha < \kappa : \alpha \text{ is a successor ordinal}\}$.

**Lemma 7 (Folklore).** *If $\kappa > \omega$ is a regular cardinal, then $\mathfrak{r}(\kappa) \geq \kappa^+$.*

**Proof.** Let $F \subset [\kappa]^\kappa$ be a family with $|F| \leq \kappa$. We must find a $B \in \mathcal{P}(\kappa)$ which reaps $F$. If $F$ is empty, then $B = \kappa$ will work. So assume $F$ is non-empty. Let $\{A_\alpha : \alpha < \kappa\}$ enumerate $F$, possibly with repetitions. For each $\alpha < \kappa$, let $C_\alpha = \{\delta < \kappa : \delta \text{ is closed under } s_{A_\alpha}\}$. Then $C = \{\delta < \kappa : \forall \alpha < \delta\, [\delta \in C_\alpha]\}$ is a club in $\kappa$. For each $\xi \in \kappa$, let $B_\xi = \{\zeta < e_C(\xi + 1) : e_C(\xi) \leq \zeta\}$. Note that for all $\alpha < e_C(\xi + 1)$, $A_\alpha \cap B_\xi \neq 0$. Also for any distinct $\xi, \xi' \in \kappa$, $B_\xi \cap B_{\xi'} = 0$. Put $B = \bigcup\left\{B_\xi : \xi \in \lim(\kappa)\right\}$. Then $B \in \mathcal{P}(\kappa)$ and since for each $\alpha < \kappa$ and each $\xi \in \lim(\kappa) \setminus \alpha$, $A_\alpha \cap B_\xi \neq 0$, $|A_\alpha \cap B| = \kappa$, for all $\alpha < \kappa$. Furthermore, $\bigcup\left\{B_{\xi'} : \xi' \in \mathrm{succ}(\kappa)\right\} \subset \kappa \setminus B$, and since for each $\alpha < \kappa$ and for each $\xi' \in \mathrm{succ}(\kappa) \setminus \alpha$, $A_\alpha \cap B_{\xi'} \neq 0$, $|A_\alpha \cap (\kappa \setminus B)| = \kappa$, for all $\alpha < \kappa$. Thus $B$ reaps $F$. □

The above proof really shows that $\mathfrak{r}(\kappa) \geq \mathfrak{b}(\kappa)$. However we will not need this in what follows. The proof of the main theorem is broken into two cases. For the remainder of this section, let $\kappa > \omega$ be a fixed regular cardinal. The crucial definition is the following.

**Definition 8.** Let $E_2 \subset E_1$ both be clubs in $\kappa$. For each $\xi \in \kappa$, define $\mathrm{set}(E_1, \xi) = \{\zeta < s_{E_1}(\xi) : \xi \leq \zeta\}$. Define $\mathrm{set}(E_2, E_1) = \bigcup\{\mathrm{set}(E_1, \xi) : \xi \in E_2\}$.

**Lemma 9.** *Suppose that $F \subset [\kappa]^\kappa$ is an unreaped family with $|F| = \mathfrak{r}(\kappa)$. Assume there is a club $E_1 \subset \kappa$ such that for each club $E \subset E_1$, there exists $A \in F$ with $A \subset^* \mathrm{set}(E, E_1)$. Then $\mathfrak{d}(\kappa) \leq \mathfrak{r}(\kappa)$.*

**Proof.** For each $A \in F$ define a function $g_A : \kappa \to \kappa$ as follows. Given $\beta \in \kappa$, $g_A(\beta) = s_A(s_{E_1}(\beta))$. Then $|\{g_A : A \in F\}| \leq |F| = \mathfrak{r}(\kappa)$, and we will check that this is a dominating family of functions. To this end, fix any $f \in \kappa^\kappa$. Put

$$E_f = \{\xi \in E_1 : \xi \text{ is closed under } f\}.$$

Then $E_f \subset E_1$ and it is a club in $\kappa$. By hypothesis there exist $A \in F$ and $\delta \in \kappa$ with $A \setminus \delta \subset \mathrm{set}(E_f, E_1)$. We claim that for any $\zeta \in \kappa$, if $\zeta \geq \delta$, then $f(\zeta) < g_A(\zeta)$. Indeed suppose $\delta \leq \zeta < \kappa$ is given. Let $\gamma = s_{E_1}(\zeta) > \zeta$ and let $g_A(\zeta) = \beta = s_A(s_{E_1}(\zeta))$. Then $\beta \in A$ and $\delta \leq \zeta < s_{E_1}(\zeta) < \beta$. Thus $\beta \in \mathrm{set}(E_f, E_1)$. Let $\zeta' \in E_f$ be such that $\zeta' \leq \beta < s_{E_1}(\zeta')$. It could not be the case that $\zeta' < \gamma$, for if that were the case, then the inequality $\beta < s_{E_1}(\zeta') \leq \gamma = s_{E_1}(\zeta) < \beta$ would be true, which is impossible. Therefore $\gamma \leq \zeta'$ and since $\zeta < \gamma \leq \zeta'$ and $\zeta'$ is closed under $f$, we have $f(\zeta) < \zeta' \leq \beta = g_A(\zeta)$, as claimed. Hence $f \leq^* g_A$. As $f \in \kappa^\kappa$ was arbitrary, this proves that $\{g_A : A \in F\}$ is dominating, and so $\mathfrak{d}(\kappa) \leq |\{g_A : A \in F\}| \leq \mathfrak{r}(\kappa)$. □

The proof in the case when the hypothesis of Lemma 9 fails will make use of Shelah's Revised GCH, which is a theorem of ZFC. Let us recall the definition of various notions that are relevant to the revised GCH.

**Definition 10.** Let $\kappa$ and $\lambda$ be cardinals. Define $\lambda^{[\kappa]}$ to be

$$\min\left\{|\mathcal{P}| : \mathcal{P} \subset [\lambda]^{\leq\kappa} \text{ and } \forall u \in [\lambda]^{\kappa} \exists \mathcal{P}_0 \subset \mathcal{P}\left[|\mathcal{P}_0| < \kappa \text{ and } u = \bigcup \mathcal{P}_0\right]\right\}.$$

The operation $\lambda^{[\kappa]}$ is sometimes referred to as the *weak power*.

The following remarkable ZFC result was obtained by Shelah in[8] as one of the many fruits of his PCF theory. A nice exposition of its proof may also be found in Abraham and Magidor[10]. Another relevant reference is Shelah[11].

**Theorem 11 (The Revised GCH).** *If $\theta$ is a strong limit uncountable cardinal, then for every $\lambda \geq \theta$, there exists $\sigma < \theta$ such that for every $\sigma \leq \kappa < \theta$, $\lambda^{[\kappa]} = \lambda$.*

**Corollary 12.** *Let $\mu \geq \beth_\omega$ be any cardinal. There exists an uncountable regular cardinal $\theta < \beth_\omega$ and a family $\mathcal{P} \subset [\mu]^{\leq\theta}$ such that $|\mathcal{P}| \leq \mu$ and for each $u \in [\mu]^\theta$, there exists $v \in \mathcal{P}$ with the property that $v \subset u$ and $|v| \geq \aleph_0$.*

**Proof.** $\beth_\omega$ is a strong limit uncountable cardinal. Therefore Theorem 11 applies and implies that there exists $\sigma < \beth_\omega$ such that for every $\sigma \leq \theta < \beth_\omega$, $\mu^{[\theta]} = \mu$. It is possible to choose an uncountable regular cardinal $\theta$ satisfying $\sigma \leq \theta < \beth_\omega$. Since $\mu^{[\theta]} = \mu$, there exists $\mathcal{P} \subset [\mu]^{\leq\theta}$ such that $|\mathcal{P}| = \mu$ and for each $u \in [\mu]^\theta$, there exists $\mathcal{P}_0 \subset \mathcal{P}$ with the property that $|\mathcal{P}_0| < \theta$ and $u = \bigcup \mathcal{P}_0$. Now suppose that $u \in [\mu]^\theta$ is given. Let $\mathcal{P}_0 \subset \mathcal{P}$ be such that $|\mathcal{P}_0| < \theta$ and $u = \bigcup \mathcal{P}_0$. Since $\theta$ is a regular cardinal and $|u| = \theta$, it follows that $|v| = \theta \geq \aleph_0$, for some $v \in \mathcal{P}_0$. This is as required because $v \in \mathcal{P}$ and $v \subset u$. □

The proof of the following theorem is similar to the proof of Cummings and Shelah's theorem from[3] that if $\kappa \geq \beth_\omega$, then $\mathfrak{d}(\kappa) = \mathfrak{d}_{\mathrm{cl}}(\kappa)$.

**Theorem 13.** *If $\kappa \geq \beth_\omega$, then $\mathfrak{d}(\kappa) \leq \mathfrak{r}(\kappa)$.*

**Proof.** Write $\mu = \mathfrak{r}(\kappa)$. Let $F \subset [\kappa]^\kappa$ be such that $F$ is unreaped and $|F| = \mu$. Then $\beth_\omega \leq \kappa < \kappa^+ \leq \mathfrak{r}(\kappa) = \mu$. So applying Corollary 12, fix an uncountable regular cardinal $\theta < \beth_\omega$ satisfying the conclusion of Corollary 12. Note that $|\theta \times \mu| = \mu$ because $\theta < \beth_\omega < \mu$. So $|\theta \times F| = \mu$. Therefore applying Corollary 12, find a family $\mathcal{P} \subset [\theta \times F]^{\leq\theta}$ such that $|\mathcal{P}| \leq \mu$ and $\mathcal{P}$ has the property that for each $u \in [\theta \times F]^\theta$, there exists $v \in \mathcal{P}$ satisfying $v \subset u$ and $|v| \geq \aleph_0$. Put $X = F \cup \mu \cup \mathcal{P} \cup \{\theta, \mu, \kappa, \kappa^\kappa, \mathcal{P}(\kappa)\}$. Then $|X| = \mu$, and so if $\chi$ is a sufficiently large regular cardinal, then there exists $M \prec H(\chi)$ with $|M| = \mu$ and $X \subset M$. We will aim to prove that $M \cap \kappa^\kappa$ is a dominating family.

In view of Lemma 9 it may be assumed that for any club $E_1 \subset \kappa$, there exists a club $E_2 \subset E_1$ such that for all $B \in F$, $B \not\subset^* \mathrm{set}(E_2, E_1)$. Since $F$ is an unreaped family and since $\mathrm{set}(E_2, E_1) \in \mathcal{P}(\kappa)$ whenever $E_2 \subset E_1$ are both clubs in $\kappa$, it follows that for each club $E_1 \subset \kappa$, there exist a club $E_2 \subset E_1$ and a $B \in F$ such that $B \subset^* \kappa \setminus \mathrm{set}(E_2, E_1)$. Let $f \in \kappa^\kappa$ be a fixed function. Construct a sequence $\langle\langle E_i, E_i^!, B_i\rangle : i < \theta\rangle$ by induction on $i < \theta$ so that the following conditions are satisfied at each $i < \theta$:

(1) $E_i$ and $E_i^!$ are both clubs in $\kappa$, $E_i^! \subset E_i$, and $\forall j < i\left[E_i \subset E_j^!\right]$;
(2) $B_i \in F$ and $B_i \subset^* \kappa \setminus \mathrm{set}(E_i^!, E_i)$;
(3) if $i = 0$, then $E_i = \{\alpha < \kappa : \alpha \text{ is closed under } f\}$.

We first show how to construct such a sequence. When $i = 0$, put $E_i = \{\alpha < \kappa : \alpha$ is closed under $f\}$. Then $E_i$ is a club in $\kappa$, and so there exist a club $E_i^! \subset E_i$ and a $B_i \in F$ with $B_i \subset^* \kappa \setminus \mathrm{set}(E_i^!, E_i)$. Next suppose that $\theta > i > 0$ and that $\langle\langle E_j, E_j^!, B_j\rangle : j < i\rangle$ satisfying (1)–(3) is given. Then $\{E_j^! : j < i\}$ is a collection of $\leq |i| \leq i < \theta < \beth_\omega \leq \kappa$ many clubs in $\kappa$. Therefore $E_i = \bigcap_{j<i} E_j^!$ is a club in $\kappa$. We have $\forall j < i\left[E_i \subset E_j^!\right]$ and moreover there exist a club $E_i^! \subset E_i$ and a $B_i \in F$ such that $B_i \subset^* \kappa \setminus \mathrm{set}(E_i^!, E_i)$. It is clear that $E_i, E_i^!$, and $B_i$ are as required. This completes the construction of the sequence $\langle\langle E_i, E_i^!, B_i\rangle : i < \theta\rangle$.

Now define a function $u : \theta \to F$ by setting $u(i) = B_i$ for all $i \in \theta$. Then $u \subset \theta \times F$ and $|u| = |\mathrm{dom}(u)| = \theta$. Hence by the choice of $\mathcal{P}$ and $M$, there exists $v \in \mathcal{P} \subset X \subset M$ such that $v \subset u$ and $|v| \geq \aleph_0$. $v$ is a function and $c = \mathrm{dom}(v) \subset \mathrm{dom}(u) = \theta$. Moreover, $\aleph_0 \leq |v| = |c|$ and $c \in M$. Hence we can find $d \in M$ so that $d \subset c$ and $\mathrm{otp}(d) = \omega$. Let $w = v \restriction d \in M$. Since $\kappa > \omega$ is regular, there exists a function $g \in \kappa^\kappa$ with the property that for each $\alpha \in \kappa$, $\forall i \in d \exists \beta \in w(i) = B_i\,[\alpha < \beta < g(\alpha)]$. We may further assume that $g \in M$ because all of the relevant parameters belong to $M$. Let $\langle i_n : n \in \omega \rangle$ be the strictly increasing enumeration of $d$. Recall that for each $n \in \omega$, $E^1_{i_n} \subset E_{i_n} \subset \kappa$ are both clubs in $\kappa$ and that $B_{i_n} \subset^* \kappa \setminus \mathrm{set}(E^1_{i_n}, E_{i_n})$. In particular, for each $n \in \omega$, there exists $\delta_n \in \kappa$ so that $B_{i_n} \setminus \delta_n \subset \kappa \setminus \mathrm{set}(E^1_{i_n}, E_{i_n})$, and also $\min(E_{i_n}) \in \kappa$. Hence $\{\delta_n : n \in \omega\} \cup \{\min(E_{i_n}) : n \in \omega\}$ is a countable subset of $\kappa$, whence $\{\delta_n : n \in \omega\} \cup \{\min(E_{i_n}) : n \in \omega\} \subset \delta$, for some $\delta \in \kappa$. We will argue that for each $\alpha \in \kappa$, if $\alpha \geq \delta$, then $f(\alpha) < g(\alpha)$. To this end, let $\alpha \in \kappa$ be fixed, and assume that $\delta \leq \alpha$. For each $n \in \omega$, since $E_{i_n} \subset \kappa$ is a club in $\kappa$ and since $\min(E_{i_n}) < \delta \leq \alpha < \alpha + 1 < \kappa$, it follows that $\xi_n = \sup(E_{i_n} \cap (\alpha + 1)) \in E_{i_n}$. Also $\forall n \in \omega\,[\xi_{n+1} \leq \xi_n]$ because $\forall n \in \omega\,[E_{i_{n+1}} \subset E_{i_n}]$. It follows that there exist $\xi$ and $N \in \omega$ such that $\forall n \geq N\,[\xi_n = \xi]$. Note that $\xi \in E_{i_{N+1}} \subset E^1_{i_N}$. Consider $s_{E_{i_N}}(\xi)$. $s_{E_{i_N}}(\xi) \in E_{i_N}$ and $s_{E_{i_N}}(\xi) > \xi = \xi_N = \sup(E_{i_N} \cap (\alpha + 1))$. Therefore $s_{E_{i_N}}(\xi) \geq \alpha + 1 > \alpha$. Since $s_{E_{i_N}}(\xi) \in E_{i_N} \subset E_0$, $s_{E_{i_N}}(\xi)$ is closed under $f$. Therefore $f(\alpha) < s_{E_{i_N}}(\xi)$. Next by the choice of $g$, there exists $\beta \in B_{i_N}$ with $\alpha < \beta < g(\alpha)$. Note that $\delta_N < \delta \leq \alpha < \beta$. Hence $\beta \in B_{i_N} \setminus \delta_N \subset \kappa \setminus \mathrm{set}(E^1_{i_N}, E_{i_N})$, in other words, $\beta \notin \mathrm{set}(E^1_{i_N}, E_{i_N})$. Note that $\xi = \sup(E_{i_N} \cap (\alpha + 1)) \leq \alpha < \beta$. Since $\xi \in E^1_{i_N}$, $\beta \geq s_{E_{i_N}}(\xi)$. Putting all this information together, we have $f(\alpha) < s_{E_{i_N}}(\xi) \leq \beta < g(\alpha)$, as required.

Thus we have proved that $f \leq^* g$. Since $f \in \kappa^\kappa$ was arbitrary and since $g \in M \cap \kappa^\kappa$, we have proved that $M \cap \kappa^\kappa$ is a dominating family. Therefore $\mathfrak{d}(\kappa) \leq |M \cap \kappa^\kappa| \leq |M| = \mu = \mathfrak{r}(\kappa)$. □

## 4. Questions

Raghavan and Shelah [12] introduced the method of forcing with a carefully chosen Boolean ultrapower of a forcing iteration to obtain the following result.

**Theorem 14 ([12]).** *Let $\kappa \geq \omega$ be any regular cardinal. If there is a supercompact cardinal $\theta > \kappa$, then there is a cardinal preserving forcing extension in which $\theta < \mathfrak{b}(\kappa) = \mathfrak{d}(\kappa) < \mathfrak{a}(\kappa)$. There is also a cardinal preserving forcing extension in which $\theta < \mathfrak{b}(\kappa) < \mathfrak{d}(\kappa) < \mathfrak{a}(\kappa)$.*

In the models of $\mathfrak{b}(\kappa) < \mathfrak{a}(\kappa)$ obtained in [12], the value of $\mathfrak{b}(\kappa)$ is much larger than $\kappa$. It is unknown how large $\mathfrak{b}(\kappa)$ needs to be for the configuration $\mathfrak{b}(\kappa) < \mathfrak{a}(\kappa)$ to be consistent. So we ask

**Question 15.** Does $\mathfrak{b}(\kappa) = \kappa^{++}$ imply that $\mathfrak{a}(\kappa) = \kappa^{++}$, for every regular cardinal $\kappa > \omega$?

It is not possible to step-up the proof of Theorem 5 in any straightforward way. If Question 15 has a positive answer, then the proof is likely to involve quite a different argument.

Theorem 13 of course gives no information about the relationship between $\mathfrak{d}(\kappa)$ and $\mathfrak{r}(\kappa)$ when $\kappa < \beth_\omega$.

**Question 16.** If $\omega < \kappa < \beth_\omega$ is a regular cardinal, then does $\mathfrak{d}(\kappa) \leq \mathfrak{r}(\kappa)$ hold? In particular, is $\mathfrak{d}(\aleph_n) \leq \mathfrak{r}(\aleph_n)$, for all $1 \leq n < \omega$?

In trying to tackle this problem, it may seem reasonable to first try to produce a model where $\mathfrak{r}(\aleph_n) < 2^{\aleph_n}$, for if $\mathfrak{r}(\aleph_n)$ is provably equal to $2^{\aleph_n}$, then of course $\mathfrak{d}(\aleph_n) \leq \mathfrak{r}(\aleph_n)$. This is closely related to a well-known question of Kunen about the minimal size of a base for a uniform ultrafilter on $\aleph_1$

**Question 17.** Is $\mathfrak{r}(\aleph_1) < 2^{\aleph_1}$ consistent? Is $\mathfrak{u}(\aleph_1) < 2^{\aleph_1}$ consistent?

## References


1. A. Blass, T. Hyttinen and Y. Zhang, Mad families and their neighbours (Preprint).
2. D. Raghavan and S. Shelah, Two inequalities between cardinal invariants, *Fund. Math.* **237**, 187 (2017).
3. J. Cummings and S. Shelah, Cardinal invariants above the continuum, *Ann. Pure Appl. Logic* **75**, 251 (1995).
4. T. Suzuki, About splitting numbers, *Proc. Japan Acad. Ser. A Math. Sci.* **74**, 33 (1998).
5. J. Zapletal, Splitting number at uncountable cardinals, *J. Symbolic Logic* **62**, 35 (1997).
6. A. Blass, Combinatorial cardinal characteristics of the continuum, in *Handbook of set theory. Vols. 1, 2, 3*, (Springer, Dordrecht, 2010) pp. 395–489.
7. S. Shelah, Two cardinal invariants of the continuum ($\mathfrak{d} < \mathfrak{a}$) and FS linearly ordered iterated forcing, *Acta Math.* **192**, 187 (2004).
8. S. Shelah, The generalized continuum hypothesis revisited, *Israel J. Math.* **116**, 285 (2000).
9. K. Kunen, *Set theory: An introduction to independence proofs*, Studies in Logic and the Foundations of Mathematics, Vol. 102 (North-Holland Publishing Co., Amsterdam, 1980).
10. U. Abraham and M. Magidor, Cardinal arithmetic, in *Handbook of set theory. Vols. 1, 2, 3*, (Springer, Dordrecht, 2010) pp. 1149–1227.
11. S. Shelah, More on the revised GCH and the black box, *Ann. Pure Appl. Logic* **140**, 133 (2006).
12. D. Raghavan and S. Shelah, Boolean ultrapowers and iterated forcing, *Preprint*.

# PART B

# The 15$^{th}$ Asian Logic Conference

# Logical Revision by Counterexamples: A Case Study of the Paraconsistent Counterexample to *Ex Contradictione Quodlibet*

Seungrak Choi

*Department of Philosophy, Korea University,*
*Seoul, 02841, Republic of Korea*
*E-mail: choi.seungrak.eddy@gmail.com*

It is often said that a correct logical system should have no counterexample to its logical rules and the system must be revised if its rules have a counterexample. If a logical system (or theory) has a counterexample to its logical rules, do we have to revise the system? In this paper, focussing on the role of counterexamples to logical rules, we deal with the question.

We investigate two mutually exclusive theories of arithmetic - intuitionistic and paraconsistent theories. The paraconsistent theory provides a (strong) counterexample to *Ex Contradiction Quodlibet* (*ECQ*). On the other hand, the intuitionistic theory gives a (weak) counterexample to the *Double Negation Elimination* (*DNE*) of the paraconsistent theory. If any counterexample undermines the legitimate use of logical rules, both theories must be revised.

After we investigate a paraconsistent counterexample to *ECQ* and the intuitionist's answer against it, we arrive at the unwelcome conclusion that *ECQ* has both a justification and a counterexample. Moreover, we argue that if a logical rule were abolished whenever it has a counterexample, a promising conclusion would be logical nihilism which is the view that there is no valid logical inference, and so a correct logical system does not exist. Provided that the logical revisionist is not a logical nihilist, we claim that not every counterexample is the ground for logical revision. While logical rules of a given system have a justification, the existence of a counterexample loses its role for logical revision unless the rules and the counterexample share the same structure.

*Keywords*: *Ex Contradictione Quodlibet*; Counterexample; Logical revision; Logical nihilism.

## 1. Introduction

Mathematical practice is directed toward two major goals – the formulation of proofs and the construction of counterexamples. Logical practice shares the major goals. Especially, the construction of counterexamples often causes a logical revision for the mathematical practice. In logic, a counterexample is an exceptional case to a valid form of argument which preserves the truth value of the premises to the conclusion. A traditional picture of logic takes it for granted that a valid argument has a *general* truth-preserving relation, no matter what case it applies.

Therefore, a form of argument is valid if it does not have any counterexample which makes it be failed to keep the truth-preserving relation. A correct logical system has logical rules which reflect the valid forms of argument. When there exists a counterexample to the logical rules of the logical system, the counterexample undermines the legitimate use of the rules. The counterexample gives a reason for a revision of the logical system.

An intuitionist has requested a revision of classical logic due to the fact that there are counterexamples to classical rules. One of such is the *Double Negation Elimination* (*DNE*), which derives a sentence from its double negative form. Unlike the classicist, the intuitionist has a different conception of truth. (S)he claims that all truths are knowable and provides an interpretation of logical connectives and quantifiers in terms of provability. An intuitionistic weak counterexample to *DNE* has been given in an intuitionistic theory of real analysis. The weak counterexample has been the main ground for a revision of classical logic.[a] If an argument has a case that its premises are recognizably true, but its conclusion is recognizably not true, though it is not recognizably false, it has a *weak* counterexample. When there is a case that exemplifies an argument whose premises are true but whose conclusion is false, the argument has a *strong* counterexample. *DNE* does not have a strong counterexample but has a weak counterexample.

If a classicist rejects the intuitionistic interpretation, the weak counterexample to *DNE* does not matter for classical logic. As the intuitionist and the classicist have different (mathematical) structures for logic, only in the intuitionistic structure, *DNE* has the weak counterexample. However, if a valid logical rule generally applies to all legitimate mathematical structures and both the classical/intuitionistic structures are legitimate, *DNE* is not valid. In this perspective, the intuitionist considers that the classical theory of mathematics uses an incorrect logical rule, and so it should be revised. Should the classicist accept the weak counterexample to *DNE*? If one finds a counterexample to logical rules, ought we to request a revision of logic?

Focussing on the role of a counterexample to logical rules, the present paper deals with the suggested questions, and the answer is 'no' because not every counterexample has the main role for logical revision. When the theory (or system) has a justification of its logical rules, the justification cancels out the role of counterexamples for reforming the theory. A case study is a paraconsistent counterexample to *Ex Contradictione Quodlibet* (*ECQ*) which states that, from contradictory premises, anything can be inferred. A logical consequence relation is said to be *paracon-*

[a]For a detailed explanation of the weak/strong counterexample, the reader can consult page 189 of Michael Dummett[9] and page from 8 to 16 of A.S. Troelstra and Dirk Van Dalen[43].

*sistent* if contradictory premises do not imply every sentence. Paraconsistentists including Robert Meyer and Chris Mortensen have claimed that *ECQ* is not valid because *ECQ* has a strong counterexample in their inconsistent theory of arithmetic. The situation is different from the tension between the classicist and the intuitionist since the classicist does not provide any counterexample to logical rules of intuitionistic logic. The paraconsistentist gives a counterexample to *ECQ* which is the main reasoning for contradiction (or absurdity) in the intuitionistic theory of arithmetic, called *Heyting Arithmetic* (*HA*). In addition, the paraconsistent theory has *DNE* which has a weak counterexample in the intuitionistic theory of real analysis. Both theories are mutually exclusive in the sense that they provide counterexamples to logical rules of each other's theory. Also, both are incorrect if a correct theory must have logical rules which immune to counterexamples in any legitimate structures.

From Sections 2 to 5, we will introduce the paraconsistent counterexample to *ECQ* and see the answers from the intuitionist that *ECQ* has its justification. It is an unwelcome conclusion that *ECQ* has both the justification and the counterexample. If logical revisionists – the intuitionist and the paraconsistentist – believe that any sort of counterexample is prior to all kinds of justifications, they will fail to claim that their logical system (or theory) is correct. For instance, if the intuitionist claims that the paraconsistent system has to be revised due to the fact that it includes an incorrect rule, i.e. *DNE*, which has a weak counterexample, then (s)he should accept that either the intuitionistic theory is to be revised since *ECQ* has a counterexample. The paraconsistentist will suffer the similar situation. The situation seems to raise the question of whether every counterexample is a ground for logical revision.

As yet, though, the issue of the role of counterexamples for logical revision does not seem to have been addressed. In Section 6, we shall argue that a counterexample to logical rules is dependent on a mathematical structure as the validity of logical rules is on the structure. Not every counterexample is to be a ground for logical revision. If a logical rule were abolished whenever we find a structure that constructs its counterexample, the promising conclusion would be logical nihilism which is the view that there is no valid logical inference and no correct logical system. Since the revisionists, like the intuitionist and the paraconsistentist, hope to find a correct logical system, they should not accept the view that every counterexample is a ground for logical revision. Rather, for them, it is better to focus on the justification of truth-preserving relation unless they agree with the legitimate scope of *all* mathematical structures. The consideration of the justification would be a good alternative against logical nihilism.

## 2. *ECQ* on Meyer-Mortensen's Inconsistent Arithmetic

From Sections 2 to 5, we shall investigate the paraconsistent counterexamples to *ECQ* and the intuitionist's defense against it. While Sections 2 and 3 are about the semantic counterexample, Sections 4 and 5 deal with the syntactic (or proof-theoretic) counterexample to *ECQ*, such as Mortensen's non-primeness counterexample to *Disjunctive Syllogism* (*DS*). To put it roughly, if *DS* were inferentially equivalent to *ECQ* in *HA*, *ECQ* has an indirect non-primeness counterexample. Section 5 argues that Mortensen's non-primeness objection does not work in the standard structure for *HA* since *HA* satisfies the disjunction property. Even were any theory of *HA* non-prime, the non-primeness counterexample to *DS* could not be the indirect counterexample to *ECQ* because *DS* is not inferentially equivalent to *ECQ* in any non-prime theory of *HA*. The conflict between the classicist and the intuitionist have been well introduced, but the tension between intuitionistic and paraconsistent structures has not been explained. Especially, it has not been well discussed that intuitionistic and paraconsistent structures are mutually exclusive. Hence, the discussions from Sections 2 to 5 will be helpful to understand both structures and to reconsider the role of counterexamples for logical revision.

Meyer[14] seems to have firstly proposed an inconsistent arithmetic. Meyer and Mortensen[18] has further developed the inconsistent models of $R^{\#}$ and $R^{\#\#}$. Their inconsistent model is called '$RM3$', and it assigns a truth-value to the implication with respect to the three-valued Sugihara matrix $\{+1, 0, -1\}$ which appears in Anderson and Belnap[1]. Simply put, their inconsistent arithmetic has a three-valued system taking the set of values – true ($+1$), both ($0$), and false ($-1$). They consider that an identity equation and its negation have the both-value. For example, $0 = 0$ and its negation have the both-value. $0 = 1$ is false in $RM3$, and $RM3$ regards true and both-value as the value of the true sentence. When $0 = 0$ and its negation imply $0 = 1$, $RM3$ assigns the false value to the implication because the premise is true, but the conclusion is false. Not all instances of *ECQ* preserve the truth-value of premises to a conclusion in all valuations of $RM3$. *ECQ* becomes an invalid inference in the inconsistent arithmetic. In this section, we shall introduce Meyer-Mortensen's inconsistent model, $RM3$, and its counterexample to *ECQ*.

In a general case of our discussion, we shall use $\wedge$, $\vee$, $\rightarrow$, $\leftrightarrow$, $\perp$, $\neg$, $\exists$, $\forall$ for an intuitionistic conjunction, disjunction, implication, equivalence, absurdity, negation, existential and universal quantification respectively. For a paraconsistent counterpart, we will use $\&$, $\sqcup$, $\supset$, $\equiv$, $\perp^P$, $\sim$, $\exists^P$, and $\forall^P$. Let $\varphi$, $\psi$ be any formula. We call any instances of the form $\varphi \wedge \neg\varphi$ (or $\varphi \& \sim \varphi$), 'contradiction.' An intuitionistic negation formula $\neg\varphi$ is defined by $\varphi \rightarrow \perp$ whereas a paraconsistent negation formula $\sim \varphi$ is not. As is usual, for any theory $\Gamma$, we write '$\Gamma \vdash \varphi$' to mean that $\Gamma$ derives $\varphi$ and '$\Gamma \nvdash \varphi$' means that $\Gamma$ does not derive $\varphi$. The definition of 'consistent' and 'trivial' runs as follows:

**Definition 2.1.** Let $\Gamma$ be any theory and $\mathfrak{L}$ be a given language of $\Gamma$. (1) $\Gamma$ is *consistent* iff there exists no formula $\varphi$ such that $\Gamma \vdash \varphi$ and $\Gamma \vdash \neg\varphi$; otherwise, *inconsistent*. (2) $\Gamma$ is *trivial* iff for any formula $\varphi$ in $\mathfrak{L}$, $\Gamma \vdash \varphi$; otherwise, *non-trivial*.

Mortensen[22] and Meyer and Mortensen[18] construct the inconsistent model $RM3^i$ for arithmetic. For any sentence $\varphi$, $v$ be a valuation from the sentences to truth values where $v(\varphi) \in \{+1, 0, -1\}$, say *true*, *both*, and *false*.[b] A sentence $\varphi$ is $RM3^i$-true under the valuation $v$ iff $v(\varphi) \in \{+1, 0\}$. $\varphi$ is true in $RM3^i$ iff $\varphi$ is $RM3^i$-true in all valuations $v$. For any model $\mathfrak{M}$, we call $Th(\mathfrak{M})$ a set of sentences true in $\mathfrak{M}$. $Th(RM3^i)$ is the theory of Meyer-Mortensen's relevant arithmetic. Meyer[14] and Meyer and Mortensen[18] show that $Th(RM3^i)$ is inconsistent but non-trivial.

**Theorem 2.1.** *$Th(RM3^i)$ is inconsistent and non-trivial.*

**Proof.** See Appendix A. □

Theorem 2.1 shows that $Th(RM3^i)$ is inconsistent and non-trivial. Though the issue of triviality of $Th(RM3^i)$ is different from that of the validity of *ECQ*, $RM3^i$ gives a case that $v(\varphi \& \sim \varphi) = 0$ and $v(\psi) = -1$ for the sentence $(\varphi \& \sim \varphi) \supset \psi$. When we regard *ECQ* as any form of the inference from $(\varphi \& \sim \varphi)$ to $\psi$, *ECQ* fails to preserve the *true* or *both* value from the premises to the conclusion in all valuations of $RM3^i$. *ECQ* is not valid in $Th(RM3^i)$. $RM3^i$ provides a semantic counterexample to *ECQ*. In Section 3, we will investigate the intuitionist's response to the semantic counterexample.

## 3. An Intuitionist's Answer Against the Semantic Counterexample to *ECQ*

The semantic counterexample provides the case that *ECQ* fails to preserve the both-value from the premises to the conclusion in all valuations of the inconsistent model $RM3^i$. The intuitionist has three answers for the semantic counterexample. First, there is no reason to accept the both value. Second, the counterexample to *ECQ* is the counterexample to the *paraconsistent ECQ* but not to the *intuitionistic ECQ*. Third, there exists a construction from an intuitionistic absurdity to all (atomic) sentences in *HA*.

[b] A brief introduction to Meyer-Mortensen's semantics is given in the Appendix A.

### 3.1. *The Both-Value and Dialetheism*

To begin with, the intuitionist rejects the inconsistent model and the both-value. Without the both-value, there is no case that a contradictory premise, e.g. $0 \neq 0$, is true, but a conclusion, e.g. $0 = 1$, is false. Traditionally, *ECQ* is conceived as *Ex Falso Quodlibet* which means that falsity implies everything. Even following $RM3^i$, falsity ($\perp^P$) implies every sentence. Only because, for any term $t$, $t \neq t$ has the both-value in $RM3^i$, there exist true contradictory premises and a counterexample to *ECQ*. The inconsistent models for arithmetic can be supported by dialetheism which is the view that there exists a true contradiction. However, it is unclear whether there is a true contradiction in intuitionistic theories of arithmetic.

Graham Priest[34] argues that true contradictions are derivable from the self-referential paradoxes and Gödel's incompleteness theorem. In addition, at page 66 of Priest[34], he offers a prospect of an intuitionistic dialetheism.

> It would be equally possible to have an "intuitionistic dialetheism," which took a constructive stance on negation (so that a proof of the impossibility of a proof of [$\varphi$] was required for the truth of [$\sim \varphi$]) and the other logical constants. (We noted ... that the proofs of many logical paradoxes do not require the law of excluded middle or other intuitionistically invalid principles).

Unlike the prospect of Priest[34], Tennant[41] has argued that there is no true contradiction in intuitionistic logic. Priest[34] replies to Tennant in the footnote 6 at page 286.

> In the final section of [Tennant[41]], Tennant also criticizes my account of the paradoxes of self-reference by giving his own. But he does not address the arguments of the 1st edn that would appear to apply to his account. For example, he says that the liar sentence is 'radically truth-valueless' ... but he does not address [the strengthened liar paradox]: this sentence is false or radically truth-valueless. ... Nor does he address the paradoxes that do not use the [law of excluded middle], such as Berry's. Similarly, he claims that the "Gödel Paradox" shows that the notion of naive proof cannot be formali[z]ed. He does not address the consideration ... as to why this is false or irrelevant.

Truly, Tennant did not answer, but the reason why the intuitionist would not accept the intuitionistic dialetheism is enough.

For the issue of Berry's paradox, Priest[30] and at page from 25 to 27 of Priest[34] have attempted to show that the law of excluded middle (*LEM*) is unnecessary for the derivation of a contradiction from Berry's paradox. Unfortunately, Ross Brady[4] explains how Priest implicitly assumes *LEM*. Priest[34] suggests a different argument from that of Priest[30] to derive a contradiction from Berry's paradox, but his proof uses *DNE* which is provably equivalent to *LEM*. Therefore, he fails to show that Berry's paradox leads to a contradiction without classical inference.

With regard to Gödel's incompleteness theorem, Chapter 3 of Priest[34] maintains that any correct formalization of our naive proof procedure is inconsistent and it is tantamount to establish that a true contradiction exists. On the other hand, Tennant[41] asserts that Gödel's theorem merely shows that one cannot have the complete characterization of our naive proof procedure. Tennant's interpretation of Gödel's theorem may be a consistent counterpart whereas Priest's view may presume that our inconsistent linguistic practice leads to an inconsistency of our naive proof procedure. They may have a different conception of 'naive proof procedure.' The tension between them is based on their different intuitions of the naive proof procedure, so it seems to be hard to find any ways to ease the tension. One way to ease the tension is to claim that the lesson of Gödel's proof is that any sufficiently strong and intuitively correct arithmetic cannot be both complete and consistent. As Seungrak Choi[5] notes, a contradiction is derivable from Gödel sentence only in the complete system. It is not necessary for the intuitionist that only the complete system is correct. The intuitionist does not have to suppose the completeness of the system. Likewise, for a given natural deduction systems that the prooflessness is expressible, Choi[6] shows that $\perp$ is derivable from the strengthened liar sentence in $S$ only when $S$ is complete. Thus, Priest's arguments for dialetheism does not support intuitionistic dialetheism. It is unconvincing that there is an intuitionistic system which derives a true contradiction. Since the intuitionist has no ground for both-value, (s)he will reject the inconsistent model for arithmetic and the paraconsistent counterexample to *ECQ*.

### 3.2. *Different ECQs and the Proof of the Intuitionistic ECQ*

The second answer against the semantic counterexample to *ECQ* is that a paraconsistent *ECQ* is not an intuitionistic *ECQ*. The meaning of 'contradiction' relies on mathematical structures. Even if the $RM3^i$-model gives the semantic counterexample to *ECQ*, it is the counterexample to the paraconsistent *ECQ*, but not to the intuitionistic *ECQ*. While the paraconsistent contradiction has the both-value, the intuitionistic contradiction has the false value.[c] It is often said that to grasp a meaning of a sentence is to know the truth-(or assertion-)condition. Since the

[c]It is a controversial issue on whether $\perp$ has a truth-value or not. Following the intuitionistic interpretation of $\perp$, if $\perp$ has no proof, it follows that $\perp$ has no truth-value. Tennant[40] accepts the view that $\perp$ has

paraconsistent and the intuitionistic contradiction have different truth-conditions, they have different meanings. The paraconsistentist and the intuitionist talk about different contradictions. The semantic counterexample to the paraconsistent *ECQ* does not work for the intuitionistic *ECQ*.

A paraconsistent *DNE* which has the form $\sim\sim \varphi \supset \varphi$, has the similar case. The paraconsistent *DNE* is valid in $RM3^i$, but an intuitionistic *DNE*, $\neg\neg\varphi \rightarrow \varphi$, has a weak counterexample. The same meaning of a logical connective has the same inference relation. As the paraconsistent *DNE* and the intuitionistic *DNE* have different consequences, each *DNE* have a separate meaning of negation. Hence, the intuitionist and the paraconsistentist talk about different things. If we are to avoid elementary fallacies generated from ambiguities, we have to say that the intuitionist's weak counterexample to *DNE* does not work for the paraconsistent *DNE*. Likewise, the paraconsistent semantic counterexample to *ECQ* does not affect the validity of the intuitionistic *ECQ* as it is the counterexample to the paraconsistent *ECQ* but not to the intuitionistic *ECQ*.

The last reason to reject the semantic counterexample is that *ECQ* has a construction that makes it valid in *HA*. For any given formula $\varphi$, the intuitionistic *ECQ* has the form $\bot \rightarrow \varphi$. $\bot$ is regarded as $0 = 1$ in *HA*, so *ECQ* has the form $0 = 1 \rightarrow \varphi$. All sentences in *HA* can be constructed by the composition of atomic sentences, and an arithmetic atomic sentence has the form of an equation. When *HA* derives a conjunction of all atomic sentences, every sentence is derivable in *HA*. If it is shown that $0 = 1$ implies the conjunction of all atomic sentences, *ECQ* can be intuitionistically valid. We define in *HA* the conjunction of all atomic sentences in terms of an intuitionistic conjunction connective $\wedge$.

**Definition 3.1.** Let $\varphi(x, y)$ be any binary atomic formula and $i, j, m, n$ be natural numbers. The conjunction of all $\varphi(i, j)$ is recursively defined as follows:

$$\bigwedge_{i,j\leq 0} \varphi(i, j) =_{df} \varphi(0, 0) \tag{1}$$

$$\bigwedge_{\substack{i\leq m+1\\ j\leq n+1}} \varphi(i, j) =_{df} \bigwedge_{\substack{i\leq m\\ j\leq n}} \varphi(i, j) \wedge \varphi(m + 1, n + 1) \tag{2}$$

When we regard $\varphi(x, y)$ as a binary atomic formula stating $x = y$. Then, we easily prove the following theorem and corollary.

no meaning in the same way that the ancient Hindus used '0' for emptiness in arithmetic. However, the fact that $\bot$ has no proof does not exclude the case that $\bot$ has a disprove of it. If we regard $\bot$ as $0 = 1$ in *HA*, since $0 = 1$ has a disproof of it, $\bot$ is false in *HA*. We consider $\bot$ to be $0 = 1$ in this paper, and the standard structure of *HA* gives a false value to $0 = 1$. We set aside the view that $\bot$ has no truth-value.

**Theorem 3.1.** *Let $\varphi(x, y)$ be a binary atomic formula which describes $x = y$ and $i, j, m, n$ be any natural number.*

$$HA \vdash 0 = 1 \text{ iff } HA \vdash \bigwedge\!\!\!\bigwedge_{\substack{i \le m+1 \\ j \le n+1}} \varphi(i, j)$$

**Proof.** See Appendix B. □

**Corollary 3.1.** *Let $\psi(x, y)$ be a binary formula which states $x \neq y$ and $i, j, m, n$ be any natural number.*

$$HA \vdash 0 = 1 \text{ iff } HA \vdash \bigwedge\!\!\!\bigwedge_{\substack{i \le m+1 \\ j \le n+1}} \psi(i, j)$$

**Proof.** See Appendix B. □

The sketch of the result can be found in Appendix B. Theorem 3.1 and Corollary 3.1 provide a justification of the validity of *ECQ* in *HA*.[d]

Three objections to the semantic counterexample are based on the view that $RM3^i$ is an alien semantic for *HA*. Even though *ECQ* is invalid in $RM3^i$, the intuitionist can ask why does (s)he use an alien semantic for *HA*, and if one does, why trust the result.

Another paraconsistent counterexample to *ECQ* can be given by showing that *DS* is invalid. *ECQ* is justified by *DS*, and *ECQ* might be inferentially equivalent to *DS* in the sense that they derive the same consequence when they have the same premise. If *ECQ* and *DS* are inferentially equivalent, a counterexample to *DS* can be an indirect counterexample to *ECQ*. Mortensen[19,20] suggests non-primeness counterexample to an extensional (or truth-functional) *DS*. Although he did not discuss that his non-primeness counterexample to *DS* applies to *ECQ*, the rejection of *DS* naturally is connected to the rejection of *ECQ*. We consider that the non-primeness counterexample to *DS* is the indirect counterexample to *ECQ* if they are inferentially equivalent. In Section 4, we shall investigate how *ECQ* and *DS* can be inferentially equivalent, and how the counterexample to *DS* can be the indirect counterexample to *ECQ*.

---

[d]The suggested argument for the justification of the intuitionistic *ECQ* does not work in the paraconsistent arithmetic $R^{\#}$. In $R^{\#}$, $0 = 1$ may be provably equivalent to $\forall^P x \forall^P y(x = y)$. However, $\sim (0 = 0)$ does not imply every sentence and so $R^{\#}$ would not be trivial. Define $\varphi \supset \psi$ as $\sim \varphi \sqcup \psi$, then $R^{\#} \vdash \sim (0 = 0) \supset 0 = 1$. $\sim (0 = 0)$ is true in the model of $R^{\#}$. The application of modus ponens derives $0 = 1$ and so all sentences. However, as Meyer and Friedman[17] notes that modus ponens described by Ackermann's rule $\gamma$ is inadmissible in $R^{\#}$. Thus, $\sim (0 = 0)$ does not imply all sentences and $R^{\#}$ is not trivial. Meyer[16] admits modus ponens in $R^{\#\#}$ and Meyer[15] claims that since $R^{\#\#} \nvdash 0 = 1$, by modus tollens, $R^{\#\#} \nvdash \sim (0 = 0)$. Even if $R^{\#\#} \vdash 0 = 1 \equiv \forall^P x \forall^P y(x = y)$, $R^{\#\#}$ does not imply every sentence and it is non-trivial because $R^{\#\#}$ is consistent.

## 4. An Indirect Counterexample to *ECQ*

Much argument for rejecting *ECQ* consists of two statements. The first statement is that there is no proper (or relevant) relation between $\varphi \wedge \neg\varphi$ and $\psi$. (Cf. Anderson and Belnap[1].) As we have discussed in Section 3, the intuitionistic absurdity as $0 = 1$ and the contradiction $0 = 1 \wedge 0 \neq 1$ have the proper relation to the conjunction of all sentences in *HA*. *ECQ* in *HA* has a relevant relation between $\perp$ and every sentence, so the intuitionist can reject the first statement. The second statement is that *ECQ* is justified by *DS*, but *DS* is invalid, and so is *ECQ*. We however already have a direct argument for justifying the intuitionistic *ECQ* in *HA*, the rejection of *DS* does not imply the rejection of *ECQ*. The other argument against *ECQ* is to claim that *DS* and *ECQ* are inferentially equivalent. If *DS* is inferentially equivalent to *ECQ*, a counterexample to *DS* can be an indirect counterexample to *ECQ*.

The intuitionist may reject the indirect counterexample if an inferential equivalence is conceived in a model-theoretic way. Some intuitionists might object to the use of the model-theoretic notions by arguing that *HA* identifies the mathematical truth with provability but not truth in a classical model. As Dummett[9], Prawitz[27], and Tennant[42] have argued, a proof-theoretic interpretation of the inferential equivalence may be well-suited to the intuitionistic standpoint. We first introduce a standard interpretation of the intuitionistic logical connectives with some variations of our purpose and give logical rules for *HA* in the natural deduction style proposed by Prawitz[24].

Let $\mathfrak{L}$ be a language of *HA* and $\mathbb{P}$ be a set of proofs of atomic sentences of $\mathfrak{L}$ and an individual domain $\mathbb{N}$ of natural numbers. We assume, for convenience, that each number $n$ in $\mathbb{N}$ has a numeral. For any formula $\varphi(x)$, we understand by $\varphi(n)$ the closed sentence obtained by substituting in $\varphi(x)$ the numeral of $n$ for $x$. Let $\mathfrak{D}$ be a sequence of proofs, say 'derivation.' We use $\begin{matrix}\mathfrak{D}\\ \varphi\end{matrix}$ for a derivation for $\varphi$ – i.e. a sequence of the proofs of $\varphi$. $\begin{matrix}\varphi\\ \mathfrak{D}\\ \psi\end{matrix}$ means a derivation from $\varphi$ to $\psi$. The intuitionistic interpretation through a proof over $\mathbb{P}$ of a closed compound sentence $\varphi$ in $\mathfrak{L}$ is then proposed recursively as follows:

For any sentence $\varphi$ and $\psi$,

(1) A proof over $\mathbb{P}$ of $\varphi \wedge \psi$ is a pair $(\mathfrak{D}_g, \mathfrak{D}_y)$ of derivations such that $\begin{matrix}\mathfrak{D}_g\\ \varphi\end{matrix}$ and $\begin{matrix}\mathfrak{D}_y\\ \psi\end{matrix}$.

(2) A proof over $\mathbb{P}$ of $\varphi \vee \psi$ is a pair $(\mathfrak{D}_g, \mathfrak{D}_y)$ of derivations such that $\begin{matrix}\mathfrak{D}_g\\ \varphi\end{matrix}$ or $\begin{matrix}\mathfrak{D}_y\\ \psi\end{matrix}$.

(3) A proof over $\mathbb{P}$ of $\varphi \rightarrow \psi$ is a derivation $\mathfrak{D}$ which converts any proof of $\varphi$ into a proof of $\psi$. (i.e. $\begin{matrix}\varphi\\ \mathfrak{D}\\ \psi\end{matrix}$).

(4) Nothing is a proof of $\bot$.

(5) A proof over $\mathbb{P}$ of $\exists x\varphi(x)$ is a derivation $\mathfrak{D}$ such that $\begin{matrix}\mathfrak{D}\\ \varphi(n)\end{matrix}$ where $n$ in $\mathbb{N}$.

(6) A proof over $\mathbb{P}$ of $\forall x\varphi(x)$ is a derivation $\mathfrak{D}$ such that for any $n$ in $\mathbb{N}$, $\mathfrak{D}$ is a proof over $\mathbb{P}$ of the instance $\varphi(n)$.

Let $\mathfrak{D}_g, \ldots, \mathfrak{D}_n$ be an arbitrary derivation with respect to $\mathbb{P}$ and $\varphi$, $\psi$, $\sigma$ be any formulas in $\mathfrak{L}$. Say $x, y$ be any frce variables and $t$ be any term which is not free. We consider an intuitionistic natural deduction system $S_I$ which has the following rules:

$$\dfrac{\begin{matrix}\mathfrak{D}_g & \mathfrak{D}_y\\ \varphi & \psi\end{matrix}}{\varphi \wedge \psi}\wedge I \qquad \dfrac{\begin{matrix}\mathfrak{D}\\ \varphi \wedge \psi\end{matrix}}{\varphi}\wedge E_1 \qquad \dfrac{\begin{matrix}\mathfrak{D}\\ \varphi \wedge \psi\end{matrix}}{\psi}\wedge E_2 \qquad \dfrac{\begin{matrix}[\varphi]^1\\ \mathfrak{D}\\ \psi\end{matrix}}{\varphi \rightarrow \psi}\rightarrow I_{,1} \qquad \dfrac{\begin{matrix}\mathfrak{D}_g & \mathfrak{D}_y\\ \varphi \rightarrow \psi & \varphi\end{matrix}}{\psi}\rightarrow E$$

$$\dfrac{\begin{matrix}\mathfrak{D}_g\\ \varphi\end{matrix}}{\varphi \vee \psi}\vee I_1 \qquad \dfrac{\begin{matrix}\mathfrak{D}_y\\ \psi\end{matrix}}{\psi \vee \varphi}\vee I_2 \qquad \dfrac{\begin{matrix} & [\varphi]^1 & [\psi]^1\\ \mathfrak{D} & \mathfrak{D}_v & \mathfrak{D}_w\\ \varphi \vee \psi & \sigma & \sigma\end{matrix}}{\sigma}\vee E_{,1} \qquad \dfrac{\begin{matrix}\mathfrak{D}\\ \bot\end{matrix}}{\varphi}\bot_I$$

$$\dfrac{\begin{matrix}\mathfrak{D}\\ \varphi(y)\end{matrix}}{\forall x\varphi[x/y]}\forall I \qquad \dfrac{\begin{matrix}\mathfrak{D}\\ \forall x\varphi(x)\end{matrix}}{\varphi[t/x]}\forall E \qquad \dfrac{\begin{matrix}\mathfrak{D}\\ \varphi(t)\end{matrix}}{\exists\varphi(x)}\exists I \qquad \dfrac{\begin{matrix} & [\varphi(y)]^1\\ \mathfrak{D}_g & \mathfrak{D}_y\\ \exists x\varphi(x) & \sigma\end{matrix}}{\sigma}\exists E_{,1}$$

$[x/y]$ means the substitution of $x$ for $y$ in $\varphi$. $\forall I$-rule has the variable restriction that $y$ must not occur free in any assumption that $\varphi(y)$ depends on nor in $\forall x\varphi$. Also, $\exists E$-rule has the restriction that $y$ must not occur free in $\exists x\varphi$, $\sigma$ nor in any assumption that $\sigma$ depends on except $\varphi(y)$. A derivation is *open* when it depends on assumptions and *closed* when all assumptions are discharged or bounded. We have a minimal natural deduction system $S_M$ by dropping $\bot_I$-rule from $S_I$, and an addition of the rule for *DNE* to $S_I$ gives a classical natural deduction system $S_C$.

The intuitionistic interpretation determines a meaning of an intuitionistic logical connective. It may be desirable to understand ‘inferential equivalence’ in terms of the natural deduction system.[e]

It is easily proved in $S_M$ that $(((\varphi \vee \psi) \wedge \neg\varphi) \to \psi) \leftrightarrow ((\varphi \wedge \neg\varphi) \to \psi)$. (See Appendix C.) It is not sufficient, however, to show the inferential equivalence between *DS* and *ECQ* because an intuitionistic inferential equivalence is correspond to interdeducibility *salva* provability. Two different but inferentially equivalent rules must have the same conclusion when they have the same premise. The interdeducibility can be defined by the admissibility of logical rules.

**Definition 4.1.** Let $R$ be a rule with the premises $\varphi_1, ..., \varphi_n$ and the conclusion $\psi$. $S$ be a system of rules. $R$ is said to be *admissible* for $S$ if $S \vdash \varphi_1, ..., \varphi_n$ implies $S \vdash \psi$.

The definition of the *interdeducibility* of logical rules runs as follows:

**Definition 4.2.** Let $R_1$ and $R_2$ be logical rules, and $S$ be a system of logical rules which does not have both $R_1$ and $R_2$. Say $S_{R_1}$ and $S_{R_2}$ be systems obtained by adding $R_1$ to $S$ and $R_2$ to $S$ respectively. $R_1$ is said to be $S$ – *interdeducible* for $R_2$ if $R_1$ is admissible for $S_{R_2}$ and $R_2$ is admissible for $S_{R_1}$.

The interdeducibility in a specific system may be a promising conception of the inferential equivalence.[f] *DS* and *ECQ* have the following form of elimination rule:

$$\dfrac{\begin{matrix}\mathfrak{D}_g & \mathfrak{D}_y \\ \varphi \vee \psi & \neg\varphi\end{matrix}}{\psi}\,DS \qquad\qquad \dfrac{\begin{matrix}\mathfrak{D} \\ \varphi \wedge \neg\varphi\end{matrix}}{\psi}\,ECQ$$

---

[e]Dummett[9] as an intuitionist borrows an idea to establish a validity of argument from Gerhard Gentzen[11] and Prawitz[25,26] such that the meaning of logical connectives is implicitly defined by its introduction rules, while the elimination rules are justified by respecting the stipulation made by the introduction rule. The intuitionistic interpretation of logical connectives and Gentzen-Prawitz’s approach to the meaning are independently given in their works. Also, they are different in the sense that the intuitionistic interpretation does not concern with what justifies inferences and what makes something a valid form of reasoning, but Gentzen-Prawitz’s approach does. As Prawitz[28] has argued, however, two approaches are concerned with the meaning of what it is to prove something, and it can be seen that the intuitionistic interpretation is extensionally equivalent to Gentzen-Prawitz’s approaches. In this sense, one may claim that ‘inferential equivalence’ on the intuitionistic perspective is to be conceived in terms of the natural deduction system.

[f]There is another candidate for the interdeducibility, such as the *uniqueness* of logical rules suggested by Nuel Belnap[3], Dummett[9], and Došen and Schroeder-Heister[7]. The inferential equivalence may be regarded as the uniqueness of logical rules, but if the inferential equivalence has to be conceived as the uniqueness, *DS* and *ECQ* are not inferentially equivalent and so there is no indirect counterexample to *ECQ* via the counterexample to *DS*.

Let $S_{DS}$ and $S_{ECQ}$ be a system given by adding *DS*-rule to $S_M$ and *ECQ*-rule to $S_M$ respectively. If *DS*-rule is admissible for $S_{ECQ}$ and *ECQ*-rule is admissible for $S_{DS}$, then *DS*-rule is $S_M$ – *interdeducible* for *ECQ*-rule.

**Theorem 4.1.** *DS-rule is* $S_M$ *– interdeducible for ECQ-rule.*

**Proof.** Since $S_{DS}$ and $S_{ECQ}$ share the logical rules, except *DS* and *ECQ*-rules, we only consider the consequences of *DS* and *ECQ*-rules.

Claim. 1: *DS*-rule is admissible for $S_{ECQ}$.

**Proof.** Assume that the derivations $\begin{array}{c}\mathfrak{D}_g\\ \varphi\vee\psi\end{array}$ and $\begin{array}{c}\mathfrak{D}_y\\ \neg\varphi\end{array}$ of *DS*-rule are derivable in $S_{ECQ}$. Then, the following process in $S_{ECQ}$ derives the same conclusion, $\psi$, of *DS*-rule.

$$\dfrac{\begin{array}{c}\mathfrak{D}_g\\ \varphi\vee\psi\end{array} \qquad \dfrac{\dfrac{[\varphi]^1 \quad \begin{array}{c}\mathfrak{D}_y\\ \neg\varphi\end{array}}{\varphi\wedge\neg\varphi}\,\wedge I}{\psi}\,ECQ \qquad [\psi]^2}{\psi}\,\vee E_{,1,2}$$

□

Claim. 2: *ECQ*-rule is admissible for $S_{DS}$.

**Proof.** Assume that the derivations $\begin{array}{c}\mathfrak{D}_v\\ \varphi\wedge\neg\varphi\end{array}$ of *ECQ*-rule is derivable in $S_{DS}$. The following process in $S_{DS}$ derives the same conclusion, $\psi$, of *ECQ*-rule.

$$\dfrac{\dfrac{\dfrac{\begin{array}{c}\mathfrak{D}_v\\ \varphi\wedge\neg\varphi\end{array}}{\varphi}\,\wedge E}{\varphi\vee\psi}\,\vee I \qquad \dfrac{\begin{array}{c}\mathfrak{D}_v\\ \varphi\wedge\neg\varphi\end{array}}{\neg\varphi}\,\wedge E}{\psi}\,DS$$

□

Therefore, by the claim 1 and 2, *DS*-rule is $S_M$ – *interdeducible* for *ECQ*-rule. □

Provided that $S_M$–*interdeducibility* be a correct conception of the inferential equivalence, *ECQ* has the indirect counterexample when there exists a counterexample to *DS*. In Section 5, we shall examine a Mortensen's non-primeness counterexample to *DS* and argue that it cannot be an indirect counterexample to *ECQ*.

## 5. Mortensen's Non-Primeness Counterexample to *DS*

About any quantifier-free sentence $\varphi$ and $\psi$, $\varphi \vee \psi$ is explicitly defined as $\exists x((x = 0 \rightarrow \varphi) \wedge (x \neq 0 \rightarrow \psi))$ in *HA* and either the $\vee I$ and $\vee E$-rule are. (Cf. Troelstra and Dalen[43].) Similarly, for any quantifier-free sentence $\varphi$ and $\psi$, *DS* is explicitly definable. (See Appendix D.) The proof of the definability of *DS* uses *DNE* which holds in any quantifier-free sentences in *HA*. However, there is a quantified sentence, such as Gödel sentence, which is not provable in a consistent theory of *HA*. One may consider that, for a Gödel sentence $\varphi$, $\neg\neg\varphi \rightarrow \varphi$ is not provable in the consistent theory of *HA*. If *DNE* does not work in *HA*, *DS* would not be definable in *HA*.[g] Mortensen has proposed a counterexample to *DS* in a similar point

Mortensen[19] has given two counterexamples to *DS* – the inconsistent and the non-primeness counterexample. As we have seen in Section 2, there is an inconsistent and non-trivial theory. Suppose that *DS* holds in an inconsistent and non-trivial theory $\Gamma$ which has all logical rules of $S_M$. For some $\varphi$, $\Gamma \vdash \varphi \wedge \neg\varphi$ since $\Gamma$ is inconsistent. Applying $\wedge E$-rule to $\varphi \wedge \neg\varphi$, we have $\Gamma \vdash \neg\varphi$ and, by $\vee I$-rule, $\Gamma \vdash \neg\varphi \vee \psi$ for arbitrary $\psi$. Also, by $\wedge E$-rule, $\Gamma \vdash \varphi$. If *DS* holds in $\Gamma$, $\Gamma \vdash \psi$ and so $\Gamma$ is trivial. *Ex hypothesi*, $\Gamma$ is non-trivial. Hence, *DS* does not hold in $\Gamma$. However, the inconsistent and non-trivial theory is grounded on $RM3^i$. As argued in Section 3, the intuitionist does not have to accept $RM3^i$.

For the non-primeness case, we have the following definition.

**Definition 5.1.** Let $\Gamma$ be any theory. (1) $\Gamma$ is *non-prime with respect to* $\varphi \vee \psi$ iff $\Gamma \vdash \varphi \vee \psi$ but $\Gamma \nvdash \varphi$ and $\Gamma \nvdash \psi$. (2) $\Gamma$ is *non-prime* if $\Gamma$ is non-prime with respect to some disjunction; otherwise, $\Gamma$ is prime.

Consider $\Gamma \vdash \neg\varphi$ and $\Gamma \vdash \varphi \vee \psi$. If $\Gamma$ is non-prime with respect to any disjunction $\varphi \vee \psi$ while $\Gamma \vdash \neg\varphi$, then $\Gamma \nvdash \psi$. Therefore, *DS* does not hold in $\Gamma$. If the intuitionist must accept the non-primeness counterexample and *DS* is inferentially equivalent to *ECQ*, the non-primeness counterexample can be an indirect counterexample to *ECQ*. Unfortunately, three problems occur in the non-primeness counterexample.

Firstly, though *DS* is not fully definable in *HA*, it is not shown that there exists a non-prime theory for *HA*. Mortensen[19] already has noted that *DS* holds for any consistent and prime theory, such as intuitionistic logic, but he claimed that a classical *Peano Arithmetic*, *PA*, is non-prime if it is consistent. Let $\varphi$ be a Gödel sentence. Then, certainly $PA \vdash \varphi \sqcup \sim \varphi$, but if *PA* is consistent, $PA \nvdash \varphi$ and $PA \nvdash \sim \varphi$ by the Gödel's first incompleteness theorem. Hence, the consistent *PA*

[g]Even though *DS* would not be definable in *HA*, an intuitionist does not have a problem for *ECQ* because, as we have seen in Section 3, there exists a direct proof of the validity of *ECQ*.

is non-prime with respect to $\varphi \sqcup \sim \varphi$. *DS* does not hold in any consistent theory for arithmetic extended from *PA*. Unlike *PA*, Gödel's theorem does not support the non-primeness of *HA* since $\varphi \vee \neg\varphi$ is not a theorem of *HA*. It is not shown that *HA* is non-prime with respect to $\varphi \vee \neg\varphi$.

In addition, an intuitionistic interpretation for $\vee$ suggested in Section 4 says that if there exists a proof of $\varphi \vee \psi$, there is a proof of $\varphi$ or of $\psi$. For a given theory $\Gamma$, if, whenever $\Gamma \vdash \varphi \vee \psi$, either $\Gamma \vdash \varphi$ or $\Gamma \vdash \psi$, $\Gamma$ satisfies the *disjunction property*. Stephen Kleene[13] already proved the disjunction and the existence property of *HA* with the method of the realizability. Also, Jan von Plato[23] proved the same result in the natural deduction system. The results show that *HA* is prime. Since the non-primeness of *HA* is not established, it is unconvincing that *DS* has a non-primeness counterexample in *HA*.

Mortensen's inconsistent and non-primeness counterexample to *DS* does not work in *HA* if the suggested two arguments are correct. Though any theory of *HA* is non-prime, the non-primeness counterexample to *DS* could not be an indirect counterexample to *ECQ*. A non-primeness case says that we can derive $\varphi \vee \psi$ without any derivations of $\varphi$ and of $\psi$. It means that $\vee I$-rule is not an appropriate rule for $\vee$ which governs a meaning of 'or.' If we do not use $\vee I$-rule, then Theorem 4.1 in Section 4 fails. *DS* is not inferentially equivalent to *ECQ* in any non-prime theory of *HA*. Hence, any counterexample to *DS* is unable to be an indirect counterexample to *ECQ*. For these reasons, Mortensen's non-primeness counterexample to *DS* is not to be the indirect counterexample to *ECQ* in *HA*.

## 6. Logical Nihilism and the Role of Counterexamples

Were it the case that the arguments from Sections 3 to 5 are correct, the paraconsistent counterexamples to *ECQ* do not matter in the intuitionistic theory of arithmetic. An intuitionistic logical system for *HA* does not need to be reformed because the paraconsistent counterexamples to *DS* and *ECQ* are not to be a reason for logical revision. Likewise, the paraconsistentist can cope with the weak counterexample to *DNE*. The arguments from Sections 3 to 5 seem to show that a counterexample to logical rules is relative to the structure of a given system of the rules. In this section, we shall argue that the *mere* existence of counterexamples does not support logical revision because the counterexamples work only in their suitable structures. If the mere existence of counterexamples were the main ground for logical revision, the promising view would be logical nihilism which claims that there is no valid logical inference and no correct logic. However, logical nihilism is not a promising view for the logical revisionists since they seek to find a correct logical system. Therefore, we will argue that the revisionists do not have to overestimate the role of the counterexamples for logical revision.

If one follows the slogan that a valid logical rule generally applies all legitimate (mathematical) structures, (s)he denies that *DNE*, *DS*, and *ECQ* are valid inferences. The slogan forces us to accept the view that a logical rule is valid iff, in every legitimate structure, it preserves the truth-value of the premises to the conclusion, without having any counterexamples. Following the chapter 4 of Stewart Shapiro[37], we call this notion of validity, 'super-validity.' The problem is that truth-preserving relation is relative to the structures. Each logic has a separate interpretation of implication in its structure and provides a distinct truth-preserving relation. As the notion of validity varies across the structures, either does a counterexample. The issue of the paraconsistent counterexample may show that a counterexample works for logical revision only in its legitimate structure which shares the same interpretation of 'truth-preserving relation.' Furthermore, if one claims that a correct and a revision-free logic must have super-valid rules, (s)he would be faced with the problem of logical nihilism.

If a correct logic must have super-valid logical rules, what logic would be correct? *DNE* has a weak counterexample in an intuitionistic structure for real analysis, and either does *DS* and *ECQ* in an inconsistent structure for arithmetic. If we regard a structure for quantum logic as the legitimate structure for mathematics, we drop the standard elimination rule for $\vee$ and the distributive laws for $\wedge$ and $\vee$. (Cf. Hillary Putnam[35].) As Meyer and Friedman[17] notes that modus ponens in terms of Ackermann's rule $\gamma$ is inadmissible in $R^{\#}$, we do not have modus ponens. More seriously, one may construct a mathematical structure that formulates a sentence $\varphi$ which expresses "$\varphi$' is a premise.' Once the structure restricts that a conclusion is not, at the same time, to be a premise, $\varphi$ is true as the premise but not as the conclusion. The structure may give a counterexample to an inference from $\varphi$ to $\varphi$. For instance, Mortensen[21] argues for two relatives of logical nihilism. The one is based on the idea that no sentence is necessarily true and the other is on the idea that no sentence is true in all mathematical structures. He suggests that there is no reason why the structures, which makes $\varphi \supset \varphi$ false, should not be realized. Also, Gillian Russell[36] regards Mortensen's view as nihilism about logical truth and proposes the structure that some instances of the form $\varphi \supset \varphi$ are false.[h] Therefore, in a worst case, the logical revisionists have a quick argument for logical nihilism if they follow the super-validity.

[h]I would like to thank an anonymous reviewer for the comment that Mortensen may not be a logical nihilist. He calls his view 'possibilism' which is the view that all sentences are contingent, or that anything is possible. Truly, one may claim that possibilism is closer to relativism about logical truth but not to nihilism about logical truth. However, the main issue of this paper is not to suggest an argument for logical nihilism nor to discuss whether his view can be interpreted as nihilism or not. We set aside the issue about the interpretation of his possibilism.

If logical nihilism supports the idea that there is no correct logical system, the revisionists are reluctant to accept logical nihilism since they seek to find a correct and revision-free system. There are two ways to avoid logical nihilism. The one is to reject the super-validity and the other is to restrict the legitimate scope of mathematical structures. Unfortunately, the second option is hopeless because it is hard to find any agreement on the scope. Philosophers have often claimed that a correct ordinary language semantics verifies a correct logical rule. Any suitable structure for right logical rules seems to have a correct ordinary language semantics. The correct semantics may be a key to limit the legitimate scope of mathematical structures. We find at least three different views on the correct ordinary language semantics from Dummett[8,9] for intuitionistic logic, Priest[34] for paraconsistent logic, and Tennant[39] for intuitionistic relevant logic.

Dummett[9] believes that 'meaning' and 'truth' are the core concept of logic and a correct meaning theory provides a correct semantic for logic. Dummett[8,9] gives a meaning-theoretic argument, called 'manifestation argument,' for supporting that intuitionistic logic is a proper logic for our ordinary linguistic use. He denies *DNE* because it often assumes the principle of bivalence that all sentences are either provable or disprovable. For there are many scientific and mathematical conjectures that are not yet proved or disproved, so no one yet knows which truth-value it has. While each sentence of the disputed domain possesses an objective truth-value, independently of our means to know it, we are unable to recognize the procedure for catching that truth value. From the rejection of bivalence, he rejects *DNE* as a correct inference rule and so either classical logic. Dummett's ordinary language semantics may verify *ECQ* but not *DNE*.

On the other hand, Priest[34] keeps the view that any formalization of ordinary language semantics needs to take paradox into account. Considering a liar-type sentence $\varphi$ which expresses that $\varphi$ is false. While assuming bivalence, the case that $\sim \varphi$ is true means that $\varphi$ is false and vice versa. Then, we have a relation $\varphi \equiv \sim \varphi$, and so $\varphi \& \sim \varphi$ is true by *DNE*. He claims that this phenomenon appeals to us that there exists a true contradiction. He applies a both-value for the liar-type sentence in Priest[29,34]. $\varphi \& \sim \varphi$ is true if $\varphi$ has the both-value. In his paraconsistent logic, called the *Logic of Paradox*(*LP*), *DNE* is a correct inference, but *ECQ* may not. There exists a liar-type sentence $\varphi$ which makes $\varphi \& \sim \varphi$ true, but all sentences are not derivable in *LP*. Therefore, Priest's ordinary language semantics verify *DNE* but *ECQ* may not correct.[i]

---

[i]More precisely, *ECQ* is valid in Priest's logic *LP*, but from *ECQ* all sentences are not derivable since modus ponens is not valid. *ECQ* is valid in Priest's inconsistent model for arithmetic setting out in *LP*. Priest[29,31] gives the *LP* model, and Priest[32,33] introduces his inconsistent arithmetic. *LP* defines the

Tennant[40] agrees that *ECQ* is invalid, but at page 110 of Tennant[38] he claims that nice(or correct) logic is adequate for uncovering all inconsistencies and any intuitionistic consequences of any consistent set of axioms. Especially, in Tennant[41], he denies that there exists a true contradiction in intuitionistic (relevant) logic. His meaning theory in Tennant[39] has founded on Dummett's philosophical framework, but his expected ordinary language semantic would reject both *DNE* and *ECQ*.

In sum, Dummett's ordinary language semantics may verify *ECQ* but give a counterexample to *DNE*. Priest's semantics justifies *DNE* but provides a counterexample to *ECQ*. Tennant's semantics may give weak counterexamples to *DNE* and *ECQ*. There is not yet clear agreement on which ordinary language semantics are legitimate and which scope of the class of mathematical structures are legitimate for truth-preserving relation. Hence, for the purpose of avoiding logical nihilism, the revisionists cannot accept the second option which restricts the legitimate scope of mathematical structures.

The revisionists have the first option which rejects the super-validity. If they reject the super-validity, any interpretation of truth-preserving relation relies on (mathematical) structures. Since the counterexample works in its suitable interpretation of truth-preserving relation, it is relative to the structures. In this respect, the revisionists should not overestimate the effect of the counterexamples for logical revision, until they find an agreement on the scope of the legitimate structures for validity.

## 7. Conclusion: Is Every Counterexample a Ground for Logical Revision?

As we have seen in Section from 3 to 5, *ECQ* is not valid in $RM3^i$ but valid in the standard structure for *HA*. It is an undesirable conclusion that *ECQ* has both the counterexample and the justification of it. The counterexample to *ECQ* in $RM3^i$ is not applicable to the intuitionistic *ECQ*, and the intuitionistic weak counterexample to *DNE* does not work for the classical and paraconsistent *DNE*. Since the construction of a counterexample generates a new mathematical structure, the counterexample relies on the interpretation of truth-preservation of that structure. Moreover, if logical nihilism is not a desirable conclusion to the logical revisionists, they must not consider that every counterexample is a ground for logical revision. Rather, it is better for them to focus on a justification of truth-preserving relation in a given structure unless they have an agreement on the scope of the legitimate math-

---

implication from the disjunction and negation, i.e. $\varphi \supset \psi$ as $\sim \varphi \sqcup \psi$. *ECQ* is valid in every valuation of *LP*, but from the application of *ECQ*, all sentences are not derivable since modus ponens is not valid in *LP*. (Cf. Beall and Foster and Seligman[2].)

ematical structures for validity. For example, even in an intuitionistic perspective, an intuitionistic *DNE* has a weak counterexample in some structures, such as the structures of real analysis. Yet, it is valid in some structures of arithmetic, such as the structures of primitive recursive arithmetic because there exists a construction which transfers any proof of $\neg\neg\varphi$ to a proof of $\varphi$ in those structures. It may be the case that the existence of a proof of the truth-preserving relation has the main role of validity rather than the existence of a counterexample. If the revisionists can accept a revision-free logical system in its suitable structure, they may accept that, for a certain specific structure $\mathbb{M}$, a logical rule is valid in $\mathbb{M}$ if it has a legitimate proof of the truth-preserving relation from the premises to the conclusion and has no counterexample in $\mathbb{M}$. Although the notion of a *legitimate proof* may be relative to the structures and should be well defined, the consideration of a proof rather than a counterexample would be a good alternative against logical nihilism.

**Acknowledgments**

The topics of Sections 3 and 6 were presented at Pluralism Week on June 2016, held by the Veritas Research Center at Underwood International College, Yonsei University. An early version of this paper was presented at Spring Regular Conference of Korean Association for Logic on April 2017. I would like to thank Colin Caret, Inkyo Chung, Byungduk Lee, Teresa Kouri, Stella Moon, Nikolaj Pedersen, Andy Yu, Norbert Preining, Eunsuk Yang for helpful discussions. I especially thank Jinhee Lee for his various suggestions on the early version of this paper, which helped a great deal of improvement.

## Appendix A. Meyer-Mortensen's Semantics for Inconsistent Arithmetic and the Proof of Theorem 2.1

We begin with a language $\mathfrak{L}$ with the connectives &, $\sqcup$, $\supset$, $\equiv$, $\sim$, the quantifiers $\exists^P$, $\forall^P$, and a single binary relation =, constant 0, variables $x, y, z, ...$, basic functions +, $\cdot$, $s$ for addition, multiplication, and successor respectively. The constants, (basic) functions, and variables are terms. Open/closed terms and well-formed formulas are defined in the usual way. Positive integers 1, 2, 3, ... are defined by $S(0), S(S(0)), S(S(S(0))), ...$ For any formula $\varphi, \psi, \chi$, the relevant logic $R$ is given by the following axiom schemata and rules:

Logical Axioms:

(1) $(\varphi \supset \psi) \supset ((\psi \supset \chi) \supset (\psi \supset \chi))$,
(2) $\varphi \supset ((\varphi \supset \psi) \supset \psi)$,

(3) $(\varphi_1 \& \varphi_2) \supset \varphi_i$ where $i = 1, 2$,
(4) $((\varphi \supset \psi)\&(\varphi \supset \chi)) \supset (\varphi \supset (\psi \& \chi))$,
(5) $\varphi_i \supset (\varphi_1 \sqcup \varphi_2)$ where $i = 1, 2$,
(6) $((\varphi \supset \chi)\&(\psi \supset \chi)) \supset ((\varphi \sqcup \psi) \supset \chi)$,
(7) $(\varphi\&(\psi \sqcup \chi)) \supset ((\varphi\&\psi) \sqcup (\varphi\&\chi))$,
(8) $\sim\sim \varphi \supset \varphi$,
(9) $(\varphi \supset\sim \varphi) \supset\sim \varphi)$,
(10) $\forall x\varphi(x) \supset \varphi(t)$ for any term $t$ in $\mathfrak{L}$,
(11) $\forall x(\varphi \supset \psi) \supset (\forall x\varphi \supset \forall x\psi)$,
(12) $\varphi \supset \forall x\varphi(x)$ where $x$ is not free in $\varphi$,
(13) $\forall x(\varphi \sqcup \psi) \supset (\varphi \sqcup \forall x\psi)$ where $x$ is not free in $\varphi$,
(14) $(\forall x\varphi\&\forall x\psi) \supset \forall x(\varphi\&\psi)$.

Logical Rules:

(1) If $\varphi$ and $\psi$ are true so is $\varphi\&\psi$.
(2) If $\varphi$ and $\varphi \supset \psi$ are true so is $\psi$.

The addition of the axiom scheme $\varphi \supset (\varphi \supset \varphi)$ makes $R$ to be $RM$. The relevant arithmetic $R^\#$ and $RM^\#$ in $\mathfrak{L}$ have the following axioms with the logical axioms and rules of $R$ and $RM$.

Arithmetical Axioms:

(1) $\forall x(\sim (S(x) = 0))$,
(2) $(\forall x\forall y(S(x) = S(y)) \equiv (x = y))$,
(3) $\forall x(x + 0 = x)$,
(4) $\forall x\forall y(x + S(y) = S(x + y))$,
(5) $\forall x(x \cdot 0 = 0)$,
(6) $\forall x\forall y(x \cdot S(y) = (x \cdot y) + y)$,
(7) $\forall x\forall y\forall z(x = y \supset (x = z \supset y = z))$.

$R^\#$ and $RM^\#$ have an arithmetical rule of mathematical induction: if $\varphi(0)$ and $\forall x(\varphi(x) \supset \varphi(S(x))$ are true, so is $\forall x\varphi(x)$. The relevant arithmetic $R^{\#\#}$ and $RM^{\#\#}$ has an additional rule, called 'Hilbert's rule $\Omega$': if, for each $n$, $\varphi(0), \varphi(1), \varphi(2), ..., \varphi(n)$ are true, so is $\forall x\varphi(x)$. We now have an $RM3^i$-model for the relevant arithmetic $R^\#$ and $R^{\#\#}$. The $RM3^i$-model is an ordered pair $< \mathfrak{D}^i, I >$ where $\mathfrak{D}^i$ is the set of integers modulo $i$, and $I$ is a function which assigns denotations to non-logical symbols of $\mathfrak{L}$ in the following way:

- For any constant symbol $d$, $I(d)$ is a member of $\mathfrak{D}^i$.
- For every $n$-ary function symbol $f$, $I(f)$ is an $n$-ary function on $\mathfrak{D}^i$.

- For any terms $t_1$, $t_2$, $I(+(t_1, t_2)) = I(+(I(t_1), I(t_2)))$, $I(\cdot(t_1, t_2)) = I(\cdot(I(t_1), I(t_2)))$, $I(S(t_1)) = I(S(I(t_1)))$.

Finally, let $\varphi$, $\psi$ be a formula and $v$ be a valuation from the formulas to truth values where $v(\varphi) \in \{+1, 0, -1\}$. Let us define $x \neq y$ as $\sim (x = y)$, $x < y$ as $\exists z(x+S(z) = y)$, and $x \leq y$ as $(x < y) \sqcup (x = y)$. We introduce two characteristic functions, *min* and *max*: $min(x, y) = y$ if $y \leq x$; otherwise $x$, and $max(x, y) = y$ if $x \leq y$; otherwise $x$.

- For any terms $t_1$, $t_2$ and any atomic sentence $t_1 = t_2$,
  $v(t_1 = t_2) = 0$ iff $I(t_1) = I(t_2)$,
  $v(t_1 = t_2) = -1$ iff $I(t_1) \neq I(t_2)$.
- For any formula $\varphi$, $\psi$ of $\mathfrak{D}$,
  $v(\sim \varphi) = -1$ iff $v(\varphi) = 1$,
  $v(\sim \varphi) = 0$ iff $v(\varphi) = 0$,
  $v(\sim \varphi) = 1$ iff $v(\varphi) = -1$,
  $v(\varphi \& \psi) = min(v(\varphi), v(\psi))$,
  $v(\varphi \sqcup \psi) = max(v(\varphi), v(\psi))$,
  $v(\varphi \supset \psi) = 1$ iff $v(\varphi) \leq v(\psi)$ where $v(\varphi) \neq 0 \neq v(\psi)$,
  $v(\varphi \supset \psi) = -1$ iff $v(\psi) < v(\varphi)$,
  $v(\varphi \supset \psi) = 0$ iff $v(\psi) = v(\varphi) = 0$,
  For any term $t_1, ..., t_n$,
  $v(\forall x\varphi(x)) = v(\varphi(t_1)\&\varphi(t_2)\& \cdots \&\varphi(t_n))$,
  $v(\exists x\varphi(x)) = v(\varphi(t_1) \sqcup \varphi(t_2) \sqcup \cdots \sqcup \varphi(t_n))$.

A formula $\varphi$ is $RM3^i$-true under the valuation $v$ iff $v(\varphi) \in \{+1, 0\}$. $\varphi$ is true in $RM3^i$ iff $\varphi$ is $RM3^i$-true in all valuations $v$. Meyer-Mortensen's relevant arithmetic is the set of formulas true in $RM3^i$, say $Th(RM3^i)$. We now show that $Th(RM3^i)$ is inconsistent and non-trivial.

**Theorem** 2.1 *$Th(RM3^i)$ is inconsistent and non-trivial.*

**Proof.**

*Claim 1:* $Th(RM3^i)$ is non-trivial.

*Proof.* All theorems of $Th(RM3^i)$ take one of values $\{+1, 0\}$. Since $I(0) \neq I(1)$, $v(0 = 1) = -1$ and so $0 = 1$ is not a theorem of $Th(RM3^i)$. Hence, $Th(RM3^i)$ is non-trivial.

*Claim 2:* $Th(RM3^i)$ is inconsistent.

*Proof.* For any term $t_1, t_2$, $v(t_1 = t_2) = 0$ iff $I(t_1) = I(t_2)$ in any model $RM3^i$. Since, for any term $t$, $v(t = t) = 0 = v(t \neq t)$, $(t = t)\&(t \neq t)$ is true. Thus, $RM3^i$ is inconsistent. □

## Appendix B. The Proof of Theorem 3.1

In this section, we will show that *ECQ* has a constructive operation that converts any proof of $\bot$(or $0 = 1$) into a proof of all (atomic) sentences in *HA*. In order to show it, we need to have two steps. The first step is to show that $HA \vdash 0 = 1$ implies $HA \vdash \forall x \forall y(x = y)$ and vice versa. The second step is to establish that $\forall x \forall y(x = y)$ is equivalent to the conjunction of all atomic sentences in *HA*.

An intuitionistic negation in *HA* is defined by the intuitionistic implication and the absurdity. (Cf. Page 24 of Dummett[10].) For any formula $\varphi$, '$\neg\varphi$' is explained by '$\varphi \to \bot$' and especially in *HA* '$\bot$' is interpreted as '$0 = 1$.' Also, we have the axioms of *HA* from the arithmetical axioms of Appendix A by replacing $\sim$ by $\neg$ in the arithmetical axiom 1 and the axiom 2 by $\forall x \forall y(S(x) = S(y) \to x = y)$, with the intuitionistic interpretation of quantifications. For the two steps, we need the following lemmas.

**Lemma B.1.** *Let $\varphi(x, y)$ be a binary atomic formula that describes $x = y$ and $i, j, m, n$ be any natural number. Then we have*

$$HA \vdash \bigwedge\!\!\!\bigwedge_{\substack{i \leq m+1 \\ j \leq n+1}} \varphi(i, j) \leftrightarrow \forall x \forall y(x = y)$$

**Proof.** Since $\bigwedge\!\!\!\bigwedge_{\substack{i \leq m+1 \\ j \leq n+1}} \varphi(i, j)$ is recursively defined by Definition 3.1, we readily have $\forall x \forall y(x = y)$. Also, as we have all instances of $\varphi(x, y)$ that are constructible in *HA* from $\forall x \forall y(x = y)$, by Definition 3.1, we have $\bigwedge\!\!\!\bigwedge_{\substack{i \leq m+1 \\ j \leq n+1}} \varphi(i, j)$ from $\forall x \forall y(x = y)$. □

**Lemma B.2.** *Let $\psi(x, y)$ be a binary formula that describes $x \neq y$ and $i, j, m, n$ be any natural number. Then we have*

$$HA \vdash \bigwedge\!\!\!\bigwedge_{\substack{i \leq m+1 \\ j \leq n+1}} \psi(i, j) \leftrightarrow \forall x \forall y(x \neq y)$$

**Proof.** Immediate. □

Lemma B.1 says that $\bigwedge\!\!\!\bigwedge_{\substack{i \leq m+1 \\ j \leq n+1}} \varphi(i, j)$ is equivalent to $\forall x \forall y(x = y)$ in *HA*. It is proved in *HA* that $0 = 1$ is equivalent to $\forall x \forall y(x = y)$ as well as to $\forall x \forall y(x \neq y)$.

**Lemma B.3.** *$HA \vdash 0 = 1$ iff $HA \vdash \forall x \forall y(x = y)$.*

**Proof.** Let $\mathbb{N}$ be a set of natural numbers. To prove the relation from the right side to the left is trivial. Only the proof from the left side to the right will be suggested. We will use double induction on each variable $v_i$ and $v_j$ of the equation $v_i = v_j$ where $i, j \in \mathbb{N}$. Assuming that $HA \vdash 0 = 1$, we begin with induction basis on $v_i$.

(1) Induction Basis: $v_i = 0$.

Sub-induction Basis: Suppose $v_j = 0$. Then, trivially we have $0 = 0$.
Sub-induction Hypothesis: Suppose $v_j = n$ for any $n \in \mathbb{N}$. Then, we have $0 = n$. If $HA \vdash 0 = S(0)$, we have $0 = S(n)$. By induction, we have $\forall y(0 = y)$.

(2) Induction Hypothesis: $v_i = n$ for any $n \in \mathbb{N}$.

Sub-induction Basis: Suppose $v_j = 0$. Then, we have $n = 0$. If $HA \vdash S(0) = 0$, $S(n) = 0$.
Sub-induction Hypothesis: Suppose $v_j = m$ for any $m \in \mathbb{N}$. Then, we have $n = m$. If $S(0) = 0$ and, by (1), $\forall y(0 = y)$, $S(0) = 0 = m$ and $0 = n$. Hence, $S(n) = m$.

Therefore, by induction with (1) and (2), $HA \vdash \forall x \forall y(x = y)$. □

**Lemma B.4.** *$HA \vdash 0 = 1$ iff $HA \vdash \forall x \forall y(x \neq y)$.*

**Proof.** For the relation from the left to the right, we suppose $HA \vdash 0 = 1$. Then, by Lemma B.3, $HA \vdash \forall x \forall y(x = y)$. With the arithmetical axiom 1 of $HA$, $\forall x(S(x) \neq 0)$, we have $HA \vdash \forall x \forall y(x \neq y)$. From the right to the left, if $HA \vdash \forall x \forall y(x \neq y)$, then $HA \vdash 0 = 0 \rightarrow 0 = 1$. Hence, $HA \vdash 0 = 1$. □

**Theorem** 3.1 *Let $\varphi(x, y)$ be a binary atomic formula which describes $x = y$ and $i, j, m, n$ be any natural number.*

$$HA \vdash 0 = 1 \text{ iff } HA \vdash \bigwedge\!\!\!\bigwedge_{\substack{i \leq m+1 \\ j \leq n+1}} \varphi(i, j)$$

**Proof.** By Lemmas B.1 and B.3. □

**Corollary** 3.1 *Let $\psi(x, y)$ be a binary formula which states $x \neq y$ and $i, j, m, n$ be any natural number.*

$$HA \vdash 0 = 1 \text{ iff } HA \vdash \bigwedge\!\!\!\bigwedge_{\substack{i \leq m+1 \\ j \leq n+1}} \psi(i, j)$$

**Proof.** By Lemmas B.2 and B.4. □

Theorem 3.1 and Corollary 3.1 show that *ECQ* has a justification of its validity.

## Appendix C. The Proof of $S_M \vdash (((\varphi \vee \psi) \wedge \neg\varphi) \to \psi) \leftrightarrow ((\varphi \wedge \neg\varphi) \to \psi)$.

We give a proof of $S_M \vdash (((\varphi \vee \psi) \wedge \neg\varphi) \to \psi) \leftrightarrow ((\varphi \wedge \neg\varphi) \to \psi)$. The following tree gives a proof of the relation from the left sentence to the right.

$$\dfrac{\dfrac{\dfrac{\dfrac{\dfrac{\dfrac{[\varphi \wedge \neg\varphi]^1}{\varphi}\wedge E}{\varphi \vee \psi}\vee I \quad \dfrac{[\varphi \wedge \neg\varphi]^1}{\neg\varphi}\wedge E}{(\varphi \vee \psi) \wedge \neg\varphi}\wedge I \qquad [((\varphi \vee \psi) \wedge \neg\varphi) \to \psi]^2}{\psi}\to E}{(\varphi \wedge \neg\varphi) \to \psi}\to I_{,1}}{(((\varphi \vee \psi) \wedge \neg\varphi) \to \psi) \to ((\varphi \wedge \neg\varphi) \to \psi)}\to I_{,2}$$

The following tree shows the relation from the right to the left.

$$\dfrac{\dfrac{\dfrac{\dfrac{[((\varphi \vee \psi) \wedge \neg\varphi)]^1}{\varphi \vee \psi}\wedge E \quad \dfrac{\dfrac{[\varphi]^2 \quad \dfrac{[((\varphi \vee \psi) \wedge \neg\varphi)]^1}{\neg\varphi}\wedge E}{\varphi \wedge \neg\varphi}\wedge I \qquad [(\varphi \wedge \neg\varphi) \to \psi]^3}{\psi}\to E \quad [\psi]^4}{\psi}\wedge E_{,2,4}}{((\varphi \vee \psi) \wedge \neg\varphi) \to \psi}\to I_{,1}}{((\varphi \wedge \neg\varphi) \to \psi) \to (((\varphi \vee \psi) \wedge \neg\varphi) \to \psi)}\to I_{,3}$$

## Appendix D. The Proof of the Definability of *DS* in *HA*

We will show that *DS* is explicitly definable in *HA*, for any quantifier-free sentence $\varphi$ and $\psi$. To prove it, we need the following lemma.

**Lemma D.1.** $HA \vdash \forall x(((x = 0 \to 0 = 1) \to 0 = 1) \to x = 0)$.

**Proof.** We use mathematical induction on $v$. When $v = 0$, it is trivial. For an induction hypothesis, we consider that $v = n$, i.e. $HA \vdash ((n = 0 \to 0 = 1) \to 0 = 1) \to n = 0)$. From the axiom 1 of *HA* and Lemma B.3, $HA \vdash ((S(n) = 0 \to 0 = 1) \to 0 = 1) \to S(n) = 0)$. □

**Theorem D.1.** *For any quantifier-free sentence $\varphi$ and $\psi$. Let us define $\varphi \vee \psi$ as $\exists x((x = 0 \to \varphi) \wedge (x \neq 0 \to \psi))$. Then DS is explicitly definable in HA.*

**Proof.** There are two forms of rules for *DS*

$$\dfrac{\begin{matrix}\mathfrak{D}_g & \mathfrak{D}_y \\ \varphi \vee \psi & \neg\varphi\end{matrix}}{\psi}DS_1 \qquad \dfrac{\begin{matrix}\mathfrak{D}_g & \mathfrak{D}_v \\ \varphi \vee \psi & \neg\psi\end{matrix}}{\varphi}DS_2$$

We firstly show that $DS_1$–rule is explicitly definable in $HA$. Suppose that, for any given $\varphi$ and $\psi$, there exist closed derivations $\mathfrak{D}_g$ of $\varphi \vee \psi$ (i.e. $\exists x((x = 0 \to \varphi) \wedge (x \neq 0 \to \psi))$) and $\mathfrak{D}_y$ of $\neg\varphi$. Then, we have the following open derivation $\mathfrak{D}_w$ of $\psi$ from $[(n = 0 \to \varphi) \wedge (n \neq 0 \to \psi)]$.

$$\dfrac{\dfrac{[(n = 0 \to \varphi) \wedge (n \neq 0 \to \psi)]^1}{n \neq 0 \to \psi}\wedge E \qquad \dfrac{\dfrac{\dfrac{\dfrac{[(n = 0 \to \varphi) \wedge (n \neq 0 \to \psi)]^1}{n = 0 \to \varphi}\wedge E \quad [n = 0]^2}{\varphi}\to E \quad \begin{matrix}\mathfrak{D}_y \\ \neg\varphi\end{matrix}}{0 = 1}\to E}{n = 0 \to 0 = 1}\to I_{,2}}{\psi}\to E$$

Having the open derivation $\mathfrak{D}_w$, we have a derivation of $\psi$.

$$\dfrac{\begin{matrix}\mathfrak{D}_g \\ \exists x((x = 0 \to \varphi) \wedge (x \neq 0 \to \psi))\end{matrix} \qquad \begin{matrix}[(n = 0 \to \varphi) \wedge (n \neq 0 \to \psi)]^1 \\ \mathfrak{D}_w \\ \psi\end{matrix}}{\psi}\exists E_{,1}$$

At Second, we consider $DS_2$–rule. Suppose that, for any given $\varphi$ and $\psi$, there exist closed derivations $\mathfrak{D}_g$ of $\varphi \vee \psi$ and $\mathfrak{D}_v$ of $\neg\psi$. Then, with Lemma D.1 and $\mathfrak{D}_v$, we have an open derivation $\mathfrak{D}$ of $\varphi$ from $[(n = 0 \to \varphi) \wedge (n \neq 0 \to \psi)]$.

$$\dfrac{\dfrac{[(n = 0 \to \varphi) \wedge (n \neq 0 \to \psi)]^1}{n = 0 \to \varphi}\wedge E \qquad \dfrac{\dfrac{\dfrac{}{\forall x(((x = 0 \to 0 = 1) \to 0 = 1) \to x = 0)}\text{lemma D.1}}{(n \neq 0 \to 0 = 1) \to n = 0}\forall E \qquad \dfrac{\dfrac{\dfrac{\dfrac{\dfrac{[(n = 0 \to \varphi) \wedge (n \neq 0 \to \psi)]^1}{n \neq 0 \to \psi}\wedge E \quad [n \neq 0]^2}{\psi}\to E \quad \begin{matrix}\mathfrak{D}_v \\ \neg\psi\end{matrix}}{0 = 1}\to E}{n \neq 0 \to 0 = 1}\to I_{,2}}{n = 0}\to E}{\varphi}\to E$$

Having the open derivation $\mathfrak{D}$, we have a derivation of $\varphi$.

$$\dfrac{\begin{matrix}\mathfrak{D}_g \\ \exists((x = 0 \to \varphi) \wedge (x \neq 0 \to \psi))\end{matrix} \qquad \begin{matrix}[(n = 0 \to \varphi) \wedge (n \neq 0 \to \psi)]]^1 \\ \mathfrak{D} \\ \varphi\end{matrix}}{\varphi}\exists E_{,1}$$

□

## References


1. A. R. Anderson and N. D. Belnap, *Entailment (vol. I)* (Princeton New Jersey: Princeton University Press, 1975).
2. J. Beall, T. Foster, and J. Seligman, A note on freedom from detachment in the logic of paradox, *Notre Dame Journal of Formal Logic*, **54**, 1 (2012), pp. 15-20.
3. N. D. Belnap, Tonk, plonk and plank, *Analysis*, **22** (1962), pp. 130-134.
4. R. T. Brady, Reply to Priest on Berry's paradox, *Philosophical Quarterly*, **34**, 135(1984), pp. 157-163.

5. S. Choi, Can Gödel's incompleteness theorem be a ground for dialetheism?, *Korean Journal of Logic*, **20**, 2(2017), pp. 241-271.
6. S. Choi, Liar-type paradoxes and intuitionistic natural deduction systems, *Korean Journal of Logic*, **21**, 1(2018), pp. 59-96.
7. K. Došen and P. Schroeder-Heister, Conservativeness and uniqueness, *Theoria*, **3** (1985), pp. 159-173.
8. M. Dummett, The philosophical basis of intuitionistic logic, In *Truth and Other Enigmas* (Cambridge: Harvard University Press, 1973), pp. 215-247.
9. M. Dummett, *Logical Basis of Metaphysics* (Cambridge: Harvard University Press, 1991).
10. M. Dummett, *Elements of Intuitionism* (Clarendon: Oxford University Press, 2000).
11. G. Gentzen, Investigations concerning logical deduction, In *The Collected Papers of Gerhard Gentzen*, eds. M. E. Szabo (Amsterdam and London:North-Holland, 1935), pp. 68-131.
12. A. Heyting, *Intuitionism: An Introduction* (North-Holland Publications, 1971).
13. S. C. Kleene, On the interpretation of intuitionistic number theory, *Journal of Symbolic logic*, **10**, 4 (1945), pp. 109-124.
14. R. K. Meyer, Relevant arithmetic, *Bulletin of the Section of Logic*. **5** (1976), pp. 133-137.
15. R. K. Meyer, Kurt Gödel and the consistency of $R^{\#\#}$, In *Logical Foundations of Mathematics*, ed. P. Hajek (Computer Science and Physics: Kurt Gödel's Legacy, Springer, 1996), pp. 247-256.
16. R. K. Meyer, $\supset E$ is admissible in "true" relevant arithmetic, *Journal of Philosophical Logic*. **27** (1998), pp. 327-351.
17. R. K. Meyer and J. Friedman, Whither relevant arithmetic?, *the Journal of Symbolic Logic*, **57**, 3 (1992), pp. 824-831.
18. R. K. Meyer and C. Mortensen, Inconsistent models for relevant arithmetics, *The Journal of Symbolic Logic*, **49**, 3 (1984), pp. 917-929.
19. C. Mortensen, The validity of disjunctive syllogism is not so easily proved, *Notre Dame Journal of Formal Logic*, **24**, 1 (1983), pp. 35-40.
20. C. Mortensen, Reply to Burgess and to Read, *Notre Dame Journal of Formal Logic*, **27**, 2 (1986), pp. 195-200.
21. C. Mortensen, Anything is possible, *Erkenntnis*, **30** (1989), pp. 319-337.
22. C. Mortensen, *Inconsistent Mathematics* (Kluwer Academic Publishers, 1995).
23. J. V. Plato, *Elements of Logical Reasoning* (Cambridge University Press, 2013).
24. D. Prawitz, *Natural Deduction: A Proof-Theoretical Study* (Dover Publications, 1965).
25. D. Prawitz, Towards a foundation of a general proof theory, In *Logic, Methodology, and Philosophy of Science IV*, eds. P. Suppes et al. (Amsterdam: North Holland, 1973), pp. 225-250.
26. D. Prawitz, On the idea of a general proof theory, *Synthese*, **27** (1974), pp. 63-77.
27. D. Prawitz, Inference and knowledge, In *The logica yearbook 2008*, ed. M. Pelis (London: College Publications, King's College London, 2009), pp. 175-192.
28. D. Prawitz, On the relation between Heyting's and Gentzen's approaches to meaning, In *Advances in Proof-Theoretic Semantics*, eds. T. Piecha and P. Schroeder-Heister (Springer International Publishing, 2016), pp. 5-25.

29. G. Priest, The logic of paradox, *Journal of Philosophical Logic*, **8**, 1 (1979), pp. 219-241.
30. G. Priest, The logical paradoxes and the law of excluded middle, *Philosophical Quarterly*, **33**, 131(1983), pp. 160-165.
31. G. Priest, Minimally inconsistent *LP*, *Studia Logica*, **50**, (1991), pp. 321-331.
32. G. Priest, Inconsistent models for arithmetic: I, finite models, *Journal of Philosophical Logic*, **26** (1997), pp. 223-235.
33. G. Priest, Inconsistent models for arithmetic: II, the general case, *The Journal of Symbolic Logic*, **65** (2000), pp. 1519-1529.
34. G. Priest, *In Contradiction: A Study of the Transconsistent*, expanded ed. (Clarendon: Oxford University Press, 2006).
35. H. Putnam, The logic of quantum mechanics, In *Mathematics, Matter and Method, Philosophical Papers Vol 1*, ed. H. Putnam (Cambridge University Press, 1968), pp. 174-197.
36. G. Russell, An introduction to logical nihilism, In *Logic, Methodology and Philosophy of Science – Proceedings of the 15th International Congress*, eds. H. Leitgeb, I. Niiniluoto, P. Seppala, and E. Sober (College Publications, 2017), pp. 125-135.
37. S. Shapiro, *Varieties of Logic* (Oxford University Press, 2014).
38. N. Tennant, Logic and its place in nature, In *Kant and Contemporary Epistemology*, ed. P. Parrini (Dordrecht: Kluwer Academic Press, 1994), pp. 101-113.
39. N. Tennant, *The Taming of the True* (Clarendon: Oxford University Press, 1997).
40. N. Tennant, Negation, absurdity, and contrariety, In *What is Negation?*, eds. D. Gabbay and H. Wansing (Dordrecht: Kluwer Academic Press, 1999), pp. 199-222.
41. N. Tennant, An anti-realist critique of dialetheism, In *the Law of Non-Contradiction*, eds. G. Priest, JC Beall, and B. A. Garb (Clarendon: Oxford University Press, 2004), pp. 355-384.
42. N. Tennant, Inferentialism, logicism, harmony, and a counterpoint, In *Essays for Crispin Wright: Logic, Language and Mathematics*, ed. A. Coliva (Oxford: Oxford University Press, 2008).
43. A. S. Troelstra and D. Dalen, *Constructivism in Mathematics* (Amsterdam: North Holland, 1998).

# On Stable Theories with a Special Type

Koichiro Ikeda

*Faculty of Business Administration, Hosei University, Chiyoda-ku, Tokyo 102-8160, Japan*
*E-mail: ikeda@hosei.ac.jp*

We show that any Ehrenfeucht theory has a special type. Also, we give a generic structure whose theory is $\omega$-stable and has a special type.

*Keywords*: Special type; Stable theory; Ehrenfecht theory; Generic structure.

## 1. Introduction

Let $T$ be a complete theory in a countable language. Then a type $p \in S(T)$ is called special, if there are $\bar{a}, \bar{b} \models p$ such that $\mathrm{tp}(\bar{a}/\bar{b})$ is isolated and non-algebraic, and $\mathrm{tp}(\bar{b}/\bar{a})$ is non-algebraic. In this paper, we want to explain results on theories with a special type. First, we will show that any Ehrenfeucht theory has a special type (Proposition 2.1). The proof heavily depends on Lemma 2.1, and the lemma have been obtained by Pillay[2]. Therefore Proposition 2.1 is essentially due to Pillay.

There is a well-known example with a special type (Example 2.1), but the theory is unstable. On the other hand, there are $\omega$-stable examples with a special type[3,5]. Here, using the Hrushovski amalgamation construction, we will give another example with a special type (Theorem 3.1). Our example is based on Sudoplatov's example[5].

**Notation 1.1.** $M, N, \ldots$ denote $L$-structures, and $A, B, \ldots$ subsets of structures. Elements of structures will be denoted by $a, b, \ldots$, and finite tuples of elements by $\bar{a}, \bar{b}, \ldots$. If members of the tuple $\bar{a}$ come from $A$ we sometimes write $\bar{a} \in A$. By $A \subset_{\mathrm{fin}} B$ we mean that $A$ is a finite subset of $B$. $AB$ means $A \cup B$.

Let $T$ be a complete theory and $\mathcal{M}$ a big model of $T$. Then $\mathrm{tp}(\bar{a}/A)$ denotes a type of $\bar{a}$ over $A$ in $\mathcal{M}$. $S(A)$ denotes the set of all types over $A$, and $S(T)$ denotes $S(\emptyset)$. The set of all algebraic elements over $A$ in $\mathcal{M}$ is denoted by $\mathrm{acl}(A)$.

## 2. Proposition

In what follows, $T$ is a complete theory in a countable language.

**Definition 2.1.** Let $p \in S(T)$ be nonisolated. Then $p$ is said to be special, if there are $\bar{a}, \bar{b} \models p$ such that

- $\mathrm{tp}(\bar{b}/\bar{a})$ is isolated and non-algebraic;
- $\mathrm{tp}(\bar{a}/\bar{b})$ is non-isolated.

**Example 2.1.** The following example is well-known and has a special type: Let

$$T = \mathrm{Th}(Q, <, 0, 1, 2, \ldots),$$

where $Q$ is the rationals. Then a complete type $p = \{n < x\}_{n\in\omega}$ is special. In fact, take realizations $a, b$ of $p$ in a big model satisfying $a < b$. Then $\mathrm{tp}(a/b)$ is nonisolated, and $\mathrm{tp}(b/a)$ is isolated and nonalgebraic. Hence $p$ is special.

The example stated above is an Ehrenfeucht theory (see Definition 2.6). In this section, we want to show that any Ehrenfeucht theory has a special type (Proposition 2.1). To prove the result, we need some preparation.

**Definition 2.2.**

(1) The Cantor-Bendixson rank $\mathrm{CB}(\varphi)$ of a formula $\varphi(\bar{x}) \in L$ is defined as follows:

- If $\varphi(\bar{x})$ is consistent, then $\mathrm{CB}(\varphi) \geq 0$;
- Let $\beta$ be limit. Then $\mathrm{CB}(\varphi) \geq \beta$, if $\mathrm{CB}(\varphi) \geq \alpha$ for any $\alpha < \beta$;
- $\mathrm{CB}(\varphi) \geq \alpha + 1$ if there are formulas $\varphi_i(\bar{x}) \in L$ $(i \in \omega)$ such that
  (a) $\models \neg\exists\bar{x}(\varphi_i(\bar{x}) \wedge \varphi_j(\bar{x}))$ for each $i, j \in \omega$ with $i \neq j$;
  (b) $\mathrm{CB}(\varphi \wedge \varphi_i) \geq \alpha$ for each $i \in \omega$.
- If $\mathrm{CB}(\varphi) \geq \alpha$ for all $\alpha$, then we say $\mathrm{CB}(\varphi) = \infty$;
- If $\mathrm{CB}(\varphi) \geq \alpha$ and $\mathrm{CB}(\varphi) \not\geq \alpha + 1$, then we say $\mathrm{CB}(\varphi) = \alpha$.

(2) The rank $\mathrm{CB}(p)$ of a type $p \in S(T)$ is defined to be $\min\{\mathrm{CB}(\varphi) : \varphi \in p\}$.
(3) The degree $\deg(\varphi)$ of $\varphi$ is defined to be the greatest $m \in \omega$ such that there are distinct $p_1, \ldots, p_m \in S(T)$ with $\mathrm{CB}(p_i) = \mathrm{CB}(\varphi)$ for $i = 1, \ldots, m$.
(4) Let $\mathrm{CB}(\bar{a})$ denote $\mathrm{CB}(\mathrm{tp}(\bar{a}))$.

**Definition 2.3.** A theory $T$ is said to be small, if $S(T)$ is countable.

**Remark 2.1.**

(1) If $\bar{a} \in \mathrm{acl}(\bar{b})$, then $\mathrm{CB}(\bar{b}) = \mathrm{CB}(\bar{a}\bar{b})$.
(2) If $T$ is small, then each formula has the CB-rank.

The following lemma was suggested by A. Pillay, and it can be found in his PhD thesis[2].

**Lemma 2.1.** *Suppose that $T$ is small. Let $p \in S(T)$ and $\bar{a}, \bar{b} \models p$. If* $\text{tp}(\bar{b}/\bar{a})$ *is algebraic, then* $\text{tp}(\bar{a}/\bar{b})$ *is isolated.*

**Proof.** Assume that $T$ is small. By Remark 2.1, we can take a formula $\varphi(\bar{x}, \bar{y}) \in \text{tp}(\bar{a}\bar{b})$ with $\text{CB}(\bar{a}\bar{b}) = \text{CB}(\varphi(\bar{x}, \bar{y}))$ and $\deg(\varphi(\bar{x}, \bar{y})) = 1$. Since $\text{tp}(\bar{b}/\bar{a})$ is algebraic, we can assume that $\models \varphi(\bar{a}', \bar{b}')$ implies $\bar{b}' \in \text{acl}(\bar{a}')$. We want to show that

$$\varphi(\bar{x}, \bar{b}) \vdash \text{tp}(\bar{a}/\bar{b}).$$

Take any $\bar{a}' \models \varphi(\bar{x}, \bar{b})$. Clearly we have $\text{CB}(\bar{a}'\bar{b}) \leq \text{CB}(\bar{a}\bar{b})$. Since $\bar{b} \in \text{acl}(\bar{a}')$, by Remark 2.1, we have $\text{CB}(\bar{b}) \leq \text{CB}(\bar{a}')$. Then we have $\text{CB}(\bar{b}) \leq \text{CB}(\bar{a}') \leq \text{CB}(a'\bar{b}) \leq \text{CB}(\bar{a}\bar{b}) \leq \text{CB}(\bar{a}) = \text{CB}(\bar{b})$. Hence $\text{CB}(\bar{a}'\bar{b}) = \text{CB}(\bar{a}\bar{b})$. Since $\deg(\varphi(\bar{x}, \bar{y})) = 1$, we have $\text{tp}(\bar{a}'\bar{b}) = \text{tp}(\bar{a}\bar{b})$. Therefore we have $\bar{a}' \models \text{tp}(\bar{a}/\bar{b})$. $\square$

**Definition 2.4.** Let $p \in S(T)$ be non-isolated. Then $p$ is said to be powerful, if any model realizing $p$ realizes every type over $\emptyset$.

**Remark 2.2.** It is well-known that any Ehrenfeucht theory has a poweful type.

**Definition 2.5.** $\text{tp}(b/a)$ is said to be semi-isolated, if there is a formula $\varphi(x, a) \in \text{tp}(b/a)$ with $\varphi(x, a) \vdash \text{tp}(b)$.

The following fact can be found in Pillay's paper[4].

Fact 2.1. Any non-isolated type $p \in S(T)$ has realizations $\bar{b}, \bar{b}'$ such that $\text{tp}(\bar{b}'/\bar{b})$ is not semi-isolated.

**Definition 2.6.** A theory $T$ is said to be Ehrenfeucht, if it has finitely many countable models, and is not $\omega$-categorical.

The following proposition can be directly obtained by Lemma 2.1, and it was also suggested by A. Pillay.

**Proposition 2.1.** *Any Ehrenfeucht theory has a special type.*

**Proof.** Suppose that $T$ is an Ehrenfeucht theory. By Remark 2.2, there is a powerful type $p(\bar{x})$. By Fact 2.1, we can take $\bar{b}, \bar{b}' \models p$ such that $\text{tp}(\bar{b}'/\bar{b})$ is not semi-isolated. Since $p$ is powerful, we can take $\bar{a} \models p$ such that $\text{tp}(\bar{b}\bar{b}'/\bar{a})$ is isolated. By the transitivity of semi-isolation, $\text{tp}(\bar{a}/\bar{b})$ is nonisolated. By Lemma 2.1, $\text{tp}(\bar{b}/\bar{a})$ is not algebraic. Hence $p$ is special. $\square$

## 3. Example

Proposition 2.1 says that any Ehrenfeucht theory has a special type. In fact, Example 2.1 is Ehrenfeucht and then it has a special type. However, this example is unstable. So the following question arise naturally:

**Question 3.1.** Is there a (small) stable theory with a special type?

For this question, A. Pillay suggested that he had had an $\omega$-stable example with a special type[3]. Also, S. Sudoplatov told me that he had also obtained an example satisfying the same condition[5]. In this section, we will give an $\omega$-stable theory with a special type. This example is based on Sudoplatov's one, but it is constructed by the Hrushovski amalgamation construction.

Here, by a digraph (or directed graph) we mean a graph $(A, R^A)$ satisfying

- $A \models \forall x \forall y (R(x, y) \to \neg R(y, x))$;
- $A \models \forall x \forall y (R(x, y) \to x \neq y)$,

where $R^A = \{ab \in A : A \models R(a, b)\}$, Let $Q(x, y)$ denote $R(x, y) \vee R(y, x)$.

**Definition 3.1.** Let $L = \{R(*, *), U_0(*), U_1(*), ...\}$, and $\mathbf{K}$ the class of all finite $L$-structures $A$ with the following property:

(1) $(A, R^A)$ is a digraph;
(2) $(A, R^A)$ has no cycles, i.e., there is no sequence $a_0 a_1 ... a_n$ in $A$ with $A \models Q(a_0, a_1) \wedge Q(a_1, a_2) \wedge ... \wedge Q(a_n, a_0)$ for each $n \in \omega$;
(3) $\{U_i^A\}_{i \in \omega}$ is disjoint;
(4) For any $i \in \omega$, if $A \models R(a, b) \wedge U_i(b)$ then there is some $j < i$ with $A \models U_j(a)$.

For $A \in \mathbf{K}$, a predimension of $A$ is defined by

$$\delta(A) = |A| - \alpha |R^A|,$$

where $\alpha \in (0, 1]$. In our setting, let $\alpha = 1$. Let $\delta(B/A)$ denote $\delta(B \cup A) - \delta(A)$. For $A \subset B \in \mathbf{K}$, $A$ is said to be closed in $B$ (write $A \leq B$), if

$$\delta(X/A) \geq 0 \text{ for any } X \subset B.$$

For $A, B, C$ with $A = B \cap C$, $B \perp_A C$ means $R^{B \cup C} = R^B \cup R^C$. When $B \perp_A C$, a graph $B \cup C$ is denoted by $B \oplus_A C$.

**Remark 3.1.** If $A \leq B \in \mathbf{K}$ and $b \in B - A$ is connected with $A$, then there is a unique $a \in A$ such that $bb_1...b_n a$ is a path between $a$ and $b$, i.e., $B \models Q(b, b_1) \wedge Q(b_1, b_2) \wedge ... \wedge Q(b_n, a)$ for some distinct $b_1, b_2, ..., b_n \in B - A$.

**Proof.** Suppose that there were another path $bb'_1b'_2...b'_ma'$ for some $a' \in A$ and $b'_1, b'_2, ..., b'_m \in B - A$. Then we would have $\delta(bb_1...b_nb'_1...b'_m/aa') = -1 < 0$, and hence $A \nleq B$. A contradiction. □

**Lemma 3.1.** *If* $A \leq B \in \mathbf{K}, A \subset C \in \mathbf{K}$ *and* $B \bot_A C$, *then* $D = B \oplus_A C \in \mathbf{K}$.

**Proof.** Take any $A, B, C \in \mathbf{K}$ with $A \leq B, A \subset C$ and $B \bot_A C$. Let $D = B \oplus_A C$. Clearly $D$ satisfies (1), (3) and (4) of Definition 3.1. We have to check that $D$ satisfies (2). Suppose that $D$ had a cycle $S$. Since $B$ and $C$ have no cycles, there are $b \in S \cap (B - A)$ and distinct $a, a' \in S \cap A$ such that $b$ is connected with both of $a$ and $a'$. By Remark 3.1, we have $A \nleq B$. A contradiction. Hence $D \in \mathbf{K}$. □

Let $\overline{\mathbf{K}}$ be a class of (possibly infinite) $L$-structures $M$ satisfying $F \in \mathbf{K}$ for any $F \subset_{\text{fin}} M$. Let $A \subset B \in \overline{\mathbf{K}}$, we define $A \leq B$, if $A \cap F \leq B \cap F$ for any $F \subset_{\text{fin}} B$. The closure $\text{cl}_B(A)$ of $A$ in $B$ is defined by $\text{cl}_B(A) = \bigcap\{C \subset B : A \subset C \leq B\}$.

**Remark 3.2.** For any finite $A \subset B \in \overline{\mathbf{K}}$, $\text{cl}_B(A)$ is finite, because $\alpha$ is rational.

**Definition 3.2.** A countable $L$-structure $M$ is said to be $(\mathbf{K}, \leq)$-generic, if

(1) $M \in \overline{\mathbf{K}}$;
(2) if $A \leq B \in \mathbf{K}$ and $A \leq M$ then there is a $B' \cong_A B$ with $B' \leq M$;
(3) if $A \subset_{\text{fin}} M$ then $\text{cl}_M(A)$ is finite.

By Lemma 3.1, $(\mathbf{K}, \leq)$ has the (free) amalgamation property, i.e., if $A \leq B \in \mathbf{K}$ and $A \leq C \in \mathbf{K}$ then $B \oplus_A C \in \mathbf{K}$. Then there is the $(\mathbf{K}, \leq)$-generic structure $M$.

**Remark 3.3.** By the back and forth argument, it can be seen that a generic structure $M$ is ultra-homogeneous over closed sets, i.e., if $A, B \subset_{\text{fin}} M$ satisfy $A, B \leq M$ and $A \cong B$ then $\text{tp}(A) = \text{tp}(B)$.

In what follows, $M$ is $(\mathbf{K}, \leq)$-generic, $T = \text{Th}(M)$, and $\mathcal{M}$ is a big model of $T$.

For $n \in \omega$ and $A \subset B$ we define $A \leq_n B$ by $A \leq X \cup A$ for any $X \subset B - A$ with $|X| \leq n$. Also, for $A, A'$, we define $A \cong_n A'$, if $A$ and $A'$ are isomorphic in the language $\{R, U_0, ..., U_n\}$.

**Lemma 3.2.** *If* $A \leq B \in \mathbf{K}$ *and* $A \leq \mathcal{M}$, *then there is a* $B' \cong_A B$ *with* $B' \leq \mathcal{M}$.

**Proof.** For $n \in \omega$ and $C \subset_{\text{fin}} \mathcal{M}$, let $\theta^n_C(X)$ be a formula expressing that $X \cong_n C$ and $X \leq_n \mathcal{M}$. Take any $A, B \in \mathbf{K}$ with $A \leq B$ and $A \leq \mathcal{M}$. By compactness, it is enough to show that

$$\models \forall X(\theta^n_A(X) \rightarrow \exists Y \theta^n_{AB}(XY))$$

for each sufficiently large $n \in \omega$. Take any $n \in \omega$ with $n \geq \max\{m \in \omega : U^B_m \neq \emptyset\}$. Take any $A'$ with $M \models \theta^n_A(A')$. Let $C' = \mathrm{cl}_M(A')$. Note that $C'$ is finite and $A' \leq_n C'$. It is easily checked that there is a $B^* \in \mathbf{K}$ with $B^*A' \cong_n BA$. By Lemma 3.1, we have $C' \leq B^* \oplus_{A'} C' \in \mathbf{K}$. By genericity of $M$, we can assume that $B^* \oplus_{A'} C' \leq M$, and then $M \models \theta^n_{AB}(A'B^*)$. □

**Lemma 3.3.** *M is saturated.*

**Proof.** Take any $A \subset_{\mathrm{fin}} M$ and any type $p \in S(A)$. We want to show that $p$ is realized by $M$. For simplicity, we assume that $A = \emptyset$. Take a realization $\bar{b} \models p$ in $\mathcal{M}$. By Remark 3.2, $B_0 = \mathrm{cl}(\bar{b})$ is finite. By genericty of $M$, we can take $B_0'$ with $B_0' \leq M$ and $B_0' \cong B_0$. By Lemma 3.2 and the back-and-forth argument, we have $\mathrm{tp}(B_0) = \mathrm{tp}(B_0')$. Take $\bar{b}'$ with $\mathrm{tp}(B_0\bar{b}) = \mathrm{tp}(B_0'\bar{b}')$. Hence $p$ is realized by $\bar{b}' \in M$. □

The following fact can be found in papers of Wagner[6] and Baldwin-Shi[1].

Fact 3.1. Let $M$ be a saturated generic structure. Then

(1) $\mathrm{Th}(M)$ is stable;
(2) if $\alpha$ is rational, then $\mathrm{Th}(M)$ is $\omega$-stable.

**Lemma 3.4.** *$T$ is $\omega$-stable.*

**Proof.** By Lemma 3.3, $M$ is saturated. Note that $\alpha = 1$. By Fact 3.1, $T$ is $\omega$-stable. □

**Lemma 3.5.** *$T$ has a special type.*

**Proof.** We work in the generic $M$, because $M$ is saturated by Lemma 3.3, Let $p(x) = \{\neg U_0(x), \neg U_1(x), ...\}$. Then $p$ is complete, since any 1-element is closed in $M$. Take $a, b \models p$ with $\models R(a, b)$ and $ab \leq M$.

First, we show that $\mathrm{tp}(b/a)$ is isolated and non-algebraic. In fact, it can be shown that $R(a, x)$ isolates $\mathrm{tp}(b/a)$. Take any $b' \in M$ with $\models R(a, b')$. Since $a \models p$, by (4) of Definition 3.1, we have $b' \models p$, and then we have $b'a \cong ba$. Moreover, by (2) of Definition 3.1, we have $ab' \leq M$. By Remark 3.3, we have $\mathrm{tp}(b'/a) = \mathrm{tp}(b/a)$. Hence $\mathrm{tp}(b/a)$ is isolated.

By genericity of $M$, for each $n \in \omega$ there are distinct $b_1, b_2, ..., b_n \in M$ with $R(a, b_i)$ and $ab_i \leq M$ for any $i = 1, ..., n$. Again, by Remark 3.3, we have $\mathrm{tp}(b_i/a) = \mathrm{tp}(b/a)$ for each $i = 1, ..., n$. Hence $\mathrm{tp}(b/a)$ is non-algebraic.

Next, we show that $\mathrm{tp}(a/b)$ is non-isolated. It can be easily seen that $\{R(x, b)\} \cup p(x) \vdash \mathrm{tp}(a/b)$. Suppose that $\mathrm{tp}(a/b)$ were isolated. Then there would be some $n \in \omega$ such that $R(x, b) \wedge \neg U_0(x) \wedge \cdots \wedge \neg U_n(x) \vdash \mathrm{tp}(a/b)$. On the other hand, by genericity of $M$, we can take $a' \in M$ with $\models R(a', b) \wedge \neg U_0(a') \wedge \cdots \wedge \neg U_n(a') \wedge U_{n+1}(a')$. This is a contradiction. Hence $\mathrm{tp}(a/b)$ is non-isolated. □

By Lemma 3.4 and 3.5, we have the following.

**Theorem 3.1.** *There is a generic structure whose theory is $\omega$-stable and has a special type.*

## Acknowledgments

The author wishes to express his thanks to A. Pillay and S. V. Sudoplatov for helpful comments. He also wishes to express his thanks to the referee for pointing out a mistake in the first version of this paper. The author is supported by Grants-in-Aid for Scientific Research (No. 17K05350).

## References

1. J. T. Baldwin and N. Shi, Stable generic structures, *Ann. Pure Appl. Logic* **79** (1996), 1-35.
2. A. Pillay, Gaifman operations, minimal models and the number of countable models, Ph.D thesis, Bedford College, London University, 1977.
3. A. Pillay, unpbulished note.
4. A. Pillay, Instability and theories with few models, *Proceedings of the American Mathematical Society* **80** (1980), 461-468.
5. S. V. Sudoplatov, Powerful types in small theories, *Siberian Mathematical Journal* **31** (1990), 629-638.
6. F. O. Wagner, Relational structures and dimensions, in *Automorphisms of first-order structures*, (Clarendon Press, Oxford, 1994).

# On Hrushovski's Pseudoplanes

Hirotaka Kikyo

*Graduate School of System Informatics, Kobe University,*
*1-1 Rokkodai-cho, Nada, Kobe, Hyogo 657-8501, Japan*
*E-mail: kikyo@kobe-u.ac.jp*

Shunsuke Okabe

*Graduate School of System Informatics, Kobe University,*
*1-1 Rokkodai-cho, Nada, Kobe, Hyogo 657-8501, Japan*
*E-mail: 137x606x@stu.kobe-u.ac.jp*

Let $\alpha$ be a real number with $3/5 < \alpha < 2/3$, and $M$ a Hrushovski's pseudoplane associated to $\alpha$. If $\alpha$ is rational then the theory of $M$ is countably categorical and model complete. If the boundary function is bounded then the theory of $M$ is not model complete. If $\alpha$ is a quadratic irrational then the Hrushovski's boundary function is bounded.

*Keywords*: Hrushovski's amalgamation construction, pseudoplane, model completeness.

## 1. Introduction

Generic structures constructed by the Hrushovski's amalgamation construction are known to have theories which are nearly model complete. If an amalgamation class has the full amalgamation property then its generic structure has a theory which is not model complete[2]. On the other hand, Hrushovski's strongly minimal structure constructed by the amalgamation construction, refuting a conjecture of Zilber has a model complete theory[5].

We have shown that the generic structure of $\mathbf{K}_f$ with a coefficient between 0 and 1 for the predimension function has a model complete theory under some assumption on $f$[8].

Hrushovski's original boundary function does not satisfy our assumption above. Nevertheless, we show the model completeness of the theory of the generic graph associated to $\alpha$ with $3/5 < \alpha < 2/3$.

We also show that if the boundary function $f$ is bounded then the theory of the generic structure is not model complete. For example, if $\alpha$ is a quadratic irrational then a Hrushovski's boundary function is bounded.

We essentially use notation and terminology from Baldwin-Shi[3] and Wagner[12]. We also use some terminology from graph theory[4].

For a set $X$, $[X]^n$ denotes the set of all subsets of $X$ of size $n$, and $|X|$ the cardinality of $X$.

We recall some of the basic notions in graph theory we use in this paper[4]. Let $G$ be a graph. $V(G)$ denotes the set of vertices of $G$ and $E(G)$ the set of edges of $G$. $E(G)$ is a subset of $[V(G)]^2$. $|G|$ denotes $|V(G)|$. The *degree* of a vertex $v$ is the number of edges at $v$. A vertex of degree 0 is *isolated*. A vertex of degree 1 is a *leaf*. $G$ is a *path* $x_0x_1 \dots x_k$ if $V(G) = \{x_0, x_1, \dots, x_k\}$ and $E(G) = \{x_0x_1, x_1x_2, \dots, x_{k-1}x_k\}$ where the $x_i$ are all distinct. $x_0$ and $x_k$ are *ends* of $G$. The number of edges of a path is its *length*. A path of length 0 is a single vertex. $G$ is a *cycle* $x_0x_1 \dots x_{k-1}x_0$ if $k > 3$, $V(G) = \{x_0, x_1, \dots, x_{k-1}\}$ and $E(G) = \{x_0x_1, x_1x_2, \dots, x_{k-2}x_{k-1}, x_{k-1}x_0\}$ where the $x_i$ are all distinct. The number of edges of a cycle is its *length*. A *girth* of a graph $G$ is the length of the shortest cycle in $G$. A non-empty graph $G$ is *connected* if any two of its vertices are linked by a path in $G$. A *connected component* of a graph $G$ is a maximal connected subgraph of $G$. A *forest* is a graph not containing any cycles. A *tree* is a connected forest.

To see a graph $G$ as a structure in the model theoretic sense, it is a structure in language $\{E\}$ where $E$ is a binary relation symbol. $V(G)$ will be the universe, and $E(G)$ will be the interpretation of $E$. The language $\{E\}$ will be called *the graph language*.

Suppose $A$ is a graph. If $X \subseteq V(A)$, $A|X$ denotes the substructure $B$ of $A$ such that $V(B) = X$. If there is no ambiguity, $X$ denotes $A|X$. We usually follow this convention. $B \subseteq A$ means that $B$ is a substructure of $A$. A substructure of a graph is an induced subgraph in graph theory. $A|X$ is the same as $A[X]$ in Diestel's book[4].

We say that $X$ is *connected* in $A$ if $X$ is a connected graph in the graph theoretical sense[4]. A maximal connected substructure of $A$ is a *connected component* of $A$.

Let $A$, $B$, $C$ be graphs such that $A \subseteq C$ and $B \subseteq C$. $AB$ denotes $C|(V(A) \cup V(B))$, $A \cap B$ denotes $C|(V(A) \cap V(B))$, and $A - B$ denotes $C|(V(A) - V(B))$. If $A \cap B = \emptyset$, $E(A, B)$ denotes the set of edges $xy$ such that $x \in A$ and $y \in B$. We put $e(A, B) = |E(A, B)|$. $E(A, B)$ and $e(A, B)$ depend on the graph in which we are working. When we are working in a graph $G$, we sometimes write $E_G(A, B)$ and $e_G(A, B)$ respectively.

Let $D$ be a graph and $A$, $B$, and $C$ substructures of $D$. We write $D = B \otimes_A C$ if $D = BC$, $B \cap C = A$, and $E(D) = E(B) \cup E(C)$. $E(D) = E(B) \cup E(C)$ means that there are no edges between $B - A$ and $C - A$. $D$ is called a *free amalgam of $B$ and $C$ over $A$*. If $A$ is empty, we write $D = B \otimes C$, and $D$ is also called a *free amalgam of $B$ and $C$*.

**Definition 1.1.** Let $\alpha$ be a real number such that $0 < \alpha < 1$.

(1) For a finite graph $A$, we define a predimension function $\delta$ by $\delta(A) = |A| - \alpha|E(A)|$.
(2) Let $A$ and $B$ be substructures of a common graph. Put $\delta(A/B) = \delta(AB) - \delta(B)$.

**Definition 1.2.** Let $A$ and $B$ be graphs with $A \subseteq B$, and suppose $A$ is finite.
$A \leq B$ if whenever $A \subseteq X \subseteq B$ with $X$ finite then $\delta(A) \leq \delta(X)$.
$A < B$ if whenever $A \subsetneq X \subseteq B$ with $X$ finite then $\delta(A) < \delta(X)$.
We say that $A$ is *closed* in $B$ if $A < B$.

If $\alpha$ is irrational then $\leq$ and $<$ are the same relations, but they are different if $\alpha$ is a rational number. Our relation $<$ is often denoted by $\leqslant$ in the literature and some people use $\leq^*$ for our $<$. Since we want to use the relation $\leq$ as well, we use the symbol $<$ for the closed substructure relation.

Let $\mathbf{K}_\alpha$ be the class of all finite graphs $A$ such that $\emptyset < A$.

The following facts appear in the literature[3,12,13] mostly without proofs. Some proofs are given in a first author's paper[11].

Fact 1.1. Let $A$, $B$, $C$ be finite substructures in a common graph.

(1) If $A \cap C$ is empty then $\delta(A/C) = \delta(A) - \alpha e(A, C)$.
(2) If $A \cap C$ is empty and $B \subseteq C$ then $\delta(A/B) \geq \delta(A/C)$.
(3) $A \leq B$ if and only if $\delta(X/A) \geq 0$ for any $X \subseteq B$.
(4) $A < B$ if and only if $\delta(X/A) > 0$ for any $X \subseteq B$ with $X - A$ non-empty.
(5) $A \leq A$.
(6) If $A \leq B$ then $A \cap C \leq B \cap C$.
(7) If $A \leq B$ and $B \leq C$ then $A \leq C$.
(8) If $A \leq C$ and $B \leq C$ then $A \cap B \leq C$.
(9) $A < A$.
(10) If $A < B$ then $A \cap C < B \cap C$.
(11) If $A < B$ and $B < C$ then $A < C$.
(12) If $A < C$ and $B < C$ then $A \cap B < C$.

Fact 1.2. Let $D = B \otimes_A C$.

(1) $\delta(D/A) = \delta(B/A) + \delta(C/A)$.
(2) If $A \leq C$ then $B \leq D$.
(3) If $A \leq B$ and $A \leq C$ then $A \leq D$.
(4) If $A < C$ then $B < D$.
(5) If $A < B$ and $A < C$ then $A < D$.

Fact 1.3.

(1) Let $A$, $B$, $C$ and $D$ be graphs with $D = B \otimes C$ and $A \subseteq D$. Then $\delta(D/A) = \delta(B/A \cap B) + \delta(C/A \cap C)$.
(2) Let $D$ be a graph and $A$ a substructure of $D$. Let $\{D_1, D_2, \ldots, D_k\}$ be the set of all connected components of $D$ where the $D_i$ are all distinct. Then

$$\delta(D/A) = \sum_{i=1}^{k} \delta(D_i/A \cap D_i).$$

Let $B$, $C$ be graphs and $g : B \to C$ a graph embedding. $g$ is a *closed embedding of $B$ into $C$* if $g(B) < C$. Let $A$ be a graph with $A \subseteq B$ and $A \subseteq C$. $g$ is a *closed embedding over $A$* if $g$ is a closed embedding and $g(x) = x$ for any $x \in A$.

In the rest of the paper, **K** denotes a class of finite graphs closed under isomorphisms.

**Definition 1.3.** Let **K** be a subclass of $\mathbf{K}_\alpha$. $(\mathbf{K}, <)$ has the *amalgamation property* if for any finite graphs $A, B, C \in \mathbf{K}$, whenever $g_1 : A \to B$ and $g_2 : A \to C$ are closed embeddings then there is a graph $D \in \mathbf{K}$ and closed embeddings $h_1 : B \to D$ and $g_2 : C \to D$ such that $h_1 \circ g_1 = h_2 \circ g_2$.

**K** has the *hereditary property* if for any finite graphs $A$, $B$, whenever $A \subseteq B \in \mathbf{K}$ then $A \in \mathbf{K}$.

**K** is an *amalgamation class* if $\emptyset \in \mathbf{K}$ and **K** has the hereditary property and the amalgamation property.

A countable graph $M$ is a *generic structure* of $(\mathbf{K}, <)$ if the following conditions are satisfied:

(1) If $A \subseteq M$ and $A$ is finite then there exists a finite graph $B \subseteq M$ such that $A \subseteq B < M$.
(2) If $A \subseteq M$ and $A$ is finite then $A \in \mathbf{K}$.
(3) For any $A, B \in \mathbf{K}$, if $A < M$ and $A < B$ then there is a closed embedding of $B$ into $M$ over $A$.

Let $A$ be a finite structure of $M$. By Fact 1.1 (12), there is a smallest $B$ satisfying $A \subseteq B < M$, written $\mathrm{cl}_M(A)$. The set $\mathrm{cl}_M(A)$ is called a *closure* of $A$ in $M$.

The following fact is fundamental[3,12,13].

Fact 1.4. Let $(\mathbf{K}, <)$ be an amalgamation class. Then there is a generic structure of $(\mathbf{K}, <)$. Let $M$ be a generic structure of $(\mathbf{K}, <)$. Then any isomorphism between finite closed substructures of $M$ can be extended to an automorphism of $M$.

**Definition 1.4.** Let $\mathbf{K}$ be a subclass of $\mathbf{K}_\alpha$. A graph $A \in \mathbf{K}$ is *absolutely closed* in $\mathbf{K}$ if whenever $A \subseteq B \in \mathbf{K}$ then $A < B$.

Note that the notion of being absolutely closed in $\mathbf{K}$ is invariant under isomorphisms.

A proof of the following fact appears in a paper of the first author[11].

Fact 1.5. Let $\mathbf{K}$ be a subclass of $\mathbf{K}_\alpha$ and $M$ a generic structure of $(\mathbf{K}, <)$. Assume that $M$ is countably saturated. Suppose for any $A \in \mathbf{K}$ there is $C \in \mathbf{K}$ such that $A < C$ and $C$ is absolutely closed in $\mathbf{K}$. Then the theory of $M$ is model complete.

**Definition 1.5.** Let $\mathbf{K}$ be a subclass of $\mathbf{K}_\alpha$. $(\mathbf{K}, <)$ has the *free amalgamation property* if whenever $D = B \otimes_A C$ with $B, C \in \mathbf{K}$, $A < B$ and $A < C$ then $D \in \mathbf{K}$.

By Fact 1.2 (4), we have the following.

Fact 1.6. Let $\mathbf{K}$ be a subclass of $\mathbf{K}_\alpha$. If $(\mathbf{K}, <)$ has the free amalgamation property then it has the amalgamation property.

**Definition 1.6.** Let $\mathbb{R}^+$ be the set of non-negative real numbers. Suppose $f : \mathbb{R}^+ \to \mathbb{R}^+$ is a strictly increasing concave (convex upward) unbounded function. Assume that $f(0) = 0$, and $f(1) \leq 1$. We assume that $f$ is piecewise smooth. $f'_+(x)$ denotes the right-hand derivative at $x$. We have $f(x + h) \leq f(x) + f'_+(x)h$ for $h > 0$. Define $\mathbf{K}_f$ as follows:

$$\mathbf{K}_f = \{A \in \mathbf{K}_\alpha \mid B \subseteq A\mathbb{R}ightarrow\delta(B) \geq f(|B|)\}.$$

Note that if $\mathbf{K}_f$ is an amalgamation class then the generic structure of $(\mathbf{K}_f, <)$ has a countably categorical theory[13].

**Definition 1.7.** Let $R, S$ be sets and $\mu : R \to S$ a map. For $Z \subseteq [R]^m$, put

$$\mu(Z) = \{\{\mu(x_1), \ldots, \mu(x_m)\} \mid \{x_1, \ldots, x_m\} \in Z\}.$$

Let $B$, $C$, and $D$ be graphs and $X$ a set of vertices. We write $D = B \rtimes_X C$ if $C|X$ has no edges and the following hold:

(1) $V(D) = V(B) \cup V(C)$.
(2) $X = V(B) \cap V(C)$.
(3) $E(D) = E(B) \cup E(C)$.

Since we are assuming that $C$ has no edges on $X$, $B$ is a usual substructure of $D$ but $C$ may not be a substructure of $D$ in general. If $B$ has no edges on $X$, then $D$ is the free amalgam of $B$ and $C$ over $X$.

Fact 1.7. Let $D$ be a graph with $D = B \rtimes_X C$.

(1) $\delta(D/B) = \delta(C/X)$.
(2) $\delta(D) = \delta(B) + \delta(C/X)$.

Fact 1.8. Let $D$ be a graph with $D = B \rtimes_X C$.

(1) If $C|X < C$ then $B < D$.
(2) If $C|X \leq C$ then $B \leq D$.

## 2. Zero-Extensions and Intrinsic Extensions

**Definition 2.1.** Let $A$ and $B$ be graphs. $B$ is a *zero-extension of* $A$ if $A \leq B$ and $\delta(B/A) = 0$. $B$ is a *minimal zero-extension of* $A$ if $B$ is a proper zero-extension of $A$ and minimal with this property. In this case, $A \subsetneq U \subsetneq B$ implies $A < U$.

$B$ is a *biminimal zero-extension of* $A$ if $B$ is a minimal zero-extension of $A$ and whenever $A' \subseteq A$ and $\delta(B - A/A') = 0$ then $A' = A$.

**Definition 2.2.** Let $A$ and $B$ be graphs. $B$ is a *minimal intrinsic extension of* $A$ if $B$ is an extension of $A$ with $A \not< B$ and minimal with this property. In this case, $A \subsetneq U \subsetneq B$ implies $A < U$.

$B$ is a *biminimal intrinsic extension of* $A$ if $B$ is a minimal intrinsic extension of $A$ and whenever $A' \subseteq A$ and $A' \not< B$ then $A' = A$.

A minimal zero-extension is a minimal intrinsic extension.

We will use the following facts many times.

Fact 2.1. Let $A$ be a substructure of a graph $B$. The following are equivalent:

(1) $B$ is a biminimal zero-extension of $A$.
(2) $\delta(B/A) = 0$ and whenever $D \subsetneq B$ then $A \cap D < D$.

Fact 2.2. Let $D = B \otimes_A C$ where $B$ and $C$ are zero-extensions of $A$. Then $D$ is a zero-extension of $A$.

**Proof.** We have $A \leq D$ by Fact 1.2 (3). We have $\delta(D/A) = 0$ by Fact 1.2 (1). □

**Definition 2.3 (A special sequence for a rational number $\alpha$).** *Let $\alpha = m/d$ be a rational number with coprime positive integers $m$ and $d$. Assume $1/2 < \alpha < 1$. Then $1 - \alpha > 0$ and $1 - 2\alpha < 0$. We consider a finite sequence as a function from $\{0, 1, \ldots, k-1\}$ for some integer $k$. $k$ is a* length *of the sequence. A sequence $s$ is called a* special sequence for $\alpha$ *if $s$ has length $m$, each entry of $s$ is either $1 - \alpha$ or $1 - 2\alpha$, for each $j < m - 1$ we have $0 < \sum_{i=0}^{j} s(i) < \alpha$, and $\sum_{i=0}^{m-1} s(i) = 0$.*

A special sequence for $\alpha$ is defined also for $\alpha \leq 1/2^{11}$.

Fact 2.3. If $1/2 < \alpha < 1$ and $\alpha$ is a rational number then a special sequence for $\alpha$ exists. It can be defined inductively as follows:

(1) $s(0) = 1 - \alpha$.
(2) Let $j > 0$. If $\sum_{i=0}^{j-1} s(i) + (1 - 2\alpha) \geq 0$ then put $s(j) = 1 - 2\alpha$. Otherwise put $s(j) = 1 - \alpha$.
(3) Terminate when $\sum_{i=0}^{j} s(i) = 0$.

**Definition 2.4 (A twig and a wreath).** *Let s be a special sequence for $\alpha$.*

*A graph W is called a* twig associated to *s if W can be written as $W = BF$ with substructures B and F having the following properties:*

*(1) B is a path $b_0 b_1 \cdots b_{m-1}$ of length m.*
*(2) $F = \{f_0\} \cup \{f_i \mid s(i) = 1 - 2\alpha\}$. Here, $f_i \neq f_j$ if $i \neq j$.*
*(3) Each $f_i \in F$ is adjacent to $b_i$.*
*(4) Each $f_i \in F$ is a leaf of W.*

*Figure 1 shows an example with $\alpha = 5/8$.*

*Let D be a substructure of W. F(D) denotes $F \cap D$.*

*Let $k > 0$ be an integer. A graph W is called a* wreath associated to $s^k$ *if W can be written as $W = BF$ with substructures B and F having the following properties:*

*(1) B is a cycle $b_0 b_1 \cdots b_{mk-1} b_0$ of length mk.*
*(2) $F = \{f_i \mid s^k(i) = 1 - 2\alpha\}$. Here, $f_i \neq f_j$ if $i \neq j$.*
*(3) Each $f_i \in F$ is adjacent to $b_i$.*
*(4) Each $f_i \in F$ is a leaf of W.*

*We also say that W is a wreath for $\alpha$ without referring to $s^k$. Figure 2 shows an example with $\alpha = 5/8$. Let D be a substructure of W. F(D) denotes $F \cap D$.*

Figure 1. A twig for 5/8.

Fact 2.4. Let $W$ be a twig or a wreath for $\alpha$. Then $W$ is a biminimal zero-extension of $F(W)$. In particular, if $D$ is a proper substructure of $W$ then $F(D) < D$ by Fact 2.1.

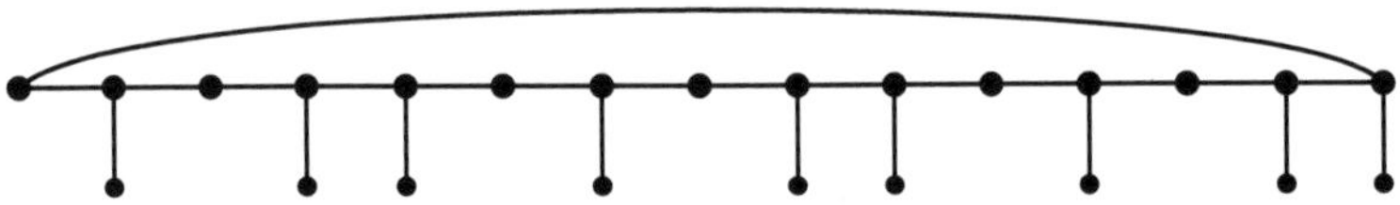

Figure 2. A wreath for 5/8.

If $B = A \rtimes_{F(W)} W$ then $B$ is a minimal zero-extension of $A$. Moreover, if $F(W) = V(A)$ then $B$ is a biminimal zero-extension of $A$.

## 3. Hrushovski's Boundary Function

Most of the definitions and facts appear in Hrushovski's paper[6].

**Definition 3.1.** Let $\alpha$ be a positive real number. We define $x_n, e_n, k_n, d_n$ for integers $n \geq 1$ by induction as follows: Put $x_1 = 2$ and $e_1 = 1$. Assume that $x_n$ and $e_n$ are defined. Let $r_n$ be a smallest rational number $r$ such that $r = k/d > \alpha$ with $d \leq e_n$ where $k$ and $d$ are positive integers. Let $k_n$ and $d_n$ be coprime positive integers with $k_n/d_n = r_n$. Finally, let $x_{n+1} = x_n + k_n$, and $e_{n+1} = e_n + d_n$.

Let $P_0 = (0, 0)$, and $P_n = (x_n, x_n - e_n\alpha)$ for $n \geq 1$. Let $f$ be a function from $\mathbb{R}^+$ to $\mathbb{R}^+$ whose graph on interval $[x_n, x_{n+1}]$ with $n \geq 0$ is a line segment connecting $P_n$ and $P_{n+1}$. We call $f$ a *Hrushovski's boundary function associated to* $\alpha$.

**Lemma 3.1.** *Assume* $3/5 < \alpha < 2/3$. *Then we have a following chart of sequences below:*

| $n$ | 1 | 2 | 3 | 4 | 5 | 6 | 7 |
|---|---|---|---|---|---|---|---|
| $x_n$ | 2 | 3 | 4 | 6 | 8 | 10 | 17 |
| $e_n$ | 1 | 2 | 3 | 6 | 9 | 12 | 23 |
| $k_n$ | 1 | 1 | 2 | 2 | 2 | 7 | 12 |
| $d_n$ | 1 | 1 | 3 | 3 | 3 | 11 | 19 |

We state a few facts known from a Hrushovski's papar[6].

Fact 3.1. Let $f$ be a Hrushovski's boundary function associated to $\alpha$. Then $f$ is strictly increasing and concave, $f'_+(x)$ is decreasing and converges to zero as $x$ tends to infinity.

The set of $\alpha$ for which $f$ is unbounded is dense in the interval (0, 1) (by Baire Categoricity Theorem).

$(\mathbf{K}_f, <)$ has the free amalgamation property. Therefore, there is a generic structure of $(\mathbf{K}_f, <)$. Any one point structure is absolutely closed in $\mathbf{K}_f$.

**Proposition 3.1.** *Let $f$ be a Hrushovski's boundary function associated to $\alpha$. If $\alpha$ is a rational number then $f$ is unbounded.*

**Proof.** Let $x_n$, $e_n$, $k_n$, $d_n$ and $P_n$ be as in Definition 3.1. Let $y_n$ be the $y$-coordinate of $P_n$. Then $y_{n+1} - y_n = k_n - d_n\alpha > 0$ since $k_n/d_n > \alpha$. Suppose $\alpha = m/d$. Then $k_n - d_n\alpha \geq 1/d$. Therefore, $f(x_n) = y_n \geq n/d$. Hence $\lim_{n\to\infty} f(x_n) = \infty$. □

## 4. Model Completeness

Let $f$ be a Hrushovski's boundary function associated to $\alpha$. Let $M$ be a generic structure of $(\mathbf{K}_f, <)$. We show that if $\alpha$ is rational with $3/5 < \alpha < 2/3$ then the theory of $M$ is model complete. In the rest of the paper, we assume that $3/5 < \alpha < 2/3$ and $\alpha = m/d$ with coprime positive integers $m$ and $d$.

In order to discuss if a given graph is in $\mathbf{K}_f$ or not, the following definition will be convenient.

**Definition 4.1.** Let $B$ be a graph and $c \geq 0$ an integer. $B$ is *normal* to $f$ if $\delta(B) \geq f(|B|)$. $B$ is *$c$-normal* to $f$ if $\delta(B) \geq f(|B| + c)$. $B$ is *$c$-critical to $f$* if $B$ is $c$-normal to $f$ and $c$ is maximal with this property.

The following three lemmas are immediate from the definitions[11].

**Lemma 4.1.** *Let $A$ be a finite graph.*

*(1) Suppose $A$ is normal to $f$ and non-empty. Then $\delta(A) > 0$.*
*(2) $A \in \mathbf{K}_f$ if and only if every substructure of $A$ is normal to $f$.*
*(3) Let $c$ and $c'$ be integers such that $0 \leq c \leq c'$. If $A$ is $c'$-normal to $f$ then $A$ is $c$-normal to $f$, and in particular, $A$ is normal to $f$.*
*(4) Let $A$ be normal to $f$. Let $n$ be an integer such that $\delta(A) \geq f(n)$ but $\delta(A) < f(n+1)$. Such an $n$ uniquely exists. Let $c = n - |A|$. Then $A$ is $c$-critical to $f$. $c$ is a unique integer $u$ such that $A$ is $u$-critical to $f$.*
*(5) Let $B$ be another graph such that $\delta(A) = \delta(B)$, $|A| \leq |B|$ and $A$ and $B$ are normal to $f$. Then $B$ is $c$-critical to $f$ if and only if $A$ is $(|B| - |A| + c)$-critical to $f$.*

**Lemma 4.2.** *There is a positive integer $x_0$ such that If $B \in \mathbf{K}_f$ with $|B| \geq x_0$, and $B$ is $c$-critical to $f$ with $0 \leq c < m$ then $B$ is absolutely closed in $\mathbf{K}_f$.*

**Proof.** Since $f'_+(x)$ is decreasing and converges to 0, there is $x_0$ such that $x \geq x_0$ implies $f'_+(x) < 1/(md)$. We can choose $x_0$ as an positive integer.

Since $\alpha = m/d$, there are no positive integers $x$, $y$ such that $x - y\alpha = 0$ with $x < m$. Hence, there are no extension $C$ of $B$ with $\delta(C/B) = 0$ with $|C - B| < m$.

Suppose $B \in \mathbf{K}_f$, $|B| \geq x_0$, and $B$ is $c$-critical to $f$ with $0 \leq c < m$. Suppose there is an extension $C$ of $B$ with $\delta(C/B) < 0$, $C \in \mathbf{K}_f$ and $|C - B| < m$. Then we have $\delta(C/B) \leq -1/d$. Then $f(|B|) < f(|C|) \leq \delta(B) - 1/d$. But since $x_0 \leq |B|$, we have $f'_+(|B|) < 1/(md)$. Therefore,

$$f(|B| + m) \leq f(|B|) + mf'_+(|B|) < \delta(B) - 1/d + m \cdot 1/(md) = \delta(B).$$

But this contradicts the fact that $B$ is $c$-critical to $f$ with $c < m$. □

**Lemma 4.3.** *Let $A$, $U$ be graphs such that $A \subsetneq U$, $\delta(A) \leq \delta(U)$, and $A$ is $|U - A|$-normal to $f$. Then $U$ is normal to $f$.*

**Proof.** $\delta(U) \geq \delta(A) \geq f(|A| + |U - A|) = f(|U|)$. □

**Definition 4.2.** We call $B$ a *special extension of $A$ over $P$* if $B = (AP) \rtimes_{F(W)} W$ where $W$ is a twig or a wreath for $\alpha$, $P$ has no edges, $AP = A \otimes P$, and $V(A) \cap F(W)$ is a proper subset of $F(W)$.

We call $C$ a *semi-special extension of $A$ over $P$* if we can write $C = B_1 \otimes_{AP} B_2 \otimes_{AP} \cdots \otimes_{AP} B_n$ where each $B_i$ is a special extension of $A$ over $P$.

**Lemma 4.4.** *Let $C$ be a semi-special extension of $A$ over $P$. Then $A < C$.*

**Proof.** Suppose $A \subsetneq U \subseteq C$. We can write $B_i = AP \rtimes_{F(W_i)} W_i$ for some twig or wreath $W_i$. So, we can write

$$U = (U \cap B_1) \otimes_{U \cap (AP)} \cdots \otimes_{U \cap (AP)} (U \cap B_n).$$

If $U \cap (AP)$ is a proper extension of $A$ then $\delta(A) < \delta(U \cap (AP))$. Since $AP \leq B_i$ for each $i$, we have $U \cap (AP) \leq U \cap B_i$. Therefore, $U \cap (AP) \leq U$. Hence, $\delta(A) < \delta(U)$.

Suppose $U \cap (AP) = A$. Then $V(U) \cap F(W_i)$ is a proper subset of $F(W_i)$. Hence, $U \cap (AP) = U \cap A < U \cap B_i$. Since $U$ is a proper extension of $A$, there is $j$ such that $U \cap B_j$ is a proper extension of $U \cap A$. Hence, $\delta(U/U \cap A) \geq \delta(U \cap B_j/U \cap A) > 0$.

We have shown that $A < C$. □

**Lemma 4.5.** *Recall that $3/5 < \alpha < 2/3$. Let $s$ be a special sequence for $\alpha$. First 4 terms of $s$ are*

$$s(0) = 1 - \alpha,\ s(1) = 1 - 2\alpha,\ s(2) = 1 - \alpha,\ s(3) = 1 - 2\alpha.$$

*Also, we have $s(m-1) = 1-2\alpha$ for the final term. If $s(j) = 1-\alpha$ then $s(j+1) = 1-2\alpha$.*

**Proof.** We use Fact 2.3. $s(0) = 1 - \alpha$ always. We have $s(0) + (1 - 2\alpha) = 2 - 3\alpha > 0$ by $\alpha < 2/3$. So, $s(1) = 1 - 2\alpha$.

We have $s(0) + s(1) + (1 - 2\alpha) = 3 - 5\alpha < 0$ by $3/5 < \alpha$. Hence, $s(2) = 1 - \alpha$.

We have $s(0) + s(1) + s(2) + (1 - 2\alpha) = 4 - 6\alpha > 0$ by $\alpha < 2/3$. Hence, $s(3) = 1 - \alpha$.

Now, we consider the final term. We have $\sum_{i=0}^{m-2} s(i) > 0$ and $1 - \alpha > 0$. Since $\sum_{i=0}^{m-1} s(i) = 0$, we must have $s(m - 1) = 1 - 2\alpha$.

Finally, suppose $s(j) = 1 - \alpha$. Then $s(j) + 1 - 2\alpha = 2 - 3\alpha > 0$. $\sum_{i=0}^{j-1} s(i) > 0$ if $j \geq 1$. Hence, $\sum_{i=0}^{j} s(i) + 1 - 2\alpha > 0$. Therefore, $s(j + 1) = 1 - 2\alpha$. □

**Lemma 4.6.**

*(1) Let $C = A \otimes_p B$ where $p$ is a single vertex and $A, B \in \mathbf{K}_f$. Then $C \in \mathbf{K}_f$.*
*(2) Any finite forest belongs to $\mathbf{K}_f$.*
*(3) Any cycle of length 6 or more belongs to $\mathbf{K}_f$.*

**Proof.** (1) Since one point structure is absolutely closed in $\mathbf{K}_f$, we have $p < A$ and $p < B$. Therefore, $C \in \mathbf{K}_f$ by the free amalgamation property.

(2) follows by induction on the number of vertices using (1).

(3) Any paths belongs to $\mathbf{K}_f$ by (2). In a path of length 3 or more, the end vertices is closed in the path by $\alpha < 2/3$. Amalgamating 2 paths of length 3 or more over its end vertices produces a cycle of length 6 or more. Hence, it belongs to $\mathbf{K}_f$ by the free amalgamation property. Any cycle of length 6 or more can be produced in this way. □

**Lemma 4.7.** *Let $B = A \rtimes_{\{x,y\}} P$ where $P = x \cdots y$ is a path. Suppose $A \in \mathbf{K}_f$.*

*(1) If the distance of $x$ and $y$ is 3 or more in $A$ and the length of $P$ is 3 then $B \in \mathbf{K}_f$.*
*(2) If the distance of $x$ and $y$ is 3 or more in $A$ and the length of $P$ is 3 or more then $B \in \mathbf{K}_f$.*
*(3) Suppose the distance of $x$ and $y$ is 2 or more in $A$ and the length of $P$ is 4 or more then $B \in \mathbf{K}_f$.*
*(4) Suppose the distance of $x$ and $y$ is 1 or more in $A$ and the length of $P$ is 5 or more then $B \in \mathbf{K}_f$.*

**Proof.** (1) Suppose $P$ has length 3. We can write $P = xuvy$. Let $U$ be a substructure of $B$. $U \cap A$ is normal to $f$ because $A \in \mathbf{K}_f$. If $U = (U \cap A) \otimes_x xu$ or $U = (U \cap A) \otimes_y vy$ then $U \in \mathbf{K}_f$ by Lemma 4.6.

Suppose $U = (U \cap A) \otimes_{\{x,y\}} xuvy$. We have

$$\delta(U) = \delta(U \cap A) + 2 - 3\alpha.$$

Let $t = |U \cap A|$. If $t \geq 4$ then $f'_+(t) \leq 1 - (3/2)\alpha$ and

$$f(|U|) = f(t + 2) \leq f(t) + 2f'_+(t) \leq \delta(U \cap A) + 2 - 3\alpha = \delta(U).$$

Suppose $t \le 3$. This means that $U \cap A = xy$ or $U \cap A = xyw$ with $w \in A$. Since $x$ and $y$ has distance 3 or more, $w$ is not connected to $x$ or $y$. Therefore, $U$ is a path or $U = xuvy \otimes w$. Hence, $U \in \mathbf{K}_f$ by Lemma 4.6.

(2) – (4) We can write $P = x \cdots x'uvy$. Let $A' = A \otimes_x x \cdots x'$. Then $A' \in \mathbf{K}_f$ by Lemma 4.6. Also, the distance between $x'$ and $y$ is 3 or more by the assumption. Now, $B = A' \otimes_{\{x',y\}} x'uvy$. $B$ belongs to $\mathbf{K}_f$ by (1). □

**Lemma 4.8.** *Let $B = A \rtimes_{\{x,y,z\}} S$ where $S = xx'c \otimes_c yy'c \otimes zz'c$. Suppose $A \in \mathbf{K}_f$ and each pair of vertices in $\{x, y, z\}$ has distance 2 or more. Then $B \in \mathbf{K}_f$ also.*

**Proof.** Let $U$ be a substructure of $B$. $U \cap A$ is normal to $f$ because $A \in \mathbf{K}_f$.

Suppose $V(S)$ is not a subset of $V(U)$. Then $U$ can be obtained from $U \cap A$ by amalgamations over 1 vertex, and connecting two points from $x, y, z$ by a path of length 4. In this case, $U$ is normal to $\mathbf{K}_f$ by Lemma 4.7.

Suppose $V(S)$ is a subset of $V(U)$. Then $U$ is an extension of $U \cap A$ by 4 vertices and 6 edges. We have $|U| - |U \cap A| = 4$ and $\delta(U) - \delta(U \cap A) = 4 - 6\alpha$.

Case $|U \cap A| = 3$. This means that $V(U \cap A) = \{x, y, z\}$. Since each pair from $\{x, y, z\}$ has distance 2 or more in $A$, there are no edges among them. So, $U = S$ is a tree and thus $U \in \mathbf{K}_f$, and therefore $U$ is normal to $f$.

Case $|U \cap A| \ge 4$. Then

$$f'_+(|U \cap A|) \le f'_+(4) = 1 - (3/2)\alpha = \frac{4 - 6\alpha}{4} = \frac{\delta(U) - \delta(U \cap A)}{|U| - |U \cap A|}.$$

Therefore, $\delta(U) \ge f(|U|)$. This means that $U$ is normal to $f$.

Now, we see that $B \in \mathbf{K}_f$. □

By Lemma 4.6, we have the following.

**Lemma 4.9.** *Let $\alpha = m/d$ with coprime integers $m$ and $d$. Note that $m \ge 5$ because $3/5 < \alpha < 2/3$.*

*Any twig for $\alpha$ belongs to $\mathbf{K}_f$. Let $W$ be a wreath for $\alpha$. If the girth of $W$ is $2m$ or more then $W$ belongs to $\mathbf{K}_f$.*

**Lemma 4.10.** *Let $A$ be a graph in $\mathbf{K}_f$ with $|A| \ge 2$, and $k$ a integer with $0 \le k \le |A|$. Suppose $\alpha = m/d$ with coprime positive integers $m$, $d$. Then there is a semi-special extension $D = C \otimes_{AP} B$ of $A$ over $P$ such that $D \in \mathbf{K}_f$, $B$ is a special extension of $A$ over $P$ by a wreath $W_1$ with girth $m|A|$ ($|B - (AP)| = m|A|$), $F(W_1) = V(AP)$, and $|C - (AP)| = mk$. Note that $|F(W_1)|$ is determined by the girth $m|A|$, and $|P| = |F(W_1)| - |A|$. Therefore, $|P|$ does not depend on $k$ and $\delta(D) = \delta(B) = \delta(C) = \delta(AP) = \delta(A) + |P|$.*

**Proof.** We prove the lemma in case that $2 \le k \le |A|$. Let $s$ be a special sequence for $\alpha = m/d$. Consider a wreath $W_1$ with girth $m|A|$ and a wreath $W_2$ with girth

$mk$. We can assume that $W_1$ has a cycle $b_0 b_1 \cdots b_{m-1} \cdots b_{m|A|-1} b_0$, and leaves $F(W_1) = \{f_i \mid s(i \bmod m) = 1 - 2\alpha,\ 0 \le i < m|A|\}$, and $b_i$ and $f_i$ are adjacent if $f_i \in F(W_1)$. We can also assume that $W_2$ has a cycle $c_0 c_1 \cdots c_{m-1} \cdots b_{mk-1} b_0$, and leaves $F(W_2) = \{f_i \mid s(i \bmod m) = 1 - 2\alpha,\ 0 \le i < mk\}$, and $c_i$ and $f_i$ are adjacent if $f_i \in F(W_2)$. We already have $F(W_2) \subseteq F(W_1)$.

By Lemma 4.5, $f_1$ and $f_3$ belong to $F(W_2)$.

Identify $f_{im+3}$ for $i = 0, 1, \ldots, |A| - 1$ with the vertices of $A$. Let $P = F(W_1) - V(A)$. $W|P$ has no edges. Let $B = A \rtimes_{V(A)} W_1$. Then $B = AP \rtimes_{F(W_1)} W_1$. Let $C = AP \rtimes_{F(W_2)} W_2$. Let $D$ be the whole structure. Then $D = C \otimes_{AP} B$.

We show that $D$ belongs to $\mathbf{K}_f$.

We explain another way of constructing $D$. First, enumerate $V(A)$ as $\{a_{im+3} \mid 0 \le i < |A|\}$. $a_{im+3}$ is identical to $f_{im+3}$.

Put a path $a_j b_j$ at $a_j$ for $j = im + 3$ with $0 \le i < |A|$. Since $A$ belongs to $\mathbf{K}_f$, The resulting graph also belongs to $\mathbf{K}_f$ by Lemma 4.6.

Connect $b_{im+3}$ and $b_{im+3+m}$ by a path $b_{im+3} b_{im+3+1} \cdots b_{im+3+m}$ of length $m$ for $i = 0, \cdots, |A| - 1$. Here, identify $b_{m|A|}, b_{m|A|+1}, b_{m|A|+2}, b_{m|A|+3}$ with $b_0, b_1, b_2, b_3$. The resulting graph belongs to $\mathbf{K}_f$ by Lemma 4.7.

Now, we construct $W_2$. Put a new path $c_3 a_3$ on $a_3$. The resulting graph belongs to $\mathbf{K}_f$ by Lemma 4.6.

Suppose that $i$ is the largest index such that $c_i$ appears in the current graph which belongs to $\mathbf{K}_f$.

Suppose that $i < mk - 1$. If $i \equiv 1 \pmod{m}$, then $i + 1 \equiv 2 \pmod{m}$ and $i + 2 \equiv 3 \pmod{m}$. Connect $c_i$ and $a_{i+2}$ with a new path $c_i c_{i+1} c_{i+2} a_{i+2}$. Since the distance between $c_i$ and $a_{i+2}$ is 3 or more, the resulting graph belongs to $\mathbf{K}_f$ by Lemma 4.7.

If $s((i + 1) \bmod m) = 1 - 2\alpha$, then connect $c_i$ and $b_{i+1}$ with a new path $c_i c_{i+1} f_{i+1} b_{i+1}$. Since the distance between $c_i$ and $b_{i+1}$ is 3 or more, The resulting graph belongs to $\mathbf{K}_f$ by Lemma 4.7.

Otherwise, $s((i + 1) \bmod m) = 1 - \alpha$. So, we have $s((i + 2) \bmod m) = 1 - 2\alpha$ by Lemma 4.5. Then connect $c_i$ and $b_{i+2}$ with a new path $c_i c_{i+1} c_{i+2} f_{i+2} b_{i+2}$. Since the distance between $c_i$ and $b_{i+2}$ is more than 2, the resulting graph belongs to $\mathbf{K}_f$ by Lemma 4.7.

Repeat this process until $i = mk - 1$ will be the largest index such that $c_i$ appears in the current graph belonging to $\mathbf{K}_f$.

Finally, connect $c_{mk-1}$, $b_1$, and $c_3$ by a structure $c_{mk-1} c_0 c_1 c_2 c_3 \otimes_{c_1} c_1 f_1 b_1$. The distance between any pair from $c_{mk-1}$, $b_1$, and $c_3$ is more than 2. Hence, the resulting graph belongs to $\mathbf{K}_f$ by Lemma 4.8.

We can see that this structure is $D$ above.

See Figure 3 for a example with $\alpha = 5/8$. White circles in the middle part are the vertices of $A$. □

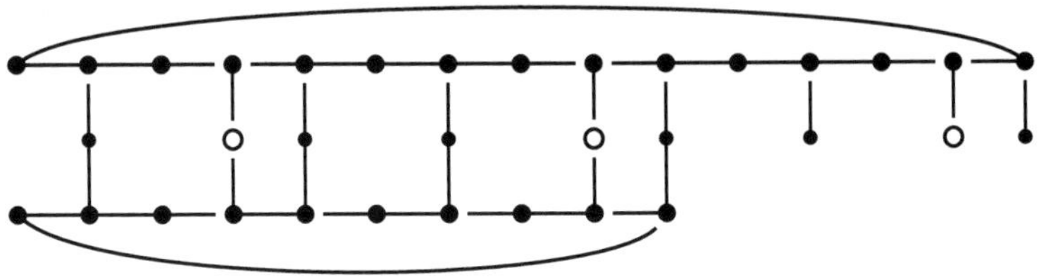

Figure 3. A construction of a semi-special extension. ($\alpha = 5/8$).

**Lemma 4.11.** *Let $D = C \otimes_{AP} B$ be a semi-special extension of $A$ over $P$. Assume that $D \in \mathbf{K}_f$ and $B$ is an extension by a wreath $W$ with $F(W) = V(AP)$. Let*

$$G = C \otimes_{AP} B_1 \otimes_{AP} B_2 \otimes_{AP} \cdots \otimes_{AP} B_n$$

*where $B_i \cong_{AP} B$ for each $i = 1, \ldots, n$. If $G$ is normal to $f$ then $G \in \mathbf{K}_f$.*

**Proof.** Note that $C \otimes_{AP} B$ and $C \otimes_{AP} B_j$ for $j \geq 1$ are isomorphic over $C$. So, $C \otimes_{AP} B_j$ belongs to $\mathbf{K}_f$ for any $j \geq 1$.

We have $B = (AP) \rtimes_{F(W)} W$ with $F(W) = V(AP)$. Let $W_i$ for $i \geq 1$ be a wreath isomorphic to $W$ such that $B_i = (AP) \rtimes_{F(W_i)} W_i$.

Suppose $U \subseteq G$.

Case $AP \subseteq U$. Since $G$ is normal to $f$, $U$ is normal to $f$ by Lemma 4.3.

Case $A \not\subseteq U$. Then $U \cap A$ is a proper subset of $A$. For each $i$ with $0 \leq i \leq n$, put $U_i = U \cap B_i$. Then for $i \geq 1$, we have $U_i = (U \cap AP) \rtimes_{F(D_i)} D_i$ where $F(D_i)$ is a proper subset of $F(W_i) = V(AP)$. Hence, $F(D_i) < D_i$ by Lemma 2.4 for each $i \geq 1$. We have $U \cap C < (U \cap C) \rtimes_{F(D_i)} D_i$ by Lemma 1.8. Put $U_i' = (U \cap C) \rtimes_{F(D_i)} D_i$. Then $U \cap C < U_i'$. Note that it is possible that $U \cap C = U_i'$. Since $(U \cap C) \rtimes_{F(D_i)} D_i = (U \cap C) \otimes_{U \cap (AP)} U_i$, we have

$$U = U_1' \otimes_{U \cap C} \cdots \otimes_{U \cap C} U_n'.$$

Since $U_i' = (U \cap C) \otimes_{U \cap A} U_i$ is a substructure of $C \otimes_A B_i \in \mathbf{K}_f$, we have $U_i' \in \mathbf{K}_f$ for $i = 1, \ldots, n$. Therefore, $U$ belongs to $\mathbf{K}_f$ by the free amalgamation property. □

**Theorem 4.1.** *Let $f$ be a Hrushovski's boundary function associated to rational number $\alpha$ with $3/5 < \alpha < 2/3$. Let $M$ be a generic structure of $(\mathbf{K}_f, <)$. Then the theory of $M$ is model complete.*

**Proof.** We shall show the following claim.

**Claim 4.1.1.** *Let $A \in \mathbf{K}_f$. Then there is a graph $G$ in $\mathbf{K}_f$ such that $A < G$ and $G$ is absolutely closed in $\mathbf{K}_f$.*

With this claim, we have the theorem by Fact 1.5.

By Lemma 4.2, we can choose a positive integer $x_0$ such that If $X \in \mathbf{K}_f$ with $|X| \geq x_0$, and $X$ is $c$-critical to $f$ with $0 \leq c < m$ then $X$ is absolutely closed in $\mathbf{K}_f$.

First, we show Claim 4.1.1 assuming $|A| \geq x_0$.

Let $B$ be a semi-special extension of $A$ over $P$ by a wreath $W_1$ for $\alpha$ such that $B \in \mathbf{K}_f$, $|B - (AP)| = m|A|$, and $F(W_1) = V(AP)$. Such a $B$ exists by Lemma 4.10 with $k = 0$.

Let $n$ be such that $\delta(AP) \geq f(n)$ but $\delta(B) < f(n)$.

Let $n - |AP| = m|A|l + t$ with $0 \leq t < m|A|$, and $t = mk + r$ with $0 \leq r < m$.

By Lemma 4.10, there is $D \in \mathbf{K}_f$ such that $D = C \otimes_{AP} B$ where $C$ is also a special extension of $A$ over $P$ with $|C - (AP)| = mk$.

Let

$$G = C \otimes_{AP} B_1 \otimes_{AP} B_2 \otimes_{AP} \cdots \otimes_{AP} B_l$$

where $B_i \cong_{AP} B$ for each $i = 1, \ldots, l$. Then $|G| = |AP| + mk + m|A|l = n - r$. Hence, $G$ is normal to $f$. By Lemma 4.11, $G$ belongs to $\mathbf{K}_f$. $G$ is $r$-critical and $0 \leq r < m$. Also, we have $|G| \geq |AP| > x_0$. Hence, $G$ is absolutely closed in $\mathbf{K}_f$ by Lemma 4.2.

$G$ is a semi-special extension of $A$ over $P$. Therefore, $A < G$ by Proposition 4.4.

Now, we prove the claim without assuming $|A| \geq x_0$.

Let $A_1$ be an extension of $A$ by isolated vertices such that $|A_1| \geq x_0$. $A_1 \in \mathbf{K}_f$ by the free amalgamation property. We have shown that Claim 4.1.1 holds assuming $|A| \geq x_0$. Hence, there is $G \in \mathbf{K}_f$ such that $A_1 < G$ and $G$ is absolutely closed. It is clear that $A < A_1$. Since $A_1 < G$, we have $A < G$. We have Claim 4.1.1. □

## 5. On $\mathbf{K}_f$ with a Bounded Boundary Function

We have only to assume $0 < \alpha < 1$ in this section. But to make our life easier, proofs will be given assuming $1/2 < \alpha < 2/3$.

**Lemma 5.1.** *Assume $\alpha$ is an irrational number. Let $n > 0$ be a large integer, and let $m/d$ be the greatest rational approximation of $\alpha$ from below with $d \leq n$. Here, we assume that $m$ and $d$ are coprime. Let $W$ be a twig for $m/d$. Then $W$ is a biminimal intrinsic extension of $F(W)$. We have $\delta(W/F(W)) > -\alpha$.*

**Proof.** Let $D$ be a proper substructure of $W$. Suppose $\delta(D/F(D)) = p - q\alpha$. We have $p \leq m \leq n$ and $q \leq d \leq n$. Since $D$ is a twig for $m/d$, we have $p - q(m/d) > 0$.

So, $p/q > m/d$. If $p - q\alpha < 0$ then $p/q < \alpha$. This violates the maximality of $m/d < \alpha$ with $d \leq n$.

Since $m/d < \alpha$, $\delta(W/F(W)) = m - d\alpha < m - d(m/d) = 0$.

Since $m/d < m/(d-1)$ and $d - 1 < d \leq n$, we have $m/(d-1) > \alpha$. So, $m - (d-1)\alpha > 0$. Therefore, $\delta(W/F(W)) = m - d\alpha > -\alpha$. □

**Lemma 5.2.** *Assume $\alpha$ is an irrational number. Let $n > 0$ be any integer. Then there is a twig $W$ for a rational number such that $|F(W)| \geq n$ and $W$ is a biminimal intrinsic extension of $F(W)$ in $\mathbf{K}_f$ (with respect to $\alpha$).*

**Proof.** First, note that if $W$ is a twig for $m/d$ with coprime positive integers $m$ and $d$, then $|F(W)| = d - m + 1$. Assuming $m/d < \alpha$, we have $(3/2)m < d$ since $\alpha < 2/3$. Hence, $|F(W)| = d - m + 1 > m/2 + 1$.

Choose an integer $q > 0$. Let $m$ and $d$ be coprime positive integers such that $m/d$ is the largest rational approximation of $\alpha$ from below with $d \leq q$.

Let $l$ be a positive integer such that $1/l < \alpha - m/d$. Then $m/d < m/d + 1/l < \alpha$. We have $m/d + 1/l = (lm+1)/(ld)$. Let $m'/d'$ be the largest rational approximation of $\alpha$ from below with $d \leq ld$. Then $m'/d' \geq (lm+1)/(ld) > m/d$. We cannot have $d' \leq q$ by the choice of $m/d$. So, $d' > q \geq d$. Since $m'/d' > m/d$ and $d' > d$, we have $m' > m$.

Now, suppose an integer $n > 0$ is given. We can choose a large enough integer $q$ such that choosing a largest rational approximation $m/d$ of $\alpha$ from below with $d \leq q$, we have $m/2 + 1 > n$.

We can assume that $m$ and $d$ are coprime. Let $W$ be a twig for $m/d$. Then $|F(W)| > m/2 + 1 > n$. Also, $W$ is a biminimal intrinsic extension of $F(W)$ in $\mathbf{K}_f$. □

**Theorem 5.1.** *Let $\alpha$ be any real number with $1/2 < \alpha < 2/3$. Let $M$ be a generic structure of $(\mathbf{K}_f, <)$. If $f$ is bounded then the theory of $M$ is not model complete.*

**Proof.** We assume that $\alpha$ is irrational. If $\alpha$ is rational, we can argue in a similar way with a wreath for $\alpha$.

Suppose $f$ is bounded. Then there is $n$ such that $n - 1 > f(x)$ for any $x \geq 0$.

By Lemma 5.2, we can choose a large twig $W$ for some rational number such that $W$ is a biminimal intrinsic extension of $F(W)$ and $|F(W)| > n$. Since $F(W)$ has no edges as a substructure of $W$. Hence, $\delta(F(W)) = |F(W)|$. Therefore, $\delta(W) = \delta(F(W)) + \delta(W/F(W)) > n - \alpha > n - 1$.

Let $W_0$ be a substructure of $M$ such that $W_0 < M$ and $W$ and $W_0$ are isomorphic. $F(W_0)$ is an isomorphic image of $F(W)$. $F(W_0)$ as a substructure of $M$ has no edges.

Let $A$ be a substructure of $M$ such that $A < M$, $A$ has no edges, and $|A| = |F(W_0)|$.

Any 1-to-1 map from $A$ to $F(W_0)$ is a graph isomorphism.

**Claim 5.1.1.** *The existential type of A in M is a subset of the existential type of $F(W_0)$ in M.*

Fix an isomorphism $\tau : A \to F(W_0)$. Assume $M \models \exists x\varphi(x, A)$ with $\varphi(x, y)$ a quantifier-free formula in the language of graphs. Let $B \subset M$ be such that $M \models \varphi(B, A)$. Then $A \subseteq BA \subseteq M$. Since $A < M$, we have $A < AB$.

Let $D = C \otimes_{F(W_0)} W$ be a graph such that there is a graph isomorphism $\tau' : BA \to C$ extending $\tau : A \to F(W_0)$.

We show that $D \in \mathbf{K}_f$. Suppose $U \subseteq D$. Then $U = (U \cap C) \otimes_{U \cap F(W_0)} (U \cap W_0)$. We have $U \cap C \in \mathbf{K}_f$ and $U \cap W_0 \in \mathbf{K}_f$ because $C \in \mathbf{K}_f$ and $W_0 \in \mathbf{K}_f$.

Since $A < AB$ we have $F(W_0) < C$. Hence, $U \cap F(W_0) < U \cap C$.

If $U \cap W$ is a proper substructure of $W$, then $U \cap F(W_0) < U \cap W$ since $W$ is a biminimal intrinsic extension of $F(W_0)$. Therefore, $U \in \mathbf{K}_f$ by the free amalgamation property.

Suppose that $U \cap W = W$. Then $U = C \otimes_{F(W_0)} W_0$. We have $\delta(C) \geq \delta(F(W_0)) > n$. We also have $\delta(W_0/F(W_0)) > \alpha$ by Lemma 5.1. Therefore, $\delta(U) = \delta(C) + \delta(W_0/F(W_0)) > n - \alpha > n - 1 > f(|U|)$. Hence, $U$ is normal to $f$. Now, we know that $D \in \mathbf{K}_f$.

Since $F(W_0) < C$, we have $W_0 < D$. With this and by the claim, $D$ can be embedded in $M$ over $F(W_0)$ because $M$ is generic. Let $B'$ be the isomorphic image of $B$ in the embedded $D$ in $M$. Then $M \models \varphi(B', F(W_0))$. We have the claim.

Suppose that the theory $Th(M)$ of $M$ is model complete. Then every formula is equivalent to an existential formula modulo $Th(M)$.

Hence, the claim implies that the full types of $A$ and $F(W_0)$ are the same. But while $A$ has no finite extension $U$ in $M$ with $A \not< U$, $F(W_0)$ has a finite extension $W_0$ with $F(W_0) \not< W_0$. This is a contradiction. □

In the case that $\alpha$ is irrational and Hrushovski's boundary function $f$ is unbounded, it is still open that whether the generic structure of $(\mathbf{K}_f, <)$ has a model complete theory or not.

Finally, we show that Hrushovski's boundary function is bounded in "usual" cases.

The following fact is well-known.

Fact 5.2. Let $\alpha$ be an irrational number with $0 < \alpha < 1$. $\alpha$ has a continued fraction representation

$$\alpha = [0; a_1, a_2, \ldots, a_n, \ldots] = \cfrac{1}{a_1 + \cfrac{1}{a_2 + \cfrac{1}{\ddots + \cfrac{1}{a_n + \ddots}}}}$$

where $a_1$, $a_2$, ..., $a_n$, ... are integers greater than or equal to 1.

Let $p_n/q_n = [0; a_1, a_2, \ldots, a_{n-1}]$ with coprime positive integers $p_n$, $q_n$. Then the following recurrence equations hold:

$$p_0 = 1, \quad p_1 = 0, \quad p_{n+1} = a_n p_n + p_{n-1},$$
$$q_0 = 0, \quad q_1 = 1, \quad q_{n+1} = a_n q_n + q_{n-1}.$$

Also, the following hold:

$$\left|\alpha - \frac{p_n}{q_n}\right| < \frac{1}{q_n^2} \quad \text{for all } n \geq 0,$$

and in particular,

$$0 < \frac{p_{2n}}{q_{2n}} - \alpha < \frac{1}{q_{2n}^2} \quad \text{for all } n \geq 0.$$

**Proposition 5.1.** *Let $\alpha = [0; a_1, a_2, \cdots]$ where $a_n \geq 1$ is an integer for any $n \geq 1$. Let $f$ be a Hrushovski's boundary function associated to $\alpha$. Recall the sequence $\{x_n\}_{n\geq 0}$ from the definition of $f$. Let $n_0$ be such that $q_4 \leq e_{n_0}$. For any $n > n_0$, there is $m$ such that*

$$f(x_n) - f(x_{n_0}) \leq \sum_{i=2}^{m} \frac{a_{2i+1}(a_{2i}+1)(a_{2i-1}(a_{2i-2}+1)+1)}{q_{2i-2}}.$$

**Proof.** Recall $x_n$, $e_n$, $k_n$, $d_n$ from the definition of $f$ and $p_n$, $q_n$ from Fact 5.2. We have

$$f(x_n) = \sum_{i=1}^{n}(k_n - d_n\alpha).$$

Define $h : \mathbb{R}^+ \to \mathbb{R}^+$ by

$$h(x) = k_n/d_n - \alpha \; (e_n \leq x < e_{n+1})$$

for each integer $n \geq 1$. Then we have

$$f(x_n) = (2 - \alpha) + \int_{e_1}^{e_{n+1}} h(x)dx.$$

Define $g : \mathbb{R}^+ \to \mathbb{R}^+$ by

$$g(x) = \frac{1}{q_{2m-2}^2} \ (q_{2m} \le x < q_{2m+2})$$

for each integer $m \ge 2$.

**Claim 5.2.1.** *If $x \ge q_4$ then $h(x) < g(x)$.*

We can choose $m, n$ such that $q_{2m} \le x < q_{2m+2}$ and $e_n \le x < e_{n+1}$.

Case $q_{2m} \le e_n$. $k_n/d_n$ is a smallest $k/d$ with $d \le e_n$. So, $k_n/d_n \le p_{2m}/q_{2m}$. Therefore, $h(x) = k_n/d_n - \alpha \le p_{2m}/q_{2m} - \alpha < 1/q_{2m}^2 < 1/q_{2m-2}^2 = g(x)$.

Case $q_{2m-2} \le e_n < q_{2m}$. We have $k_n/d_n \le p_{2m-2}/q_{2m-2}$. Hence, $h(x) = k_n/d_n - \alpha \le p_{2m-2}/q_{2m-2} - \alpha < 1/q_{2m-2}^2 = g(x)$.

Case $e_n < q_{2m-2} < q_{2m}$. Since $q_{2m} \le x < e_{n+1}$, we have $e_n < q_{2m-2} < q_{2m} < e_{n+1}$. Since $d_n \le e_n$, we have $e_{n+1} = e_n + d_n \le 2e_n$. Then $2e_n < 2q_{2m-2} \le (a_{2m-1} + 1)q_{2m-2} \le a_{2m-1}q_{2m-1} + q_{2m-2} = q_{2m}$. Hence, $e_{n+1} < q_{2m}$. This is impossible. We have the claim.

Suppose $e_n \le q_{2m}$. Then

$$\int_{e_{n_0}}^{e_n} h(x)dx < \int_{q_4}^{q_{2m}} g(x)dx = \sum_{i=2}^{m-1} \frac{q_{2i+2} - q_{2i}}{q_{2i-2}^2}. \tag{1}$$

We have

$$\begin{aligned} q_{2i+2} &= a_{2i+1}q_{2i+1} + q_{2i} \\ &= a_{2i+1}(a_{2i}q_{2i} + q_{2i-1}) + q_{2i} \\ &= (a_{2i+1}a_{2i} + 1)q_{2i} + a_{2i+1}q_{2i-1} \\ &\le (a_{2i+1}a_{2i} + 1)q_{2i} + a_{2i+1}q_{2i} \\ &= (a_{2i+1}(a_{2i} + 1) + 1)q_{2i}. \end{aligned}$$

Hence,

$$q_{2i} \le (a_{2i-1}(a_{2i-2} + 1) + 1)q_{2i-2},$$

and also,

$$q_{2i+2} - q_{2i} \le a_{2i+1}(a_{2i} + 1)q_{2i}.$$

Therefore,

$$q_{2i+2} - q_{2i} \le a_{2i+1}(a_{2i} + 1)(a_{2i-1}(a_{2i-2} + 1) + 1)q_{2i-2}.$$

With inequality (1), we have the proposition. □

**Corollary 5.1.** *Let $\alpha = [0; a_1, a_2, \cdots]$ and $a_n \geq 1$ for any $n \geq 1$. Suppose values of $a_n$ $(n \geq 1)$ are bounded from above by a single integer. Then a Hrushovski's boundary function is bounded.*

*If $\alpha$ is a quadratic irrational then a Hrushovski's boundary function is bounded.*

**Proof.** In Proposition 5.1, the denominator of each term is uniformly bounded. By the form of the recurrence equation which $q_n$ satisfies, we can see that $\sum_{i=1}^{\infty} 1/q_n$ converges. Therefore, $f(x_n)$ is bounded as $n$ tends to the infinity.

It is well-known that for any quadratic irrational, the continued fraction eventually repeats. Hence, the values of $a_n$ are bounded from above by a single integer. □

## Acknowledgments

This work is supported by JSPS KAKENHI Grant Number 17K05345.

## References

1. J.T. Baldwin and K. Holland, Constructing $\omega$-stable structures: model completeness, Ann. Pure Appl. Log. **125**, 159–172 (2004).
2. J.T. Baldwin and S. Shelah, Randomness and semigenericity, Trans. Am. Math. Soc. **349**, 1359–1376 (1997).
3. J.T. Baldwin and N. Shi, Stable generic structures, Ann. Pure Appl. Log. **79**, 1–35 (1996).
4. R. Diestel, *Graph Theory*, Fourth Edition, Springer, New York (2010).
5. K. Holland, Model completeness of the new strongly minimal sets, J. Symb. Log. **64**, 946–962 (1999).
6. E. Hrushovski, A stable $\aleph_0$-categorical pseudoplane, preprint (1988).
7. E. Hrushovski, A new strongly minimal set, Ann. Pure Appl. Log. **62**, 147–166 (1993).
8. K. Ikeda, H. Kikyo, Model complete generic structures, in *the Proceedings of the 13th Asian Logic Conference*, World Scientific, 114–123 (2015).
9. H. Kikyo, Model complete generic graphs I, RIMS Kokyuroku **1938**, 15–25 (2015).
10. H. Kikyo, Balanced Zero-Sum Sequences and Minimal Intrinsic Extensions, to appear in RIMS Kokyuroku.
11. H. Kikyo, Model Completeness of Generic Graphs in Rational Cases, Archive for Mathematical Logic, published on line (2017), doi.org/10.1007/s00153-017-0601-4.
12. F.O. Wagner, Relational structures and dimensions, in *Automorphisms of first-order structures*, Clarendon Press, Oxford, 153–181 (1994).
13. F.O. Wagner, *Simple Theories*, Kluwer, Dordrecht (2000).

# The First Homology Group of a $G$-Set

Junguk Lee

*Department of Mathematics, Yonsei University,*
*Seoul, 03722, South Korea*
*E-mail: ljw@yonsei.ac.kr*

We introduce a notion of a geometric $G$-set, which is a $G$-set equipped with a closure operation and an independence relation. For a geometric $G$-set, we define homology groups of orbits and we identify the first homology groups of orbits. These homology groups generalize the homology groups of types in model theory. Also we construct a groupoid on the set of orbit whose vertex groups are the first homology groups of orbits and from this, we associated a groupoid on the Stone space of types whose vertex groups are the first homology groups of types.

*Keywords*: Geometric $G$-set; The first homology group of a $G$-set; Groupoid on the set of orbits.

## 1. Introduction

We consider a $G$-set equipped with a closure operation and an independence relation. Such $G$-set is called *geometric*. For such geometric $G$-sets, we define homology groups on each orbit. Any $G$-set is geometric by associating a trivial closure operation and a trivial independence relation. If we consider the $G$-action on $G$ by left multiplication, then in this case, the orbit of identity is $G$ itself and the first homology group of $G$ turns out to be the abelianization of $G$. A non-trivial example of a geometric $G$-set is the automorphism group action on a monster model of a complete theory equipped with the algebraic closure and the trivial independence relation. In this case, its first homology groups are the first homology groups of types. In algebraic topology, for a topological space $X$, we associate a groupoid $\pi(X)$, called the fundamental groupoid of $X$, whose objects are the points of $X$ and morphisms are the homotopy classes of paths in $X$ so that its vertex groups are the fundamental groups. Similarly, for a geometric $G$-set $X$, we associate a groupoid on the set $X/G$ of orbits whose vertex groups are the first homology groups of orbits. As a corollary, for a given complete theory $T$, we get a groupoid on the Stone space of types whose vertex groups are the first homology groups of types.

Let $G$ be a group acting on a set $X$, called a $G$-set and dented by $(X, G)$. Our main object is a $G$-set equipped with a closure operation and an independence relation. For $A \subset X$, $G_A$ is the subgroup of elements in $G$ fixing $A$ pointwise. For a cardinal $\lambda$, let $X^\lambda$ be the set of $\lambda$-tuples of elements of $X$. Then $X^\lambda$ is also a $G$-set with the following actions: For $x = (x_i) \in X^\lambda$ and $g \in G$, $gx := (y_i)$ with $y_i = gx_i$ for $i \in \lambda$. We denote by $\mathrm{o}(a/A)$ the orbit of $a$ under the action of $G_A$ for $a \in X^\lambda$ and $A \subset X$. For $A \subset X$, if $A = \{a_i | \ i < |A|\}$, we consider $A$ as a $|A|$-tuple $(a_i)$ and conversely for a $\lambda$-tuple $a = (a_i)$, we consider $a$ as the set $\{a_i | \ i < \lambda\}$. For $a = (a_i) \in X^\lambda$ and $A \subset X$, we write $A \subset a$ or $a \subset A$ if $A \subset \{a_i\}$ or $\{a_i\} \subset A$, respectively. For $a = (a_i) \in X^\lambda$ and $b = (b_j) \in X^\kappa$ with $\lambda \leq \kappa$, we write $a \subset b$ if $a_i = b_i$ for $i \in \lambda$.

A *closure operation* on a $G$-set $X$ is a map $\mathrm{cl} : \mathcal{P}(X) \to \mathcal{P}(X)$ satisfying the following: For $A, B \subset X$ and $g \in G$,

- $A \subset \mathrm{cl}(A)$ and $\mathrm{cl}(\mathrm{cl}(A)) = \mathrm{cl}(A)$;
- $g\,\mathrm{cl}(A) = \mathrm{cl}(gA)$;
- $\mathrm{cl}(A) \subset \mathrm{cl}(B)$ if $A \subset B$.

Now we consider a $G$-set $X$ equipped with a closure operation cl, denoted by $(X, G, \mathrm{cl})$. We introduce a notion of an independence relation on $(X, G, \mathrm{cl})$. A ternary relation $\mathop{\smile\hskip-0.9em\lower0.3ex\hbox{$\vert$}}^*$ between sets of $X$ is called an *independence relation* if it satisfies the following conditions:

- invariance : for any $A, B, C$ and $g \in G$, we have $A \mathop{\smile\hskip-0.9em\lower0.3ex\hbox{$\vert$}}^*_C B$ iff $gA \mathop{\smile\hskip-0.9em\lower0.3ex\hbox{$\vert$}}^*_{gC} gB$;
- normality: for any $A, B, C$, if $A \mathop{\smile\hskip-0.9em\lower0.3ex\hbox{$\vert$}}^*_C B$, then $A \mathop{\smile\hskip-0.9em\lower0.3ex\hbox{$\vert$}}^*_C \mathrm{cl}(BC)$;
- symmetry: for any $A, B, C$, we have $A \mathop{\smile\hskip-0.9em\lower0.3ex\hbox{$\vert$}}^*_C B$ iff $B \mathop{\smile\hskip-0.9em\lower0.3ex\hbox{$\vert$}}^*_C A$;
- transitivity: for any $A, B, C, D$ with $B \subset C \subset D$, $A \mathop{\smile\hskip-0.9em\lower0.3ex\hbox{$\vert$}}^*_B D$ iff $A \mathop{\smile\hskip-0.9em\lower0.3ex\hbox{$\vert$}}^*_B C$ and $A \mathop{\smile\hskip-0.9em\lower0.3ex\hbox{$\vert$}}^*_C D$;
- extension: for any $A, B, C$ with $B \subset C$, there is $g \in G_B$ such that $gA \mathop{\smile\hskip-0.9em\lower0.3ex\hbox{$\vert$}}^*_B C$.

Throughout this paper we call the above axioms **the basic 5 axioms**. We write $(X, G, \mathrm{cl}, \mathop{\smile\hskip-0.9em\lower0.3ex\hbox{$\vert$}}^*)$ for a $G$-set $X$ with a closure operation cl and an independence relation $\mathop{\smile\hskip-0.9em\lower0.3ex\hbox{$\vert$}}^*$, and we call it a *geometric G-set*. Any $G$-set is geometric by taking the *trivial* closure operation and ternary relation, that is, $\mathrm{cl}(A) = A$ and $A \mathop{\smile\hskip-0.9em\lower0.3ex\hbox{$\vert$}}^*_C B$ for any $A, B, C$. In this case, we call this geometric $G$-set *trivial*. A typical non-trivial geometric $G$-set comes from model theory.

**Example 1.1.** Let $X = \mathcal{M}$ be a monster model of a complete theory $T$, and let $G = \mathrm{Aut}(\mathcal{M})$ be the group of automorphisms of $\mathcal{M}$. Take $\mathrm{cl} = \mathrm{acl}$ and $\mathop{\smile\hskip-0.9em\lower0.3ex\hbox{$\vert$}}^*$ as the trivial independence relation. Then we get a non-trivial geometric $G$-set $(X, G, \mathrm{cl}, \mathop{\smile\hskip-0.9em\lower0.3ex\hbox{$\vert$}}^*)$ unless $\mathrm{acl}(A) = A$ for any $A \subset X$.

From now on, we fix a geometric $G$-set $(X, G, \mathrm{cl}, \mathop{\smile\!\!\!\!\!\!\,|}^{*})$. We also fix an orbit $\mathrm{o} = \mathrm{o}(a/B)$ for $a \in X^{\lambda}$ and $B \subset X$ with $B \subset a$. We shall define the first homology group of o with respect to $\mathop{\smile\!\!\!\!\!\!\,|}^{*}$, analogously to the case of types in model theory. We first introduce some notations.

**Notation 1.1.**

(1) Let $s$ be an arbitrary finite set of natural numbers. Given any subset $X \subseteq \mathcal{P}(s)$, we may view $X$ as a category where for any $u, v \in X$, $\mathrm{Mor}(u, v)$ consists of a single morphism $\iota_{u,v}$ if $u \subseteq v$, and $\mathrm{Mor}(u, v) = \emptyset$ otherwise. If $f\colon X \to C$ is any functor into some category $C$, then for any $u, v \in X$ with $u \subseteq v$, we let $f_v^u$ denote the morphism $f(\iota_{u,v}) \in \mathrm{Mor}_C(f(u), f(v))$. We shall call $X \subseteq \mathcal{P}(s)$ a *primitive category* if $X$ is non-empty and *downward closed*; i.e., for any $u, v \in \mathcal{P}(s)$, if $u \subseteq v$ and $v \in X$ then $u \in X$. (Note that all primitive categories have the empty set $\emptyset \subset \omega$ as an object.)

(2) We use $C_B$ to denote the category whose objects are the tuples of $X$ containing $B$, and whose morphisms from $a$ to $b$ are elements $g \in G_B$ such that $ga = b$.

For a functor $f : X \to C_B$ with a primitive category $X$ and objects $u \subseteq v$ of $X$, $f_v^u(u)$ denotes the set $f_v^u(f(u))(\subseteq f(v))$.

**Definition 1.1.** By a *$*$-independent functor in* o, we mean a functor $f$ from some primitive category $X$ into $C_B$ satisfying the following:

- If $\{i\} \subset \omega$ is an object in $X$, then $f(\{i\})$ is of the form $\mathrm{cl}(Cb)$ where $b \in \mathrm{o}$, $C = \mathrm{cl}(C) = f_{\{i\}}^{\emptyset}(\emptyset) \supseteq B$, and $b \mathop{\smile\!\!\!\!\!\!\,|}^{*}_{B} C$.
- Whenever $u (\neq \emptyset) \subset \omega$ is an object in $X$, we have

$$f(u) = \mathrm{cl}\left(\bigcup_{i \in u} f_u^{\{i\}}(\{i\})\right)$$

and $\{f_u^{\{i\}}(\{i\}) \mid i \in u\}$ is $*$-independent over $f_u^{\emptyset}(\emptyset)$.

We let $\mathcal{A}_{\mathrm{o}}^{*}$ denote the family of all $*$-independent functors in o.

A $*$-independent functor $f$ is called a *$*$-independent $n$-simplex* in o if $f(\emptyset) = B$ and $\mathrm{dom}(f) = \mathcal{P}(s)$ with $s \subset \omega$ and $|s| = n + 1$. We call $s$ the *support* of $f$ and denote it by $\mathrm{supp}(f)$.

In the rest we may call a $*$-independent $n$-simplex in $p$ just an *$n$-simplex* of $p$, as far as no confusion arises. We are ready to define the first homology group $H_1^{*}(\mathrm{o})$ of o depending on our choice of the independence relation $\mathop{\smile\!\!\!\!\!\!\,|}^{*}$.

**Definition 1.2.** Let $n \geq 0$.

$$S_n(\mathcal{A}_{\mathrm{o}}^{*}) := \{ f \in \mathcal{A}_{\mathrm{o}}^{*} \mid f \text{ is an } n\text{-simplex of o} \}$$

$$C_n(\mathcal{A}^*_{\rm o}) := \text{the free abelian group generated by } S_n(\mathcal{A}^*_{\rm o}).$$

An element of $C_n(\mathcal{A}^*_{\rm o})$ is called an *n-chain* of o. The support of a chain $c$, denoted by supp($c$), is the union of the supports of all the simplices that appear in $c$ with a non-zero coefficient. Now for $n \geq 1$ and each $i = 0, \ldots, n$, we define a group homomorphism

$$\partial^i_n : C_n(\mathcal{A}^*_{\rm o}) \to C_{n-1}(\mathcal{A}^*_{\rm o})$$

by putting, for any $n$-simplex $f : \mathcal{P}(s) \to C$ in $S_n(\mathcal{A}^*_{\rm o})$ where $s = \{s_0 < \cdots < s_n\} \subset \omega$,

$$\partial^i_n(f) := f \upharpoonright \mathcal{P}(s \setminus \{s_i\})$$

and then extending linearly to all $n$-chains in $C_n(\mathcal{A}^*_{\rm o})$. Then we define the *boundary map*

$$\partial_n : C_n(\mathcal{A}^*_{\rm o}) \to C_{n-1}(\mathcal{A}^*_{\rm o})$$

by

$$\partial_n(c) := \sum_{0 \leq i \leq n} (-1)^i \partial^i_n(c).$$

We shall often refer to $\partial_n(c)$ as the *boundary of* $c$. Next, we define:

$$Z_n(\mathcal{A}^*_{\rm o}) := \ker \partial_n$$

$$B_n(\mathcal{A}^*_{\rm o}) := \Im\, \partial_{n+1}.$$

The elements of $Z_n(\mathcal{A}^*_{\rm o})$ and $B_n(\mathcal{A}^*_{\rm o})$ are called *n-cycles* and *n-boundaries* in o, respectively. It is straightforward to check that $\partial_n \circ \partial_{n+1} = 0$. Hence we can now define the group

$$H^*_n(\mathrm{o}) := Z_n(\mathcal{A}^*_{\rm o}) / B_n(\mathcal{A}^*_{\rm o})$$

called the $n$th $*$-*homology group* of o.

**Notation 1.2.**

(1) For $c \in Z_n(\mathcal{A}^*_{\rm o})$, $[c]$ denotes the homology class of $c$ in $H^*_n(\mathrm{o})$.
(2) When $n$ is clear from the context, we shall often omit $n$ in $\partial^i_n$ and in $\partial_n$, writing simply as $\partial^i$ and $\partial$.

**Definition 1.3.** A 1-chain $c \in C_1(\mathcal{A}^*_{\rm o})$ is called a 1-$*$-*shell* (or just a 1-shell) in o if it is of the form

$$c = f_0 - f_1 + f_2$$

where $f_i$'s are 1-simplices of $p$ satisfying

$$\partial^i f_j = \partial^{j-1} f_i \quad \text{whenever } 0 \leq i < j \leq 2.$$

Hence, for $\text{supp}(c) = \{n_0 < n_1 < n_2\}$ and $k \in \{0, 1, 2\}$, it follows that

$$\text{supp}(f_k) = \text{supp}(c) \setminus \{n_k\}.$$

Notice that the boundary of any 2-simplex is a 1-shell.

**Remark 1.1.**

(1) For $B \subset a$, let $\text{o} = \text{o}(a/B)$, and let $\bar{\text{o}} = \text{tp}(\text{cl}(a)/B)$. By the definitions of the $*$-independent functors and the first homology group, $H_1^*(\text{o})$ and $H_1^*(\bar{\text{o}})$ are identical.

(2) Because of the 5 axioms of $\mathop{\smile\!\!\!\!\!\!\;|}^*$, $\mathcal{A}_\text{o}^*$ forms a non-trivial *amenable* collection of functors into $C_B$, which is introduced in[4].

**Proof.** We sketch a proof of (2) briefly. We first show that $\mathcal{A}_\text{o}^*$ forms an amenable collection. To do this, we need to show that it satisfies the four conditions in[4] Definition 1.3, that is, (1) *invariance under weak isomorphisms,* (2) *closure under restrictions and unions,* (3) *closure under localizations, and* (4) *extensions of localizations are localizations of extensions.* By the definition of $\mathcal{A}_\text{o}^*$ with normality, symmetry, and transitivity, the conditions (1) and (2) automatically hold, and by transitivity, the condition (3) holds. It remains to show that the condition (4) holds. To do this, we need the following fact: For $A, B, C, D \subset X$, if each of the sets $A$, $B$, and $C$ are $\mathop{\smile\!\!\!\!\!\!\;|}^*$-independent over $D$ and $A \mathop{\smile\!\!\!\!\!\!\;|}^*_{CD} B$, then there is $g \in G_D$ such that $A \cup B \cup gC$ is $\mathop{\smile\!\!\!\!\!\!\;|}^*$-independent over $D$ (†). By extension, there is $g \in G_D$ such that $gC \mathop{\smile\!\!\!\!\!\!\;|}^*_D AB$. By invariance, $gC$ is still $\mathop{\smile\!\!\!\!\!\!\;|}^*$-independent over $D$. Some forking calculus with transitivity and symmetry show that $A \cup B \cup gC$ is $\mathop{\smile\!\!\!\!\!\!\;|}^*$-independent over $D$. Then we deduce the condition (4) from the fact (†).

Next we show that $\mathcal{A}_\text{o}^*$ is non-trivial, that is, it has 1-amalgamation and satisfies the strong 2-amalgamation(see[4] Definition 1.13). The amenable collection $\mathcal{A}_\text{o}^*$ has 1-amalgamation by extension, and satisfies the strong 2-amalgamation by invariance, transitivity, and extension. □

From Remark 1.1(2), we get the following note:

**Note.** ([3] or[4])

$$H_1^*(p) = \{[c] \mid c \text{ is a 1-}*\text{-shell with } \text{supp}(c) = \{0, 1, 2\}\}.$$

For the rest of this paper, we suppress $B$ to $\emptyset$ for notational simplicity so that $G_B = G$. Also we assume that $B = \text{cl}(B)$.

## 2. The first homology group for G-sets

Here we identify the first homology group $H_1^*(\mathrm{o})$. Since $\mathcal{A}_{\mathrm{o}}^*$ is a non-trivial amenable collection, we can use the results from[2,6,7] to describe 2-chains with 1-shell boundaries. Thus we have the following description for $H_1^*(\mathrm{o})$ in[2][Fact 3.10, Theorem 4.4].

**Theorem 2.1.**

(1) *There is a canonical homomorphism $\Psi_{\mathrm{o}}^* : G \to H_1^*(\mathrm{o})$.*
(2) *The kernel of $\Psi_{\mathrm{o}}^*$ is the normal subgroup of $G$ consisting of all elements of $G$ fixing all orbits of elements of $\bar{\mathrm{o}}$ under the action of $G'$.*

**Remark 2.1.** Due to above Theorem 2.1, $H_1^*(\mathrm{o})$ does not depend on the choice of independence $\mathop{\smile\hskip-0.9em\lower.2ex\hbox{$\mid$}}^*$ satisfying the 6 basic axioms and depends on the choice of closure cl. So we denote it by $H_1^{\mathrm{cl}}(\mathrm{o})$.

We give two examples for computing $H_1(\mathrm{o})$.

**Example 2.1.**

(1) Let $X = \mathcal{M}$ be a monster model of a complete theory $T = T^{\mathrm{eq}}$ with $\emptyset = \mathrm{acl}(\emptyset)$, and let $G = \mathrm{Aut}(\mathcal{M})$ be the group of automorphisms of $\mathcal{M}$. Consider a geometric $G$-set $(X, G, \mathrm{cl}, \mathop{\smile\hskip-0.9em\lower.2ex\hbox{$\mid$}}^*)$ with the trivial independence $\mathop{\smile\hskip-0.9em\lower.2ex\hbox{$\mid$}}^*$ and $\mathrm{cl} = \mathrm{acl}$. Let $p = \mathrm{tp}(a)$ be a strong type over $\emptyset$. Then $p(\mathcal{M})$ is an orbit o of $a$ under the action of $G$. Then $H_1^{\mathrm{acl}}(\mathrm{o}) = H_1(p)$ is the first homology group of the type $p$. If $T = T^{\mathrm{heq}}$ is simple and we take $\mathrm{cl} = \mathrm{bdd}$ as an closure operation, then $H_1^{\mathrm{bdd}}(\mathrm{o}) = 0$ by the Independence Theorem.
(2) Let $G$ be an arbitrary group. Consider a $G$-set $X = G$ by acting on $G$ as the left multiplication, $(g, x) \mapsto gx$. Consider a trivial geometric $G$-set $(X, G, \mathrm{cl}, \mathop{\smile\hskip-0.9em\lower.2ex\hbox{$\mid$}}^*)$. Let $\mathrm{o} = \mathrm{o}(\mathrm{id}) = G$. Then $H_1^{\mathrm{cl}}(\mathrm{o}) = G/G'$. Really, in this case, $\mathrm{o} = \bar{\mathrm{o}}$ and the kernel of the canonical map $\Psi : G \to H_1(\mathrm{o})$ is exactly $G'$ because $G$ acts on $X$ freely, i.e., for $g, h \in G$, $gx = hx$ for some $x \in X$ implies $g = h$.

**Note.** For a discrete group $G$, there is a path-connected topological space $BG$ called *the classifying space* of $G$ whose fundamental group is isomorphic to $G$. So the first homology group of $BG$ is isomorphic to $G/G'$. From Example 2.1(2), we see that the first homology group of the trivial geometric $G$-set $G$ is same as the first homology group of the classifying space of $G$.

**Question 2.1.** Let $G$ be a discrete group. Let $(G, G, \mathrm{cl}, \mathop{\smile\hskip-0.9em\lower.2ex\hbox{$\mid$}}^*)$ be the trivial geometric $G$-set where $G$ acts on $G$ by left multiplication, and let $BG$ be the classifying space.

Then for each $n \geq 2$, do we have that

$$H_n^{\mathrm{cl}}(G) \cong H_n(BG)?$$

## 3. Groupoid on the set of 1-orbits

In [3,5], John Goodrick, Byunghan Kim, and Alexei Kolesnikov introduced some fundamental groupoids and proved the Hurewicz correspondence in stable theories. Namely, for each $n \geq 1$, the $n$-th fundamental group is isomorphic to the $(n+1)$-th homology group under the assumption of $(n+2)$-complete amalgamation in stable theories. By the mismatch of degrees of the fundamental groups and the homology groups, there are no fundamental groups, constructed by John Goodrick, Byunghan Kim, and Alexei Kolesnkov, corresponding to the first homology groups. In this section, given a geometric $G$-set $(X, G, \mathrm{cl}, \mathrel{\raise0.2ex\hbox{$\smile$}\kern-0.6em\lower0.1ex\hbox{$|$}}^{*})$, we construct a groupoid whose objects are the elements in $X$ and vertex group on each $x \in X$ is isomorphic to the first homology group $H_1^{\mathrm{cl}}(\mathrm{o}(x))$. As a result, we construct a groupoid on the space of strong types whose vertex groups are exactly the first homology groups of each types.

For a groupoid $\mathcal{G}$, we denote the set of objects by $\mathrm{Ob}(\mathcal{G})$ and for each $x, y, z \in \mathrm{Ob}(\mathcal{G})$, the set of morphisms from $x$ to $y$ by $\mathcal{G}(x, y)$ and the composition map between morphisms by $\circ : \mathcal{G}(y, z) \times \mathcal{G}(x, y) \to \mathcal{G}(x, z)$. And each vertex group is denoted by $\mathcal{G}(x) = \mathcal{G}(x, x)$.

Consider a $G$-set $(X, G)$. Let $X/G$ be the set of orbits of $x \in X$. Denote the orbit of $x \in X$ by $[x]_G$. Fix representatives $x_i \in X$ of each orbits in $X/G$ for $i < |X/G|$.

**Theorem 3.1.** *Assign a normal subgroup $N_i$ of $G$ to each $[x_i]_G$ for each $i < |X/G|$ such that $N_i$ contains a stabilizer $G_{x_i}$ of $x_i$ and its conjugations, i.e., $gG_{x_i}g^{-1}(= G_{gx_i}) \subset N_i$ for all $g \in G$. Then there is a groupoid $\mathcal{G}_0 := \mathcal{G}_0(X, G, (N_i)_{i<|X/G|})$ such that*

*(1)* $\mathrm{Ob}(\mathcal{G}_0) = X/G$;
*(2) for* $i < j < |X/G|$, $\mathcal{G}_0([x_i]_G, [x_j]_G) \neq \emptyset$ *if and only if* $N_i = N_j$; *and*
*(3) for* $i < |X/G|$, $\mathcal{G}_0([x_i]_G) = G/N_i$.

**Proof.** Here, we denote by $[x]$ the class $[x]_G$. We define morphisms between elements in $X/G$ satisfying (2) and (3). At first, we define an equivalence relation $\sim^{\mathrm{init}}_{\mathrm{ter}}$ on $[x_i] \times [x_j]$ with $N_i = N_j (=: N)$. For this, we first define two relations $\sim^{\mathrm{init}}$ and $\sim_{\mathrm{ter}}$ on $[x_i] \times [x_j]$ as follows: For $y_{i0}, y_{i1} \in [x_i]$ and $z_{j0}, z_{j1} \in [x_j]$,

- $(y_{i0}, z_{j0}) \sim^{\mathrm{init}} (y_{i1}, z_{j1})$ if for $g \in G$ with $gz_{j0} = z_{j1}$, there is $h_g \in N$ such that $gy_{i0} = h_g y_{i1}$; and

- $(y_{i0}, z_{j0}) \sim_{\text{ter}} (y_{i1}, z_{j1})$ if for $g \in G$ with $gy_{i0} = y_{i1}$, there is $h'_g \in N$ such that $gz_{j0} = h'_g z_{j1}$.

**Claim 3.1.** $\sim^{\text{init}}$ *and* $\sim_{\text{ter}}$ *are equivalence relations.*

**Proof.** We first show that $\sim^{\text{init}}$ is an equivalence relation.

Reflexivity) Consider $(y, z) \in [x_i] \times [x_j]$ with $N_i = N_j = N$. Since the stabilizer $G_z$ of $z$ is a subset of $N_j$ and $N_i = N_j$, $(y, z) \sim^{\text{init}} (y, z)$.

Symmetry) Consider $(y_0, z_0), (y_1, z_1) \in [x_i] \times [x_j]$ with $N_i = N_j = N$, and suppose $(y_0, z_0) \sim^{\text{init}} (y_1, z_1)$. We want to show that $(y_1, z_1) \sim^{\text{init}} (y_0, z_0)$. Consider $g \in G$ such that $g^{-1}z_1 = z_0$. So $gz_0 = z_1$ and there is $h_g \in N$ such that $gy_0 = h_g y_1$ since $(y_0, z_0) \sim^{\text{init}} (y_1, z_1)$. Then $y_1 = (h_g^{-1} g) y_0$ and $g^{-1} y_1 = (g^{-1} h_g^{-1} g) y_0$. By the normality of $N$, $g^{-1} h_g^{-1} g$ is in $N$. Take $h_{g^{-1}} = g^{-1} h_g^{-1} g$ and $(y_1, z_1) \sim^{\text{init}} (y_0, z_0)$.

Transitivity) Consider $y_k \in [x_i]$, $z_k \in [x_j]$ with $N_i = N_j = N$ for $k = 0, 1, 2$. Suppose $(y_0, z_0) \sim^{\text{init}} (y_1, z_1) \sim^{\text{init}} (y_2, z_2)$. We will show that $(y_0, z_0) \sim^{\text{init}} (y_2, z_2)$. Take $g \in G$ such that $gz_0 = z_2$. Since $[z_2] = [z_1]$, there is $g' \in G$ such that $g'z_2 = z_1$. So $(g'g)z_0 = z_1$, and there is $h_{g'g} \in N$ such that $(g'g)y_0 = h_{g'g}y_1$ because $(y_0, z_0) \sim^{\text{init}} (y_1, z_1)$. By symmetry, $(y_2, z_2) \sim^{\text{init}} (y_1, z_1)$ and there is $h_{g'} \in N$ such that $g'y_2 = h_{g'}y_1$. So $y_1 = (h_{g'}^{-1} g')y_2$ and $(g'g)y_0 = h_{g'g}y_1 = (h_{g'g}h_{g'}^{-1} g')y_2$. Therefore we have that $gy_0 = (g'^{-1} h_{g'g} h_{g'}^{-1} g')y_2$. Since $N$ is a normal subgroup of $G$, $g'^{-1}(h_{g'g}h_{g'}^{-1})g'$ is in $N$. Take $h_g = g'^{-1}(h_{g'g}h_{g'}^{-1})g'$ and we are done.

In a similar way, the relation $\sim_{\text{ter}}$ on $[x_i] \times [x_j]$ with $N_i = N_j$ is also an equivalence relation. □

Moreover these two equivalence relations are same.

**Claim 3.2.** *For* $(y_0, z_0), (y_1, z_1) \in [x_i] \times [x_j]$ *with* $N_i = N_j$,

$$(y_1, z_1) \sim^{\text{init}} (y_0, z_0) \text{ if and only if } (y_1, z_1) \sim_{\text{ter}} (y_0, z_0)$$

**Proof.** Choose $(y_0, z_0), (y_1, z_1) \in [x_i] \times [x_j]$ with $N_i = N_j = N$.

$\Rightarrow$) Suppose $(y_1, z_1) \sim^{\text{init}} (y_0, z_0)$. We will show that $(y_1, z_1) \sim_{\text{ter}} (y_0, z_0)$. Take $g \in G$ such that $gy_0 = y_1$. Since $[gz_0] = [z_1]$, there is $g' \in G$ such that $(g'g)z_0 = z_1$, and $gz_0 = g'^{-1}z_1$. It is enough to show that $g'^{-1}$ is in $N$. Since $(y_1, z_1) \sim^{\text{init}} (y_0, z_0)$, there is $h_{g'g} \in N$ such that $(g'g)y_0 = h_{g'g}y_1$. By the way, $(g'g)y_0 = g'(gy_0) = g'y_1$ and $g'y_1 = h_{g'g}y_1$. Thus $g'^{-1}h_{g'g} \in G_{y_1} \subset N$. Since $h_{g'g} \in N$, so is $g'^{-1}$.

$\Leftarrow$) It is similar with the case of left to right. □

Take $\sim^{\text{init}}_{\text{ter}} := \sim^{\text{init}} (= \sim_{\text{ter}})$ and we define the set of morphisms as follows: For $i, j < |X/G|$,

$$\mathcal{G}_0([x_i],[x_j]) = ([x_i] \times [x_j])/\sim^{\text{init}}_{\text{ter}} \text{ if } N_i = N_j, \text{or } = \emptyset \text{ otherwise.}$$

For $(x', y') \in [x] \times [y]$, write $[x', y']_{\sim^{\text{init}}_{\text{ter}}}$ for the $\sim^{\text{init}}_{\text{ter}}$-class of $(x', y')$. Next, we define a composition map

$$\circ : \mathcal{G}_0([x_j],[x_k]) \times \mathcal{G}_0([x_i],[x_j]) \to \mathcal{G}_0([x_i],[x_k])$$

as follows: Choose $(y, z) \in [x_i] \times [x_j]$ and $(z', w) \in [x_j] \times [x_k]$ with $N_i = N_j = N_k = N$. Since $[z] = [z']$, there is $g \in G$ such that $gz' = z$. Define $[z', w]_{\sim^{\text{init}}_{\text{ter}}} \circ [y, z]_{\sim^{\text{init}}_{\text{ter}}}$ as $[y, gw]_{\sim^{\text{init}}_{\text{ter}}}$. We have to show this composition is well-defined, that is, for two $g_0, g_1 \in G$ such that $g_0 z' = g_1 z' = z$, $(y, g_0.w) \sim^{\text{init}}_{\text{ter}} (y, g_1.w)$. Take $g_0, g_1 \in G$ such that $g_0 z' = g_1 z' = z$. The choices of $g_0$ and $g_1$ imply $g_0 g_1^{-1} \in G_{z'} \subset N$ and $g_0 w = (g_0 g_1^{-1} g_1) w$. Thus, for any $g \in G_y$, $g(g_0 w) = g(g_0 g_1^{-1} g_1) w = (g g_0 g_1^{-1})(g_1.w)$ and $g g_0 g_1^{-1} \in N$. So, this composition is well-defined.

Consider the groupoid $\mathcal{G}_0$. Let us see the vertex groups of $\mathcal{G}_0$. The vertex groups are of the form $\mathcal{G}_0([x_i],[x_i]) (=: \mathcal{G}_0([x_i]))$ for $i < |X/G|$. Each class of $[x_i]^2/\sim^{\text{init}}_{\text{ter}}$ has a representative of the form $(x_i, y)$ for some $y \in [x_i]$.

**Claim 3.3.** *For $y_0, y_1 \in [x_i]$, $(x_i, y_0) \sim^{\text{init}}_{\text{ter}} (x_i, y_1)$ if and only if there is $g \in N_i$ such that $gy_0 = y_1$.*

**Proof.** Let $N = N_i$. Choose $y_0, y_1 \in [x_i]$.

$\Rightarrow$) Suppose $(x_i, y_0) \sim^{\text{init}}_{\text{ter}} (x_i, y_1)$. Then $(x_i, y_0) \sim_{\text{ter}} (x_i, y_1)$. Thus, for $g \in G_{x_i} \subset N$, there is $h'_g \in N$ such that $gy_0 = h'_g y_1$. So, $y_1 = (h'^{-1}_g g) y_0$ and $h'^{-1}_g g$ is in $N$.

$\Leftarrow$) Suppose there is $g \in N$ such that $gy_0 = y_1$. Then for any $g' \in G_{x_i}$, $g' y_0 = (g' g^{-1}) y_1$ and $g' g^{-1}$ is in $N$. Thus $(x_i, y_0) \sim_{\text{ter}} (x_i, y_1)$ and $(x_i, y_0) \sim^{\text{init}}_{\text{ter}} (x_i, y_1)$. □

By Claim 3.3, each vertex group $\mathcal{G}_0([x_i])$ is isomorphic to $G/N_i$. Thus this groupoid $\mathcal{G}_0$ is a desired one. □

**Corollary 3.1.** *Let $(X, G, \text{cl}, \downarrow^*)$ be a geometric $G$-set with $\text{cl}(\emptyset) = \emptyset$. Let $N_x$ be the kernel of the canonical map from $G$ to $H_1^{\text{cl}}(\text{o}(x))$ for each $x \in X$. Then there is a groupoid $\mathcal{G}(X/G)$ such that*

- $\text{Ob}(\mathcal{G}(X/G)) = X/G$*; and*
- *For each* $\text{o} \in X/G$, $\text{Ob}(\mathcal{G}(X/G))(\text{o}) \cong H_1^{\text{cl}}(\text{o})$.

Let $\mathcal{M}$ be a monster model of a complete theory $T = T^{eq}$, and $A = \mathrm{acl}(A) \subset \mathcal{M}$ be a small subset. For each $\kappa < |\mathcal{M}|$, let $S_\kappa(A)$ be the Stone space of strong types of $a \in \mathcal{M}^\kappa$ over $A$. For some $\kappa < |\mathcal{M}|$, take $X = \mathcal{M}^\kappa$, $G = \mathrm{Aut}_A(\mathcal{M})$. Let $N_q$ be the kernel of the canonical epimorphism from $\mathrm{Aut}_A(\mathcal{M})$ to $H_1(q)$ for each $q$. Then by Corollary 3.1, we get a groupoid on $S_\kappa(A)$ whose vertex groups are isomorphic to the first homology groups of types in $S_\kappa(A)$.

**Corollary 3.2.** *For each $\kappa < |\mathcal{M}|$, there is a groupoid $\mathcal{G}(S_\kappa(A))$ such that*

- $\mathrm{Ob}(\mathcal{G}(S_\kappa(A))) = S_\kappa(A)$*; and*
- *For each $q \in S_\kappa(A)$, $\mathcal{G}(S_\kappa(A))(q) \cong H_1(q)$.*


## Acknowledgments

The author was supported by Samsung Science Technology Foundation under Project Number SSTF-BA1301-03.


## References


1. Enrique Casanovas, Daniel Lascar, Anand Pillay, and Martin Ziegler, Galois groups of first order theories, *Journal of Mathematical Logic*, **1** (2001), 305-319.
2. Jan Dobrowolski, Byunghan Kim, and Junguk Lee, The Lascar groups and the first homology groups in model theory, *Annals of Pure and Applied Logic*, **168** (2017), 2129-2151.
3. John Goodrick, Byunghan Kim, and Alexei Kolesnikov, Homology groups of types in model theory and the computation of $H_2(p)$, *Journal of Symbolic Logic*, **78** (2013), 1086-1114.
4. John Goodrick, Byunghan Kim, and Alexei Kolesnikov, Amalgamation functors and homology groups in model theory, *Proceedings of ICM 2014*, II (2014), 41-58.
5. John Goodrick, Byunghan Kim, and Alexei Kolesnikov, Homology groups of types in stable theories and the Hurewicz correspondence, *Annals of Pure and Applied Logic*, **168** (2017), 1710-1728.
6. Byunghan Kim, SunYoung Kim, and Junguk Lee, A classification of 2-chains having 1-shell boundaries in rosy theories, *Journal of Symbolic Logic*, **80** (2015), 322-340.
7. SunYoung Kim and Junguk Lee, More on 2-chains with 1-shell boundaries in rosy theories, *Journal of the mathematical society of Japan*, **69** (2017), 93-109.
8. Junguk Lee, On the first homology groups of strong types in rosy theories and some arithmetic properties on nonstandard rationals, PhD thesis, Yonsei University, 2016.
9. Jon Peter May, A Concise Course in Algebraic Topology(Chicago Lectures in Mathematics) 1st edn., (University of Chicago Press 1999).
10. Martin Ziegler, Introduction to the Lascar group, in *Tits buildings and the model theory of groups*, London Math. Soc. Lecture Note Series, 291 (Cambridge University Press 2002), 279-298.

# Via

Kamal Lodaya
*The Institute of Mathematical Sciences, Chennai 600113, India*

This paper makes a proposal for a logic to describe paths in graphs and suggests another one to compare paths as well.

*Keywords*: Branching temporal logic, computation tree logic, two-variable logic.

## 1. Introduction

That modal logic on Kripke structures translates to two-variable logic has been known for a long time. Lutz, Sattler and Wolter extended modal logic with boolean and converse operations to obtain expressive completeness with respect to two-variable logic on graphs [19]. Marx and de Rijke, building on earlier work on words, showed there is a temporal logic on trees which is expressively complete for two-variable first-order logic [21].

In earlier work on words, we showed that one can extend two-variable first-order logic with "between" relations, and again get an expressively complete temporal logic [15]. We also showed that this fragment has EXPSPACE satisfiability.

In the present paper, we change the setting to directed acyclic graphs. We design a modal logic which is expressively complete for two-variable first-order logic with "via" relations, inspired by our earlier work. We show that this fragment has elementary satisfiability.

We also extend this modal logic to a richer framework with costs. Defining a suitable two-variable logic, so as to extend this result to obtain expressive completeness, and proving satisfiability of decidability, is left as an open question.

### 1.1. *Setup*

Suppose I have to go from Bangalore, where I am, to China later this year. I look up various travel websites and discover various facts. For example, there may be a two-hop journey whose fare is cheaper than any one-hop journey. I can go via Delhi where I have several options but then I have to face the long queues at Delhi airport (which I don't care much for), or I can go via southeast Asia where there are longer stopovers.

Graded modal logics [10], logics with nominals (see [7] for an approach due to Arthur Prior, or the later work of Gargov and Goranko [11]), more generally description logics [5], provide facilities for knowledge representation and reasoning over graphs, and I might imagine specifying my requirements to a theorem prover in such a logic.

What if I was planning a longer journey, continuing from China via the Pacific to Canada? Now another such set of requirements comes up. Any logic which has a tree model property will fare badly with such repeated sets of requirements, because there are several branches reaching China and the entire tree of requirements from China to Canada has to be repeated at each China node. I could go to Dubai and then via the Atlantic to Canada, which is my eventual destination, satisfying my shopping urges along this alternative path.

### 1.2. *Rich modal logics*

Surely this could be done in graded CTL [2,9,23] or in dynamic logics, where there is a large literature [8,13,20] ? However, these logics all have the tree model property. We use nominals from hybrid logics to bring together paths. Indeed, our approach will be close to hybrid CTL [1,14,26]. We have to avoid enriching such logics to force the formation of unbounded-size grids, which leads to high undecidability (for an example from our early work on concurrency, see [17]). We would like our logic to be elementarily decidable.

In his article "Modality, si! modal logic, no!" [22], John McCarthy criticizes modal logic for not providing the richness required for representing human practice, where one sometimes introduces new modalities on an *ad hoc* basis. Some of his criticisms, about other kinds of knowledge than just *knowing that*, are dealt with in recent work of Yanjing Wang [28]. We view the requirement of expanding an edge to a *small* source-sink "via graph" as a similar source of richness which could be added to modal logics. Unlike coalgebraic modal logics [24], this kind of facility is not uniform, but as McCarthy says, on an *ad hoc* basis at a particular node of the model. We did not find such considerations in the literature.

What do we mean by *small* ? The main idea here is that one should not allow recursive specification of arbitrary properties of the logic inside the via graph, for then one allows the kind of generality which one is trying to avoid, such as the formation of complex structures. The way we do this below is to only allow constant structure. It is also possible to allow a hierarchy of structure, but we do not consider that here.

## 2. Logic

Our frames are directed graphs, although our examples and results will only talk about rooted acyclic ones.

$b ::= p \in Prop \mid o \in Nom \mid \neg b \mid b_1 \vee b_2$
$\pi ::= \textbf{via } b_1 \ldots \textbf{via } b_k \mid \pi_1, \pi_2$
$\alpha ::= b \mid \neg\alpha \mid \alpha \vee \beta \mid \langle\pi\rangle\alpha$

The boolean expressions $b$ are evaluated at possible worlds as in any Kripke structure. They include *nominals* which denote single worlds. We will require that if edges from two distinct sources meet at a target, that vertex is identified by a nominal. Frames of models meeting this requirement are called *via* dags.

The expressions $\pi$ evaluate to sets of (nontrivial) paths. The expression **via** $b_1 \ldots$ **via** $b_k$ collects the set of all paths between a source and target vertex having intermediate nodes in sequence (not including the source and target) satisfying $b_1, \ldots, b_k$. Thus **via** $false$ specifies a single edge.

The formulae $\alpha$ put all these together to describe graph properties. In particular sets of paths from a source node converge to a common target: we have $w \models \langle\pi_1, \pi_2\rangle\alpha$ iff there is a node $x \models \alpha$ accessible from $w$ through two paths evaluating to $\pi_1$ and $\pi_2$ respectively.

We can define unary CTL modalities $EX\alpha = \langle\textbf{via } false\rangle\alpha$, $EXEF\alpha = \langle\textbf{via } true\rangle\alpha$, $EF\alpha = \alpha \vee EXEF\alpha$. We will freely use these abbreviations, recall also that $AX\alpha = \neg EX\neg\alpha$ and $AG\alpha = \neg EF\neg\alpha$.

Note that the formula $\langle\pi,\pi\rangle\alpha$ simplifies to $\langle\pi\rangle\alpha$. We could read this as requiring two paths, but that takes us into graded CTL[2,9]. We duck the issue by not having any multiset requirements. Since there is an easy extension of the logic we will continue to use graded modalities informally.

This buys us a theorem. Let $\mathrm{FO}^2[desc, child, Nom]$ stand for two-variable first-order logic with child (edge) and descendant (transitive closure of edges) relations together with specified nominal unary predicates. By the future fragment of this logic we mean that in a formula $\exists y\alpha(x,y)$ with free variable $x$, it is required that $y$ is a descendant of $x$.

**Theorem 2.1.** *On* via *dags, our logic is exactly as expressive as the future fragment of two-variable first-order logic $FO^2[desc, child, Nom]$, with descendant and child relations and specified nominal unary predicates. Satisfiability is decidable in* EXPTIME.

**Proof.** We first translate our modalities into a hybrid unary CTL. The path modality $\langle\textbf{via } b\rangle\alpha$ is expressible in CTL as:

$$(b \equiv false \supset EX\alpha) \wedge (b \not\equiv false \supset EXEF(b \wedge EXEF\alpha)).$$

For a longer path from $x$ to $y$, the specified boolean conditions occur as a subword in sequence, again possibly but not necessarily consecutively. (We thank a referee for pointing out a mistake in our earlier formulation.)

The two-path modality $\langle\pi_1,\pi_2\rangle\alpha$ can be described using conjunction of the translations for $\langle\pi_1\rangle\alpha$ and $\langle\pi_2\rangle\alpha$, but such a CTL formula can have a tree model which is not the intention, here we want the two paths to have a common target node. By using a fresh nominal $o$, we translate to $\langle\pi_1\rangle o \wedge \langle\pi_2\rangle o \wedge AG(o \supset \alpha)$, which is a formula in hybrid CTL[14], where we continue to use unary CTL modalities.

Sattler and Vardi give a translation from hybrid mu-calculus (a richer logic than CTL) to two-way alternating tree automata and an EXPTIME decision procedure for satisfiability[26].

There is an easy translation $Tran(\alpha)$ of a unary CTL formula $\alpha$ into the future fragment of two-variable first-order logic. When we translate a nominal, the locution $AG(o \supset \alpha)$ at a world $x$ translates to the subformula $\forall y > x(o(y) \supset Tran(\alpha(y)))$. Because a nominal is interpreted as a single world, we pull out such subformulae as outermost conjuncts which can be seen as enforcing global constraints that we have a *via* dag. The formula is in the future fragment of $\mathrm{FO}^2[desc, child, Nom]$.

The converse direction is an extension of the technique of Marx and de Rijke for the descendant and child axes of Core XPath[21], which itself comes from earlier work. The extension is to specified nominals, and we recall that our formula is interpreted on *via* dags. Every formula of the future fragment of two-variable logic $\mathrm{FO}^2[desc, child, Nom]$ can be put into normal form as a set of formulae, one for the original formula and one for each nominal. Each normal form is a disjunction of conjuncts ranging over order types consisting of children, the descendant relation beyond children and not going beyond specified nominals. Intuitively each normal form describes a tree-like portion of a dag, rooted either at a node identified by a nominal, or at the root of the dag where the original formula holds. For each order type, a corresponding unary CTL formula in the logic is constructed. Finally all these formulae are put together as a single formula of our logic using the nominals which holds at the root of the *via* dag. □

The main reason we restricted ourselves to the future fragment is that using negation we can talk about two nodes, none of which is a descendant of the other, and we do not know how to extend the Marx-de Rijke proof idea to this case. Some complexity lower bounds we examined require a significant amount of hybrid logic or CTL[14,26] and do not go through for our logic. Hence it is possible that the complexity is lower, there is a PSPACE lower bound for unary CTL.

Let us now see how the examples which started this paper fare in this logic. A two-stop journey from Bangalore to China can be expressed as *Bangalore* $\wedge$

$EXEF\ EXEF\ EXEF\ China$. Having several options to go to China at Delhi can be expressed by nominals for Delhi and China (to represent several cities in China we can use a proposition), and using a graded formula to express side properties, as in $\langle \mathbf{via}\ Delhi\rangle China \wedge AG(Delhi \supset EX^{\geq 2}EF\ China)$ or directly in graded CTL as $EXEF(Delhi \wedge EX^{\geq 2}EF\ China)$. To describe my plight as a nervous flier, that the journey from Bangalore to Delhi should not go over any sea, I can use $\neg\langle \mathbf{via}\ \neg land\rangle Delhi$ which turns out equivalent, given that seas exist, to the binary CTL formula $land\ AU\ Delhi$.

## 2.1. *Costs*

To model fare amounts, lengths of queues and duration of stopovers one could use propositions, or techniques developed for weighted CTL[6,16]. For example, a simple change in the evaluation of paths is possible, by allowing the syntax to express the **cost** of a path compared to a constant $c$, which comes from an ordered abelian cancellative semigroup, for simplicity we assume the natural numbers with zero. Note that we make use of subtraction but not of negative values. Thus I can view the long queues at Delhi as a cost I have to bear when I pass through Delhi, but if I use some other intermediate airport, I do not consider the absence of long queues as a negative cost.

Our logic could be modified so that the expressions $\pi$ now have sets of paths possibly with costs attached to them. Thus $\mathbf{cost} \sim c\ \mathbf{via}\ b_1 \ldots \mathbf{via}\ b_k$ gathers paths between a source and target vertex which additionally specify the cost constraint. The cost of a path is the sum of costs of intermediate edges and vertices. The Rescher **preference** modality[25,27] compares paths, giving sets of paths in $\pi_1$ which are less costly than those in $\pi_2$. We require that the preference relation is a strict partial order. The cost is optional, if no cost is stated then we assume no cost is defined for the given path. Similarly, if no preference is provided, there is no preference relation between the two paths. We can ask whether there is a business class fare below a certain cost as well as an economy class fare below another cost.

The formulae $\alpha$ continue to model situations rather than necessarily performing some optimizing computations.

$$b ::= p \in Prop \mid o \in Nom \mid \neg b \mid b_1 \vee b_2$$
$$\pi ::= \mathbf{cost} \sim c\ \mathbf{via}\ b_1 \ldots \mathbf{via}\ b_k, \sim \in \{<, \leq, =, \geq, >\} \mid \pi_1, \pi_2 \mid \pi_1\ \mathbf{pref}\ \pi_2$$
$$\alpha ::= b \mid \neg\alpha \mid \alpha \vee \beta \mid \langle\pi\rangle\alpha$$

Before we jump into the technical details we would like to ask this question: should one consider satisfiability as not just a matter of assigning truth values to propositions, but also as a matter of assigning costs to propositions in order to make a formula with preferences come out true? This is reminiscent of satisfiability questions in fuzzy[12], deontic[3] and weighted[4] logics. We confess to ignorance of the philosophical implications of adopting such a view. Hence we have not addressed the question of where the basic costs come from, leaving it to future work.

We would like that satisfiability of this extended logic is decidable with elementary complexity. Intuitively one expects that the procedure used for propagation of CTL eventualities through children can be lifted to propagate costs through descendants and nominals as well, after one nondeterministically guesses costs associated with each proposition and adds them up to compute the costs of an intermediate state. One of the referees suggested we could also consider model checking questions for motivation.

## Acknowledgements

A first proposal in this direction was made at the end of an early paper on concurrency with Rohit Parikh, R. Ramanujam and P.S. Thiagarajan[17]. Another paper with Ramanujam considered navigation in graphs[18]. The present proposal generalizes work with Andreas Krebs, Paritosh Pandya and Howard Straubing[15], which I presented at the 15th ALC in Daejeon. I would like to thank all these collaborators for their inspiration.

Thanks also to Fenrong Liu and Byeong-uk Yi for the discussions at Daejeon. When Hans van Ditmarsch learned of my interest, he kindly made available his draft paper with Minghui Ma[20]. We thank the referees for encouraging this work.

## References

1. MARIO R. FOLHADELA BENEVIDES AND LUIS MENASCHÉ SCHECHTER. Modal expressiveness of graph properties, *Proc. 2nd LSFA*, Ouro Preto (M. AYALA-RINCÓN AND E.H. HAEUSLER, eds.), *ENTCS* **205**, 31–47 (2008).
2. ALESSANDRO BIANCO, FABIO MOGAVERO AND ANIELLO MURANO. Graded computation tree logic, *ACM Trans. Comput. Log.* **13**, 3, 25:1–25:53 (2012).
3. JOHAN VAN BENTHEM, DAVIDE GROSSI AND FENRONG LIU. Deontics = betterness + priority, *Proc. 10th Deon*, Fiesole (GUIDO GOVERNATORI AND GIOVANNI SARTOR, eds.), *LNCS* **6181**, 50–65 (2010).
4. BENEDIKT BOLLIG, PAUL GASTIN, BENJAMIN MONMEGE AND MARC ZEITOUN. Logical characterization of weighted pebble walking automata, *Proc. 23rd CSL & 29th LICS*, Vienna (THOMAS A. HENZINGER AND DALE MILLER, eds.), ACM, 19:1–19:10 (2014).
5. RONALD J. BRACHMAN AND JAMES G. SCHMOLZE. An overview of the KL-ONE knowledge representation system, *Cog. Sci.* **9**, 171–216 (1985).
6. PETER BUCHHOLZ AND PETER KEMPER. Model checking for a class of weighted automata, *Discr. Ev. Dyn. Syst.* **20**, 1, 103–137 (2010).

7. Robert A. Bull. An approach to tense logic, *Theoria* **36**, 3, 282–300 (1970).
8. Hans van Ditmarsch, Wiebe van der Hoek and Barteld Kooi. *Dynamic epistemic logic*, Springer (2008).
9. Alessandro Ferrante, Margherita Napoli and Mimmo Parente. Model checking for graded CTL, *Fund. Inform.* **96**, 3, 323–339 (2009).
10. Kit Fine. In so many possible worlds, *Notre Dame J. Formal Logic* **13**, 4, 516–520 (1972).
11. Geroge Gargov and Valentin Goranko. Modal logic with names, *J. Philos. Logic* **22**, 6, 607–636 (1993).
12. Petr Hájek and Dagmar Harmancová. A comparative fuzzy modal logic, *Proc. 8th FLAI*, Linz (Erich-Peter Klement and Wolfgang Slany, eds.), *LNCS* **695**, 27–34 (1993).
13. David Harel, Dexter Kozen and Jerzy Tiuryn. *Dynamic logic*, MIT Press (2000).
14. Ahmet Kara, Volker Weber, Martin Lange and Thomas Schwentick. On the hybrid extension of CTL and CTL+, *Proc. 34th MFCS*, Nový Smokovec (Rastislav Královič and Damian Niwiński, eds.), *LNCS* **5734**, 427–438 (2009).
15. Andreas Krebs, Kamal Lodaya, Paritosh Pandya and Howard Straubing. Two-variable logic with a between relation, *Proc. 31st LICS*, New York (Martin Grohe, Erik Koskinen and Natarajan Shankar, eds.), ACM/IEEE , 106–115 (2016).
16. Kim G. Larsen, Radu Mardare and Bingtian Xue. Alternation-free weighted mu-calculus: decidability and completeness, *Proc. 31st MFPS*, Nijmegen (Dan R. Ghica, ed.), *ENTCS* **319**, 289–313 (2015).
17. Kamal Lodaya, Rohit Parikh, R. Ramanujam and P.S. Thiagarajan. A logical study of distributed systems, *Inf. Comput.* **115**, 1, 91–115 (1995).
18. Kamal Lodaya and R. Ramanujam. An automaton model of user-controlled navigation on the web, *Proc. 5th CIAA*, London (CA) (Sheng Yu and Andrei Paun, eds.), *LNCS* **2088**, 208–216 (2001).
19. Carsten Lutz, Ulrike Sattler and Frank Wolter. Modal logic and the two-variable fragment, *Proc. 15th CSL*, Paris (Laurent Fribourg, ed.), *LNCS* **2142**, 247–261 (2001).
20. Minghui Ma and Hans van Ditmarsch. Dynamic graded epistemic logic, manuscript (2016).
21. Maarten Marx and Maarten de Rijke. Semantic characterizations of navigational XPath, *Sigmod Record* **34**, 2, 41–46 (2005).
22. John McCarthy. Modality, si! modal logic, no! *Studia Logica* **59**, 1, 29–32 (1997).
23. Faron Moller and Alexander Rabinovich. Counting on CTL*: on the expressive power of monadic path logic, *Inf. Comput.* **184**, 147–159 (2003).
24. Lawrence S. Moss. Coalgebraic logic, *Ann Pure Appl. Logic* **96**, 1-3, 277–317 (1999).
25. Nicholas Rescher. Plurality-quantification and quasi-categorical propositions, *J. Symb. Logic* **27**, 373–374 (1962).
26. Ulrike Sattler and Moshe Vardi. The hybrid mu-calculus, *Proc. 1st IJCAR*, Siena (Rajeev Goré, Alexander Leitsch and Tobias Nipkow, eds.), *LNAI* **2083**, 76–91 (2001).

27. ROBERT C. STALNAKER. A theory of conditionals, *Studies in logical theory* (NICHOLAS RESCHER, ed.), Basil Blackwell, 98–112 (1968).
28. YANJING WANG. A new modal framework for epistemic logic, *Proc. 16th TARK*, Liverpool (JÉRÔME LANG, ed.), *EPTCS* **251**, 515–534 (2017).

# Matrix Iterations with Vertical Support Restrictions

Diego A. Mejía

*Faculty of Science, Shizuoka University,*
*Ohya 836, Suruga-ku, Shizuoka City, Shizuoka 422-8529, Japan*
*E-mail: diego.mejia@shizuoka.ac.jp*
*http://www.researchgate.com/profile/Diego_Mejia2*

We use coherent systems of FS iterations on a power set, which can be seen as matrix iteration that allows restriction on arbitrary subsets of the vertical component, to prove general theorems about preservation of certain type of unbounded families on definable structures and of certain mad families (like those added by Hechler's poset for adding an a.d. family) regardless of the cofinality of their size. In particular, we define a class of posets called $\sigma$*-Frechet-linked* and show that they work well to preserve mad families, and unbounded families on $\omega^\omega$.

As applications of this method, we show that a large class of FS iterations can preserve the mad family added by Hechler's poset (regardless of the cofinality of its size), and the consistency of a constellation of Cichoń's diagram with 7 values where two of these values are singular.

*Keywords*: Coherent system of finite support iterations, Frechet-linked, preservation of unbounded families, preservation of mad families, Cichoń's diagram.

## 1. Introduction

### *Background*

In the framework of FS (finite support) iterations of ccc posets to prove consistency results with large continuum (that is, the size of the continuum $\mathfrak{c} = 2^{\aleph_0}$ larger than $\aleph_2$), very recently in Ref. 1 appeared the general notion of *coherent systems of FS iterations* that was used to construct a three-dimensional array of ccc posets to force that the cardinals in *Cichoń's diagram* are separated into 7 different values (see Figure 1). This is the first example of a 3D iteration that was used to prove a new consistency result. Moreover, the methods from Ref. 2 where used there to force, in addition, that the *almost disjointness number* $\mathfrak{a}$ is equal to the *bounding number* $\mathfrak{b}$, and to expand well-known results about preservation of mad families along FS iterations.

For quite some time, consistency results about many different values for cardinal invariants has been investigated. Some of the earliest results are due to

Brendle[3] who fixed standard techniques for FS iterations in this direction, and due to Blass and Shelah[4] who constructed the first example of a two-dimensional array of ccc posets to prove the consistency of the existence of a base for a non-principal ultrafilter in $\omega$ of size smaller than the *dominating number* $\mathfrak{d}$. The latter technique received the name *matrix iterations* in Ref. 2 and it was improved there to prove the consistency of e.g. $\aleph_1 < \mathfrak{b} = \mathfrak{a} < \mathfrak{s}$. Recent developments on matrix iterations appear in work of the author[5,6], where forcing models satisfying that several cardinals in Cichoń's diagram are pairwise different (at most 6 different values were achieved) are constructed, and of Dow and Shelah[7] where the splitting number $\mathfrak{s}$ is forced to be singular. Concerning Cichoń's diagram, a few months ago Goldstern, Kellner and Shelah[8] used Boolean ultrapowers of strongly compact cardinals applied to the iteration constructed in Ref. 9 to prove the consistency, modulo the existence of 4 strongly compact cardinals, of a division of Cichoń's diagram into 10 different values (the maximum number of values allowed in the diagram). Another example with 10 values, modulo 4 strongly compact cardinals, appears in Ref. 10. At this point, it is still unknown how to prove the consistency of 8 different values in this diagram modulo ZFC alone.

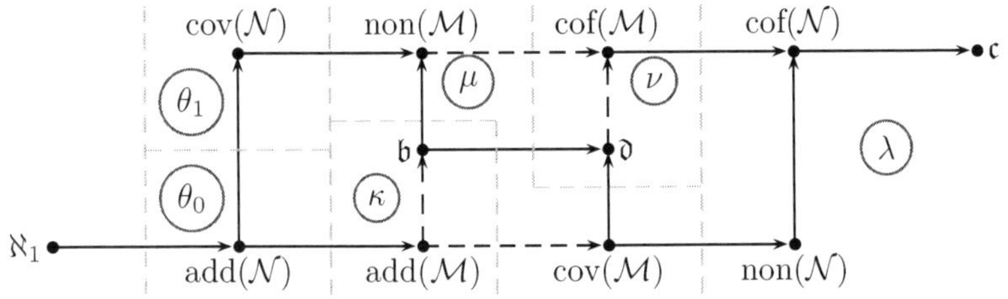


Figure 1. 7 values in Cichoń's diagram.

One drawback of the methods discussed so far is that the posets they produce force that the cardinal invariants that are not equal to $\mathfrak{c}$ must be regular. In the context of ccc forcing, one of the few exceptions is the consistency of $\mathfrak{d} < \mathfrak{c}$ with both cardinals singular, which can be obtained by a FS iteration where the last iterand is a random algebra (see e.g. Thm. 5.1(d) of Ref. 1). On the other hand, many examples can be obtained by creature forcing constructions as in Refs. 11–13, for instance, in the latter reference it is proved that the right side of Cichoń's diagram can be divided into 5 different values where 4 of them are singular. However, all these constructions are $\omega^\omega$-bounding, so they force $\mathfrak{d} = \aleph_1$ and do not allow separation of cardinal invariants below $\mathfrak{d}$.

### *Objective 1*

The main motivation of this research is to improve some of the ccc forcing methods to produce models where many cardinal invariants of the continuum are different and two or more of them are singular. As one of the main results of this paper, we show how to take advantage of the generality of coherent systems of FS iterations to produce such models where 2 cardinal invariants can be forced to be singular. In particular, we show that the 3D iteration of Ref. 1 that forces the constellation of 7 values in Cichoń's diagram can be modified so that 2 cardinals are allowed to be singular (Theorem 4.3). In addition. we modify examples from Refs. 1,5 in the same way.

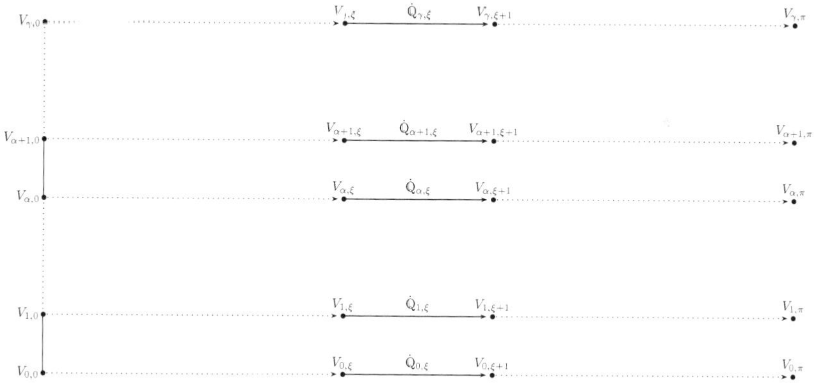

Figure 2. Matrix iteration.

### *Methods*

A coherent system of FS iterations of length $\pi$ consists of a partial order $\langle I, \leq\rangle$ and, for each $i \in I$, a FS iteration $\mathbb{P}_{i,\pi} := \langle \mathbb{P}_{i,\alpha}, \dot{\mathbb{Q}}_{i,\alpha} : \alpha < \pi\rangle$ such that any pair of such iterations are coherent in the sense that, whenever $i \leq j$ in $I$ and $\alpha \leq \pi$, the $\mathbb{P}_{i,\alpha}$-generic extension is contained in the $\mathbb{P}_{j,\alpha}$-generic extension (see details in Definition 2.3 and a picture in Figure 6). For instance, a matrix iteration is a coherent system (of FS iterations) when $\langle I, \leq\rangle$ is a well-order (see Figure 2), and a 3D iteration is a coherent system on a product of ordinals $I = \gamma \times \delta$ with the coordinate-wise order (see Figure 3).

For our applications, we construct coherent systems on partial orders of the form $\langle \mathcal{P}(\Omega), \subseteq\rangle$, which in fact look like matrix iterations, with vertical component

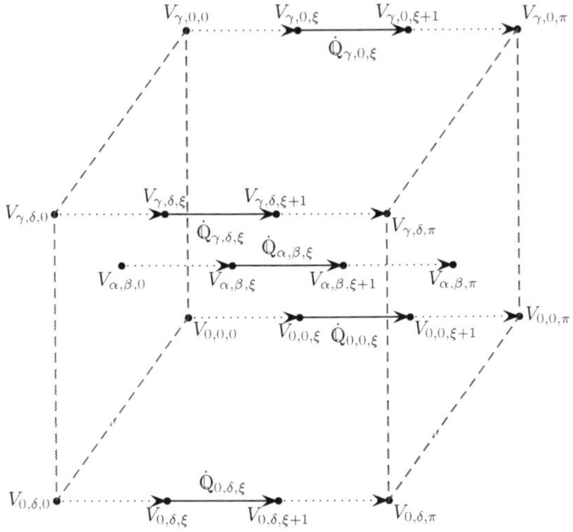


Figure 3. Three-dimensional iteration.

indexed by $\Omega$, that allow restriction on any arbitrary subset of $\Omega$. To be more precise, as the final generic extension of the forcing produced by such a system comes from the FS iteration $\langle \mathbb{P}_{\Omega,\xi}, \dot{\mathbb{Q}}_{\Omega,\xi} : \xi < \pi \rangle$, for any $A \subseteq \Omega$ the iteration $\langle \mathbb{P}_{A,\xi}, \dot{\mathbb{Q}}_{A,\xi} : \xi < \pi \rangle$ can be understood as the 'vertical' restriction on $A$ of the former FS iteration (see Figure 4). This "restriction" feature is what allows nice combinatorial arguments to force singular values for some cardinal invariants. Concretely, it allows to preserve *unbounded reals* (with respect to general structures, see Definition 3.4) that come from the vertical component (Theorem 3.15), and even maximal almost disjoint (mad) families of size of singular cardinality (Theorem 3.32). Surprisingly, the three-dimensional forcings from Ref. 1 can be reconstructed now as matrix iterations with vertical support restriction, though the real picture of the latter is the 'shape' of the lattice $\langle \mathcal{P}(\Omega), \subseteq \rangle$ plus one additional dimension (for the FS iterations).

### *Objective 2*

The theory of Brendle and Fischer[2] for preserving mad families is the cornerstone for the preservation results we propose in this paper, as well as it is in Ref. 1. In the latter reference it is proved that $\mathbb{E}$ (the standard $\sigma$-centered poset adding an eventually different real) and random forcing (thus any random algebra) behaves well in their preservation theory, which allows to prove in Thm. 4.17 of Ref. 1

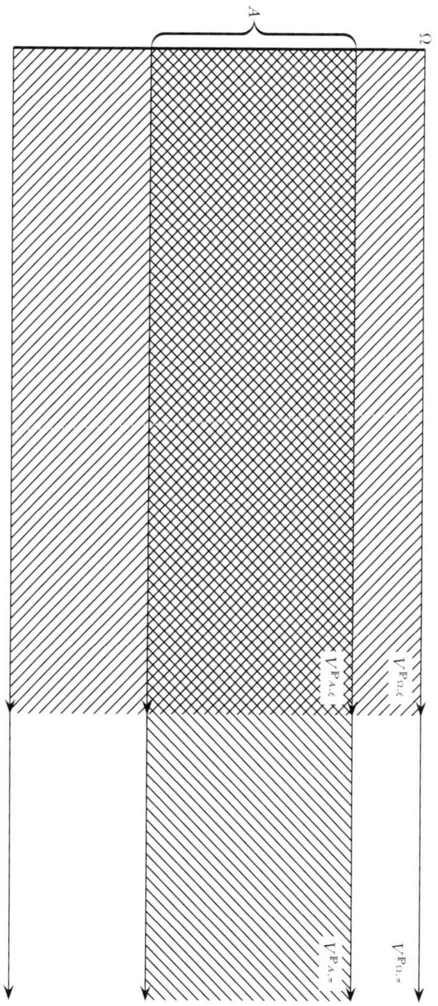

Figure 4. Matrix iteration with vertical support restriction.

that, whenever $\kappa$ is an uncountable regular cardinal, the mad family added by the Hechler poset $\mathbb{H}_\kappa$ is preserved by any further FS iteration whose iterands are either $\mathbb{E}$, a random algebra or a ccc poset of size $< \kappa$. In relation to this, we define a class of posets, which we call *$\sigma$-Frechet-linked* (see Definition 3.24), that includes $\mathbb{E}$ and random forcing, and we prove that any (definable) poset in this class behaves well with Brendle's and Fischer's preservation theory (Theorem 3.27). Moreover, by using coherent systems on a power set $\langle \mathcal{P}(\Omega), \subseteq \rangle$ we generalize Thm. 4.17 of Ref. 1 by proving that, whenever $\Omega$ is uncountable (not necessarily of regular size), the mad family added by $\mathbb{H}_\Omega$ can be preserved by a large class of FS iterations (which includes the Suslin $\sigma$-Frechet-linked posets as iterands, see Theorem 4.1). This is related to the preservation of mad families of singular size discussed in the previous paragraph. In addition we also show that, for a cardinal $\mu$, $\mu$-Frechet-linked posets behave well in the preservation theory of unbounded families (Theorem 3.30).

### *Structure of the paper*

In Section 2 we review the notion of coherent systems of FS iterations and prove general theorems about (vertical) direct limits within such a system. Section 3 is divided in two parts. In the first part, we review Judah's and Shelah's[14] and Brendle's[3] theory of preservation of strongly unbounded families (with respect to general definable structures), as well as known facts from Refs. 2,4,5 to preserve unbounded reals. At the end, a general theorem about preservation of unbounded reals through coherent systems on a power set $\langle \mathcal{P}(\Omega), \subseteq \rangle$ is proved. In the second part we review Brendle's and Fischer's theory for mad family preservation, define $\mu$-Frechet-linked posets and prove that they behave well in this preservation theory. This allows to prove at the end a general theorem about preservation of mad families through coherent systems on a power set $\langle \mathcal{P}(\Omega), \subseteq \rangle$. Afterwards, in Section 4 we show applications of the theory presented so far, namely, mad family preservation along a large class of FS iterations and consistency results about Cichoń's diagram.

The last section proposes a general framework for linkedness of subsets of posets that includes notions like $n$-linked, centered and Frechet-linked. We say that $\Gamma$ is a *linkedness property of subsets of posets* if $\Gamma(\mathbb{P})$ defines a family of subsets of $\mathbb{P}$ for each poset $\mathbb{P}$. For such a property $\Gamma$ we define its corresponding notions of *$\theta$-$\Gamma$-Knaster* (a poset $\mathbb{P}$ has this property iff any subset of size $\theta$ contains a subset in $\Gamma(\mathbb{P})$ of size $\theta$) and *$\mu$-$\Gamma$-covered* (the version of $\mu$-linked for $\Gamma$). Built on the classical FS product and iteration theorems for Knaster and $\sigma$-linked, we find sufficient conditions for $\Gamma$ to generalize these theorems for $\theta$-$\Gamma$-Knaster and $\mu$-$\Gamma$-covered.

### *Some notation*

Denote by $\mathbb{C}$ the poset that adds one Cohen real and by $\mathbb{C}_\Omega$ the poset that adds a family of Cohen reals indexed by the set $\Omega$ (which is basically a finite support product of $\mathbb{C}$). The Lebesgue measure on the Cantor space $2^\omega$ is denoted by Lb. *Random forcing*, denoted by $\mathbb{B}$, is the poset of Borel subsets of $2^\omega$ with positive Lebesgue measure, ordered by $\subseteq$. A *random algebra* on a set $\Omega$, denoted by $\mathbb{B}_\Omega$, is the poset of subsets of $2^{\Omega\times\omega}$ of the form $B\times 2^{(\Omega\smallsetminus J)\times\omega}$ for some $J\subseteq\Omega$ countable and some Borel subset $B$ of $2^{J\times\omega}$ with positive Lebesgue measure, ordered by $\subseteq$. This adds a family of random reals indexed by $\Omega$. Hechler poset for adding a dominating real is denoted by $\mathbb{D}$, and $\mathbb{E}$ is defined as the poset whose conditions are pairs $(s,\varphi)$ with $s\in\omega^{<\omega}$ and $\varphi:\omega\to[\omega]^{\leq m}$ for some $m<\omega$, ordered by $(s',\varphi')\leq(s,\varphi)$ iff $s\subseteq s'$, $\varphi(i)\subseteq\varphi'(i)$ for any $i<\omega$, and $s'(i)\notin\varphi(i)$ for any $i\in|s'|\smallsetminus|s|$. The trivial poset is denoted by $\mathbb{1}$.

Most of the cardinal invariants used in this paper are defined (or characterized) in Example 3.7. Recall that $A\subseteq[\omega]^{\aleph_0}$ is an *almost disjoint (a.d.) family* if the intersection of any two different members of $A$ is finite. A mad family is a maximal a.d. family, and $\mathfrak{a}$ is defined as the smallest size of an infinite mad family. For a set $\Omega$, Hechler's poset $\mathbb{H}_\Omega$ for adding an a.d. family (indexed by $\Omega$) is defined as the poset whose conditions are of the form $p:F_p\times n_p\to 2$ with $F_p\in[\Omega]^{<\omega}$ and $n_p<\omega$ (demand $n_p=0$ iff $F_p=\emptyset$), ordered by $q\leq p$ iff $p\subseteq q$ and $|q^{-1}[\{1\}]\cap(F_p\times\{i\})|\leq 1$ for every $i\in[n_p,n_q)$ (see Ref. 15). This poset has the Knaster property and the a.d. family it adds is maximal when $\Omega$ is uncountable. It is forcing equivalent to $\mathbb{C}$ when $\Omega$ is countable and non-empty, and it is equivalent to $\mathbb{C}_{\omega_1}$ when $|\Omega|=\aleph_1$. For any $\Omega\subseteq\Omega'$, $\mathbb{H}_\Omega\lessdot\mathbb{H}_{\Omega'}$.

## 2. Coherent systems of FS iterations

**Definition 2.1.** Let $M$ be a transitive model of ZFC. When $\mathbb{P}\in M$ and $\mathbb{Q}$ are posets, say that $\mathbb{P}$ *is a complete subposet of* $\mathbb{Q}$ *with respect to* $M$, abbreviated $\mathbb{P}\lessdot_M\mathbb{Q}$, if $\mathbb{P}$ is a subposet of $\mathbb{Q}$ and any maximal antichain of $\mathbb{P}$ that belongs to $M$ is still a maximal antichain in $\mathbb{Q}$.

If in addition $N$ is another transitive model of ZFC, $M\subseteq N$ and $\mathbb{Q}\in N$, then $\mathbb{P}\lessdot_M\mathbb{Q}$ implies that, whenever $G$ is $\mathbb{Q}$-generic over $N$, $G\cap\mathbb{P}$ is $\mathbb{P}$-generic over $M$ and $M[G\cap\mathbb{P}]\subseteq N[G]$ (see Figure 5).

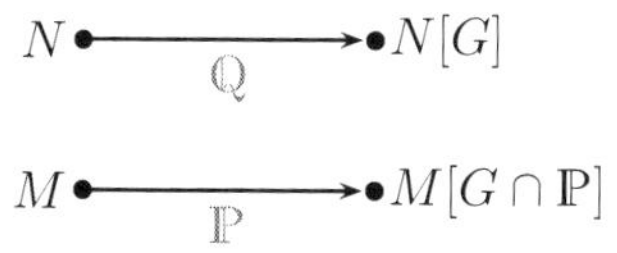


Figure 5. Generic extensions of pairs of posets ordered like $\mathbb{P}\lessdot_M\mathbb{Q}$.

**Example 2.2.** Let $M\subseteq N$ be transitive models of ZFC. When $\mathbb{P}\in M$ it is clear that $\mathbb{1}\lessdot_M\mathbb{P}$ and $\mathbb{P}\lessdot_M\mathbb{P}$. Also, if $\mathbb{S}$ is a Suslin ccc poset or a random algebra coded in $M$ then $\mathbb{S}^M\lessdot_M\mathbb{S}^N$.

**Definition 2.3 (Def. 3.2 in Ref. 1).** A *coherent system (of FS iterations)* $\mathbf{s}$ is composed of the following objects:

(I) a partially ordered set $I^{\mathbf{s}}$, an ordinal $\pi^{\mathbf{s}}$, and

(II) for each $i \in I^{\mathbf{s}}$, a FS iteration $\mathbb{P}^{\mathbf{s}}_{i,\pi^{\mathbf{s}}} = \langle \mathbb{P}^{\mathbf{s}}_{i,\xi}, \dot{\mathbb{Q}}^{\mathbf{s}}_{i,\xi} : \xi < \pi^{\mathbf{s}} \rangle$ such that, for any $i \leq j$ in $I$ and $\xi < \pi^{\mathbf{s}}$, if $\mathbb{P}^{\mathbf{s}}_{i,\xi} \lessdot \mathbb{P}^{\mathbf{s}}_{j,\xi}$ then $\mathbb{P}^{\mathbf{s}}_{j,\xi}$ forces $\dot{\mathbb{Q}}^{\mathbf{s}}_{i,\xi} \lessdot_{V^{\mathbb{P}^{\mathbf{s}}_{i,\xi}}} \dot{\mathbb{Q}}^{\mathbf{s}}_{j,\xi}$.

According to this notation, $\mathbb{P}^{\mathbf{s}}_{i,0}$ is the trivial poset and $\mathbb{P}^{\mathbf{s}}_{i,1} = \dot{\mathbb{Q}}^{\mathbf{s}}_{i,0}$. We often refer to $\langle \mathbb{P}^{\mathbf{s}}_{i,1} : i \in I^{\mathbf{s}} \rangle$ as the *base of the coherent system* $\mathbf{s}$. The condition given in (II) implies that $\mathbb{P}^{\mathbf{s}}_{i,\xi} \lessdot \mathbb{P}^{\mathbf{s}}_{j,\xi}$ whenever $i \leq j$ in $I^{\mathbf{s}}$ and $\xi \leq \pi^{\mathbf{s}}$ (see Lemma 3.14).

For $j \in I^{\mathbf{s}}$ and $\eta \leq \pi^{\mathbf{s}}$ we write $V^{\mathbf{s}}_{j,\eta}$ for the $\mathbb{P}^{\mathbf{s}}_{j,\eta}$-generic extensions. Concretely, when $G$ is $\mathbb{P}^{\mathbf{s}}_{j,\eta}$-generic over $V$, $V^{\mathbf{s}}_{j,\eta} := V[G]$ and $V^{\mathbf{s}}_{i,\xi} := V[\mathbb{P}^{\mathbf{s}}_{i,\xi} \cap G]$ for all $i \leq j$ in $I^{\mathbf{s}}$ and $\xi \leq \eta$. Note that $V^{\mathbf{s}}_{i,\xi} \subseteq V^{\mathbf{s}}_{j,\eta}$ and $V^{\mathbf{s}}_{i,0} = V$ (see Figure 6).

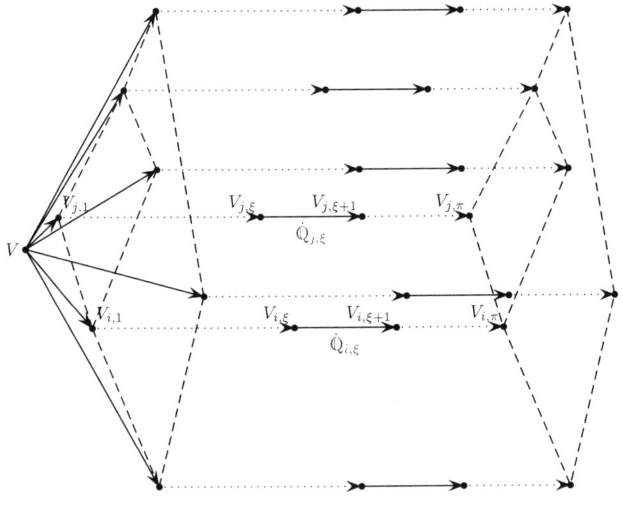


Figure 6. Coherent system of FS iterations. The figures in dashed lines represent the 'shape' of the partial order $\langle I, \leq \rangle$.

We say that the coherent system $\mathbf{s}$ has the *ccc* if, additionally, $\mathbb{P}^{\mathbf{s}}_{i,\xi}$ forces that $\dot{\mathbb{Q}}^{\mathbf{s}}_{i,\xi}$ has the ccc for each $i \in I^{\mathbf{s}}$ and $\xi < \pi^{\mathbf{s}}$. This implies that $\mathbb{P}^{\mathbf{s}}_{i,\xi}$ has the ccc for all $i \in I^{\mathbf{s}}$ and $\xi \leq \pi^{\mathbf{s}}$.

A concrete simple type of coherent system is what we call a *coherent pair (of FS iterations)*. A coherent system $\mathbf{s}$ is a coherent pair if $I^{\mathbf{s}}$ is of the form $\{i_0, i_1\}$ ordered as $i_0 < i_1$.

For a coherent system $\mathbf{s}$ and a set $J \subseteq I^{\mathbf{s}}$, $\mathbf{s}|J$ denotes the coherent system with $I^{\mathbf{s}|J} = J$, $\pi^{\mathbf{s}|J} = \pi^{\mathbf{s}}$ and the FS iterations corresponding to (II) defined as for $\mathbf{s}$; if $\eta \leq \pi^{\mathbf{s}}$, $\mathbf{s}{\restriction}\eta$ denotes the coherent system with $I^{\mathbf{s}\restriction\eta} = I^{\mathbf{s}}$, $\pi^{\mathbf{s}\restriction\eta} = \eta$ and the iterations for (II) defined up to $\eta$ as for $\mathbf{s}$. Note that, if $i_0 < i_1$ in $I^{\mathbf{s}}$, then $\mathbf{s}|\{i_0, i_1\}$ is a coherent pair and $\mathbf{s}|\{i_0\}$ is just the FS iteration $\mathbb{P}^{\mathbf{s}}_{i,\pi^{\mathbf{s}}} = \langle \mathbb{P}^{\mathbf{s}}_{i,\xi}, \dot{\mathbb{Q}}^{\mathbf{s}}_{i,\xi} : \xi < \pi^{\mathbf{s}} \rangle$.

In particular, the upper indices $\mathbf{s}$ are omitted when there is no risk of ambiguity.

The following is a generalization of Lemma 3.7 of Ref. 1.

**Lemma 2.4.** *Let $\theta$ be an uncountable regular cardinal. Assume that $\mathbf{s}$ is a coherent system that satisfies:*

*(i) $I$ has a maximum $i^*$ and $I \smallsetminus \{i^*\}$ is $< \theta$-directed,*

*(ii) each $\mathbb{P}_{i,\xi}$ forces that $\dot{\mathbb{Q}}_{i,\xi}$ is $\theta$-cc, and*

*(iii) for any $\xi < \pi$, if $\mathbb{P}_{i^*,\xi}$ is the direct limit of $\langle \mathbb{P}_{i,\xi} : i < i^* \rangle$ then $\mathbb{P}_{i^*,\xi}$ forces that $\dot{\mathbb{Q}}_{i^*,\xi} = \bigcup_{i<i^*} \dot{\mathbb{Q}}_{i,\xi}$.*

*Then, for any $\xi \leq \pi$,*

*(a) $\mathbb{P}_{i^*,\xi}$ is the direct limit of $\langle \mathbb{P}_{i,\xi} : i < i^* \rangle$ and*

*(b) if $\gamma < \theta$ and $\dot{f}$ is a $\mathbb{P}_{i^*,\xi}$-name of a function from $\gamma$ into $\bigcup_{i<i^*} V_{i,\xi}$ then $\dot{f}$ is (forced to be equal to) a $\mathbb{P}_{i,\xi}$-name for some $i < i^*$. In particular, the reals in $V_{i^*,\xi}$ are precisely the reals in $\bigcup_{i<i^*} V_{i,\xi}$.*

**Proof.** By condition (ii) it is clear that $\mathbb{P}_{i^*,\pi}$ has the $\theta$-cc. We first show that (a) implies (b). Let $\dot{f}$ be as in the condition in (b). For each $\alpha < \gamma$ choose a maximal antichain $A_\alpha$ that decides some $i < i^*$ and some $\mathbb{P}_{i,\xi}$-name to which $\dot{f}(\alpha)$ is forced to be equal. As $|A_\alpha| < \theta$ and $I \smallsetminus \{i^*\}$ is $< \theta$-directed, by (a) we can find some $i_\alpha < i^*$ and some $\mathbb{P}_{i_\alpha,\xi}$-name $\dot{x}_\alpha$ such that $\dot{f}(\alpha)$ is forced to be equal to $\dot{x}_\alpha$. Hence, as $\gamma < \theta$, there is some upper bound $j < i^*$ of $\{i_\alpha : \alpha < \gamma\}$, so $\dot{f}$ is forced to be equal to the $\mathbb{P}_{j,\xi}$-name of the function $\alpha \mapsto \dot{x}_\alpha$.

Now we prove (a) by induction on $\xi$. The case $\xi = 0$ and the limit step are clear. Assume that (a) holds for $\xi$, and we show that (a) holds for $\xi + 1$. If $p \in \mathbb{P}_{i^*,\xi+1}$ then, by (i), (iii) and (b), there is some $j_1 < i^*$ such that $p(\xi)$ is (forced to be equal to) a $\mathbb{P}_{j_1,\xi}$-name of a member of $\dot{\mathbb{Q}}_{j_1,\xi}$. On the other hand, by (a), there is some $j_0 < i^*$ such that $p{\restriction}\xi \in \mathbb{P}_{j_0,\xi}$. Hence, by (i), there is some $j < i^*$ above $j_0$ and $j_1$, so $p \in \mathbb{P}_{j,\xi}$. □

Typically, a coherent system of FS iterations is constructed by transfinite recursion. Concretely, to construct such a system $\mathbf{s}$ of FS iterations of length $\pi$, $\mathbf{s}{\restriction}\xi$ is constructed by recursion on $\xi \leq \pi$ as follows. In the step $\xi = 0$, we determine the partial order $\langle I, \leq \rangle$ that will support the base of the coherent system; in the limit step it is just enough to take direct limits of the systems constructed previously; for the successor step, assuming that $\mathbf{s}{\restriction}\xi$ has been constructed, the system $\langle \dot{\mathbb{Q}}_{i,\xi} : i \in I\rangle$ of names of posets ($\dot{\mathbb{Q}}_{i,\xi}$ is a $\mathbb{P}_{i,\xi}$-name) that will determine how it is forced in stage $\xi$ is determined, and afterwards $\mathbf{s}{\restriction}(\xi+1)$ is defined so that it extends $\mathbf{s}{\restriction}\xi$ and $\mathbb{P}_{i,\xi+1} = \mathbb{P}_{i,\xi} * \dot{\mathbb{Q}}_{i,\xi}$ for each $i \in I$. To have that $\mathbf{s}{\restriction}(\xi+1)$ is indeed a coherent system, we require that $\Vdash_{\mathbb{P}_{j,\xi}} \dot{\mathbb{Q}}_{i,\xi} \lessdot_{V_{i,\xi}} \dot{\mathbb{Q}}_{j,\xi}$ whenever $i \leq j$ in $I$.

In short, to construct a coherent system as in the previous paragraph, it is just enough to determine the partial order that will serve as base and to determine the iterands suitably. As an example (that will serve for all our applications), we define the following simple type of coherent systems. Suslin posets play an important role in such systems (see Ref. 16).

**Definition 2.5 (Def. 3.8 in Ref. 1 with variations).** A coherent system $\mathbf{s}$ is *standard* if

(I) it consists, additionally, of

- (i) a partition $\langle S^{\mathbf{s}}, C^{\mathbf{s}}\rangle$ of $\pi^{\mathbf{s}} \smallsetminus \{0\}$,
- (ii) a function $\Delta^{\mathbf{s}} : [1, \pi^{\mathbf{s}}) \to I^{\mathbf{s}}$,
- (iii) a sequence $\langle \dot{\mathbb{S}}^{\mathbf{s}}_\xi : \xi \in S^{\mathbf{s}}\rangle$ where each $\dot{\mathbb{S}}^{\mathbf{s}}_\xi$ is either a (name of a definition of a) Suslin ccc poset coded in $V^{\mathbf{s}}_{\Delta(\xi),\xi}$, or a random algebra, and
- (iv) a sequence $\langle \dot{\mathbb{Q}}^{\mathbf{s}}_\xi : \xi \in C^{\mathbf{s}}\rangle$ such that each $\dot{\mathbb{Q}}^{\mathbf{s}}_\xi$ is a $\mathbb{P}^{\mathbf{s}}_{\Delta^{\mathbf{s}}(\xi),\xi}$-name of a poset that is forced to have the ccc by $\mathbb{P}^{\mathbf{s}}_{i,\xi}$ for every $i \geq \Delta^{\mathbf{s}}(\xi)$ in $I^{\mathbf{s}}$, and

(II) for any $i \in I^{\mathbf{s}}$ it satisfies:

- (i) $\dot{\mathbb{Q}}^{\mathbf{s}}_{i,0}$ has the ccc and
- (ii) for any $0 < \xi < \pi^{\mathbf{s}}$,

$$\dot{\mathbb{Q}}^{\mathbf{s}}_{i,\xi} = \begin{cases} (\dot{\mathbb{S}}^{\mathbf{s}}_\xi)^{V^{\mathbf{s}}_{i,\xi}} & \text{if } \xi \in S^{\mathbf{s}} \text{ and } i \geq \Delta^{\mathbf{s}}(\xi), \\ \dot{\mathbb{Q}}^{\mathbf{s}}_\xi & \text{if } \xi \in C^{\mathbf{s}} \text{ and } i \geq \Delta^{\mathbf{s}}(\xi), \\ \mathbb{1} & \text{otherwise.} \end{cases}$$

As in Definition 2.3, the upper index $\mathbf{s}$ may be omitted when understood.

Note that any standard coherent system has the ccc. One of the main features of a standard coherent system is the type of generic objects that are added, which is determined in principle by the partition $\langle S^{\mathbf{s}}, C^{\mathbf{s}}\rangle$ of $\pi^{\mathbf{s}} \smallsetminus \{0\}$. When $\xi \in S^{\mathbf{s}}$, the generic object added at stage $\xi + 1$ is called *full generic*, while in the case $\xi \in C^{\mathbf{s}}$, such generic object is called *restricted generic*. The reason for this is that, in the case $\xi \in S^{\mathbf{s}}$ the generic object added by $\dot{\mathbb{S}}_\xi$ is generic over $V_{i,\xi}$ *for all* $i \in I$, while in the case $\xi \in C^{\mathbf{s}}$ the generic object added by $\dot{Q}_\xi$ is only generic over $V_{\Delta(\xi),\xi}$. For instance, this distinction is fundamental to deal with forcing constructions to separate several cardinal invariants of the continuum (as in our applications). Even more, a restricted generic can be even much more restricted, that is, not really generic over $V_{\Delta(\xi),\xi}$ but over some ZFC model $N \subseteq V_{\Delta(\xi),\xi}$ (e.g. when $\dot{\mathbb{Q}}_\xi = \mathbb{D}^N$).

The version of Lemma 2.4 for standard coherent systems requires simpler conditions.

**Corollary 2.6.** *Let* **s** *be a standard coherent system and let $\theta$ be an uncountable regular cardinal. If*

*(i) $I$ has a maximum $i^*$, $I \setminus \{i^*\}$ is $< \theta$-directed,*
*(ii) $i^* \notin \operatorname{ran}\Delta$, and*
*(iii) whenever $\pi > 0$, $\mathbb{P}_{i^*,1}$ is the direct limit of $\langle \mathbb{P}_{i,1} : i < i^* \rangle$,*

*then (a) and (b) of Lemma 2.4 hold.*

**Proof.** It is clear that hypotheses (i)-(iii) of Lemma 2.4 are satisfied. □

In our applications we use coherent systems on a power set $\langle \mathcal{P}(\Omega), \subseteq \rangle$. If **s** is a standard coherent system based in such a partial order, then $\mathbb{P}_{\Omega,\pi}$ (the largest poset in the system) can be represented as a two dimensional forcing construction supported in the plane $\Omega \times \pi$ and, for any $A \subseteq \gamma$ and $\xi \leq \pi$, $\mathbb{P}_{A,\xi}$ is seen as the restriction of the construction to the rectangle $A \times \xi$ (see Figure 4 in Section 1). The following result is a suitable consequence of Corollary 2.6 when dealing with such (standard) coherent systems.

**Lemma 2.7.** *Let $\theta$ be a cardinal of uncountable cofinality and let* **s** *be a standard coherent system where $I = I^{\mathbf{s}}$ is a suborder of $\langle \mathcal{P}(\Omega), \subseteq \rangle$. Assume that*

*(i) $I$ is closed under intersections,*
*(ii) $I \cap [\Omega]^{<\theta}$ is cofinal in $[\Omega]^{<\theta}$,*
*(iii) $\Delta(\xi) \in [\Omega]^{<\theta}$ for any $\xi \in [1, \pi)$ (see Definition 2.5(I)(ii)), and*
*(iv) whenever $\pi > 0$ and $X \in I$, $\mathbb{P}_{X,1}$ is the direct limit of $\langle \mathbb{P}_{A,1} : A \in I \cap [X]^{<\theta} \rangle$.*

*Then, for every $X \in I$ and $\xi \leq \pi$,*

*(a) $\mathbb{P}_{X,\xi}$ is the direct limit of $\langle \mathbb{P}_{A,\xi} : A \in I \cap [X]^{<\theta} \rangle$ and*
*(b) for any $\mathbb{P}_{X,\xi}$-name of a function $\dot{x}$ with domain $\gamma < cf(\theta)$ into $\bigcup_{A \in I \cap [X]^{<\theta}} V_{A,\xi}$, there is some $A \in I \cap [X]^{<\theta}$ such that $\dot{x}$ is (forced to be equal to) a $\mathbb{P}_{A,\xi}$-name.*

**Proof.** Fix $X \in I$. The lemma is trivial when $|X| < \theta$, so assume that $|X| \geq \theta$. By (i) and (ii), $I \cap [X]^{<\theta}$ is cofinal in $[X]^{<\theta}$. Put $I^* = I \cap ([X]^{<\theta} \cup \{X\})$. Hence $X$ is the maximum of $I^*$ and $I^* \setminus \{X\} = I \cap [X]^{<\theta}$ is $<$ cf($\theta$)-directed by (i) and (ii). Consider $\mathbf{s}^* = \mathbf{s}|I^*$. Note that $\mathbf{s}^*$ is a standard coherent system similar to **s** with the difference that $\Delta^* = \Delta^{\mathbf{s}^*} : [1, \pi) \to I \cap [X]^{<\theta}$ is defined as $\Delta^*(\xi) := \Delta(\xi)$ whenever $\Delta(\xi) \subseteq X$, or $\Delta^*(\xi) := \emptyset$ otherwise. Also, $\dot{\mathbb{S}}^{\mathbf{s}^*}_\xi = \dot{\mathbb{S}}^{\mathbf{s}}_\xi$ and $\dot{\mathbb{Q}}^{\mathbf{s}^*}_\xi = \dot{\mathbb{Q}}^{\mathbf{s}}_\xi$ in the first case, otherwise each one is the trivial poset. As $X \notin \mathrm{ran}\Delta^*$, the result is a direct consequence of Corollary 2.6 applied to $\mathbf{s}^*$ and $\mathrm{cf}(\theta)$. □

**Example 2.8.** Let $\Omega$ be a set.

(1) The partial order $\langle \mathcal{P}(\Omega), \subseteq \rangle$ clearly satisfies conditions (i) and (ii) of Lemma 2.7 (for any infinite cardinal $\theta$).
Assume that $\Omega = \Omega_0 \cup \Omega_1$ is a disjoint union. If $\mathbf{s}$ is a standard coherent system on $\langle \mathcal{P}(\Omega), \subseteq \rangle$ such that $\mathbb{P}_{X,1} = \mathbb{H}_{X\cap\Omega_0} \times \mathbb{C}_{X\cap\Omega_1}$ for any $X \subseteq \Omega$, then condition (iv) of Lemma 2.7 is satisfied for $\theta = \aleph_1$ (and hence for any uncountable $\theta$).
(2) If $\theta$ is a regular cardinal, $\Omega = \theta$ and $I_0$ is a cofinal subset of $\theta$, then $I := I_0 \cup \{\theta\}$ satisfies conditions (i) and (ii) of Lemma 2.7. Such a partial order $I$ (in particular $I = \theta \cup \{\theta\}$) is used to construct classical matrix iterations as in, e.g., Refs. 1,2,4,5.

## 3. Preservation properties

As mentioned in the introduction, this section is divided in two parts. For convenience with the notation fixed in the first part, we use a different notation from Refs. 1,2 for the results in the second part.

### 3.1. *Preservation theory*

A generalization of the contents of this part, as well as complete proofs and more examples, can be found in Sec. 4 of Ref. 17.

Typically, cardinal invariants of the continuum are defined through *relational systems* as follows.

**Definition 3.1.** A *relational system* is a triplet $\mathbf{R} = \langle X, Y, \sqsubset \rangle$ where $\sqsubset$ is a relation contained in $X \times Y$. For $x \in X$ and $y \in Y$, $x \sqsubset y$ is often read $y$ $\sqsubset$*-dominates* $x$. A family $F \subseteq X$ is $\mathbf{R}$*-unbounded* if there is no real in $Y$ that $\sqsubset$-dominates every member of $F$. Dually, $D \subseteq Y$ is an $\mathbf{R}$*-dominating* family if every member of $X$ is $\sqsubset$-dominated by some member of $D$. The cardinal $\mathfrak{b}(\mathbf{R})$ denotes the least size of an $\mathbf{R}$-unbounded family and $\mathfrak{d}(\mathbf{R})$ is the least size of an $\mathbf{R}$-dominating family.

Say that $x \in X$ is $\mathbf{R}$*-unbounded over a set* $M$ if $x \not\sqsubset y$ for all $y \in Y \cap M$. Given a cardinal $\lambda$ say that $F \subseteq X$ is $\lambda$-$\mathbf{R}$*-unbounded* if, for any $Z \subseteq Y$ of size $< \lambda$, there is an $x \in F$ that is $\mathbf{R}$-unbounded over $Z$. Say that $F \subseteq X$ is *strongly* $\lambda$-$\mathbf{R}$*-unbounded* if $|F| \geq \lambda$ and $|\{x \in F : x \sqsubset y\}| < \lambda$ for any $y \in Y$.

**Remark 3.2.** When $\lambda \geq 2$, any $\lambda$-$\mathbf{R}$-unbounded family is $\mathbf{R}$-unbounded. Hence, if $F$ is a $\lambda$-$\mathbf{R}$-unbounded family then $\mathfrak{b}(\mathbf{R}) \leq |F|$ and $\lambda \leq \mathfrak{d}(\mathbf{R})$. Also, if $\theta$ is regular and $F'$ is a strongly $\theta$-$\mathbf{R}$-unbounded family then it is $|F'|$-$\mathbf{R}$-unbounded, so $\mathfrak{b}(\mathbf{R}) \leq |F'| \leq \mathfrak{d}(\mathbf{R})$.

**Definition 3.3.** Let $\mathbf{R} = \langle X, Y, \sqsubset \rangle$ and $\mathbf{R}' = \langle X', Y', \sqsubset' \rangle$ be two relational systems. Say that $\mathbf{R}$ is *Tukey-Galois below* $\mathbf{R}'$ if there are two maps $F : X \to X'$ and $G : Y' \to Y$ such that, for each $x \in X$ and $b \in Y'$, if $F(x) \sqsubset' b$ then $x \sqsubset G(b)$. When, in addition, $\mathbf{R}'$ is Tukey-Galois below $\mathbf{R}$, we say that $\mathbf{R}$ and $\mathbf{R}'$ are *Tukey-Galois equivalent.*

Recall that, whenever $\mathbf{R}$ is Tukey-Galois below $\mathbf{R}'$, $\mathfrak{b}(\mathbf{R}') \leq \mathfrak{b}(\mathbf{R})$ and $\mathfrak{d}(\mathbf{R}) \leq \mathfrak{d}(\mathbf{R}')$.

**Definition 3.4.** A relational system $\mathbf{R} := \langle X, Y, \sqsubset \rangle$ is a *Polish relational system (Prs)* if the following is satisfied:

(i) $X$ is a perfect Polish space,
(ii) $Y$ is a non-empty analytic subspace of some Polish space $Z$ and
(iii) $\sqsubset = \bigcup_{n<\omega} \sqsubset_n$ for some increasing sequence $\langle \sqsubset_n \rangle_{n<\omega}$ of closed subsets of $X \times Z$ such that $(\sqsubset_n)^y = \{x \in X : x \sqsubset_n y\}$ is nwd (nowhere dense) for all $y \in Y$.

By (iii), $\langle X, \mathcal{M}(X), \in \rangle$ is Tukey-Galois below $\mathbf{R}$ where $\mathcal{M}(X)$ denotes the $\sigma$-ideal of meager subsets of $X$. Therefore, $\mathfrak{b}(\mathbf{R}) \leq \mathrm{non}(\mathcal{M})$ and $\mathrm{cov}(\mathcal{M}) \leq \mathfrak{d}(\mathbf{R})$. Moreover, (iii) implies that, whenever $c \in X$ is a Cohen real over a transitive model $M$ of ZFC and the Prs $\mathbf{R}$ is coded in $M$, $c$ is $\mathbf{R}$-unbounded over $M$.

**Definition 3.5 (Judah and Shelah[14]).** Let $\mathbf{R} = \langle X, Y, \sqsubset \rangle$ be a Prs and let $\theta$ be a cardinal. A poset $\mathbb{P}$ is $\theta$-$\mathbf{R}$-*good* if, for any $\mathbb{P}$-name $\dot{h}$ for a real in $Y$, there is a non-empty $H \subseteq Y$ of size $< \theta$ such that $\Vdash x \not\sqsubset \dot{h}$ for any $x \in X$ (in the ground model) that is $\mathbf{R}$-unbounded over $H$.

*Say that* $\mathbb{P}$ *is* $\mathbf{R}$-good *when it is* $\aleph_1$-$\mathbf{R}$-*good.*

Definition 3.5 describes a property, respected by FS iterations, to preserve specific types of $\mathbf{R}$-unbounded families. Concretely, when $\theta$ is uncountable regular,

(a) any $\theta$-$\mathbf{R}$-good poset preserves all the (strongly) $\theta$-$\mathbf{R}$-unbounded families from the ground model and
(b) FS iterations of $\theta$-cc $\theta$-$\mathbf{R}$-good posets produce $\theta$-$\mathbf{R}$-good posets.

By Remark 3.2, posets that are $\theta$-$\mathbf{R}$-good work to preserve $\mathfrak{b}(\mathbf{R})$ small and $\mathfrak{d}(\mathbf{R})$ large.

Clearly, $\theta$-$\mathbf{R}$-good implies $\theta'$-$\mathbf{R}$-good whenever $\theta \leq \theta'$, and any poset completely embedded into a $\theta$-$\mathbf{R}$-good poset is also $\theta$-$\mathbf{R}$-good.

Consider the following particular cases of interest for our applications.

**Lemma 3.6 (Lemma 4 in Ref. 5).** *If* **R** *is a Prs and* $\theta$ *is an uncountable regular cardinal, then any poset of size* $< \theta$ *is* $\theta$-**R**-*good. In particular, Cohen forcing is* **R**-*good.*

**Example 3.7.** Fix an uncountable regular cardinal $\theta$.

(1) *Preserving non-meager sets:* Consider the Polish relational system $\mathbf{Ed} := \langle \omega^\omega, \omega^\omega, \neq^* \rangle$ where $x \neq^* y$ iff $x$ and $y$ are eventually different, that is, $x(i) \neq y(i)$ for all but finitely many $i < \omega$. By Thm. 2.4.1 and 2.4.7 of Ref. 18, $\mathfrak{b}(\mathbf{Ed}) = \mathrm{non}(\mathcal{M})$ and $\mathfrak{d}(\mathbf{Ed}) = \mathrm{cov}(\mathcal{M})$.

(2) *Preserving unbounded families:* Let $\mathbf{D} := \langle \omega^\omega, \omega^\omega, \leq^* \rangle$ be the Polish relational system where $x \leq^* y$ iff $x(i) \leq y(i)$ for all but finitely many $i < \omega$. Clearly, $\mathfrak{b}(\mathbf{D}) = \mathfrak{b}$ and $\mathfrak{d}(\mathbf{D}) = \mathfrak{d}$.
Miller[19] proved that $\mathbb{E}$ is **D**-good. Furthermore, $\omega^\omega$-bounding posets, like the random algebra, are **D**-good. In Theorem 3.30 we prove that $\mu$-Frechet-linked posets are $\mu^+$-**D**-good.

(3) *Preserving null-covering families:* Define $X_n := \{a \in [2^{<\omega}]^{<\aleph_0} : \mathrm{Lb}(\bigcup_{s\in a}[s]) \leq 2^{-n}\}$ (endowed with the discrete topology) and put $X := \prod_{n<\omega} X_n$ with the product topology, which is a perfect Polish space. For every $x \in X$ denote $N^*_x := \bigcap_{n<\omega} \bigcup_{s\in x_n}[s]$, which is clearly a Borel null set in $2^\omega$.
Define the Prs $\mathbf{Cn} := \langle X, 2^\omega, \sqsubset \rangle$ where $x \sqsubset z$ iff $z \notin N^*_x$. Recall that any null set in $2^\omega$ is a subset of $N^*_x$ for some $x \in X$, so $\mathbf{Cn}$ and $\langle \mathcal{N}(2^\omega), 2^\omega, \not\ni \rangle$ are Tukey-Galois equivalent. Therefore $\mathfrak{b}(\mathbf{Cn}) = \mathrm{cov}(\mathcal{N})$ and $\mathfrak{d}(\mathbf{Cn}) = \mathrm{non}(\mathcal{N})$.
By a similar argument as in Lemma 1* of Ref. 3, any $\nu$-centered poset is $\theta$-**Cn**-good for any $\nu < \theta$ infinite. In particular, $\sigma$-centered posets are **Cn**-good.

(4) *Preserving "union of null sets is not null":* For each $k < \omega$ let $\mathrm{id}^k : \omega \to \omega$ such that $\mathrm{id}^k(i) = i^k$ for all $i < \omega$ and put $\mathcal{H} := \{\mathrm{id}^{k+1} : k < \omega\}$. Let $\mathbf{Lc} := \langle \omega^\omega, \mathcal{S}(\omega, \mathcal{H}), \in^* \rangle$ be the Polish relational system where

$$\mathcal{S}(\omega, \mathcal{H}) := \{\varphi : \omega \to [\omega]^{<\aleph_0} : \exists h \in \mathcal{H} \forall i < \omega(|\varphi(i)| \leq h(i))\},$$

and $x \in^* \varphi$ iff $\exists n < \omega \forall i \geq n(x(i) \in \varphi(i))$, which is read *$x$ is localized by $\varphi$*. As a consequence of Bartoszyński's characterization (see Thm. 2.3.9 of Ref. 18), $\mathfrak{b}(\mathbf{Lc}) = \mathrm{add}(\mathcal{N})$ and $\mathfrak{d}(\mathbf{Lc}) = \mathrm{cof}(\mathcal{N})$.
Any $\nu$-centered poset is $\theta$-**Lc**-good for any $\nu < \theta$ infinite (see Ref. 14) so, in particular, $\sigma$-centered posets are **Lc**-good. Moreover, Kamburelis[20] proved that any Boolean algebra with a strictly positive finitely additive measure is **Lc**-good. As a consequence, subalgebras (not necessarily complete) of random forcing are **Lc**-good.

**Lemma 3.8.** *If* $\mathbf{R} = \langle X, Y, \sqsubset \rangle$ *is a Prs then, for any set* $\Omega$, $\mathbb{H}_\Omega$ *is* **R**-*good.*

**Proof.** Let $\dot{y}$ be a $\mathbb{H}_\Omega$-name of a member of $Y$. Then there is some countable $A \subseteq \Omega$ such that $\dot{y}$ is a $\mathbb{H}_A$-name. As $\mathbb{H}_A$ is countable, it is **R**-good by Lemma 3.6, so there is some non-empty countable $H \subseteq Y$ witnessing this. The same $H$ witnesses goodness for $\dot{y}$ and $\mathbb{H}_\Omega$. □

In a similar way, it can be proved that any random algebra is **R**-good iff random forcing is **R**-good.

The following results indicate that (strongly) $\nu$-unbounded families can be added with Cohen reals, and the effect on $\mathfrak{b}(\mathbf{R})$ and $\mathfrak{d}(\mathbf{R})$ by a FS iteration of good posets.

**Lemma 3.9.** *Let $\nu$ be a cardinal of uncountable cofinality, $\mathbf{R} = \langle X, Y, \sqsubset\rangle$ a Prs and let $\langle \mathbb{P}_\alpha\rangle_{\alpha<\nu}$ be a $\lessdot$-increasing sequence of cf($\nu$)-cc posets such that $\mathbb{P}_\nu = \text{limdir}_{\alpha<\nu}\mathbb{P}_\alpha$. If $\mathbb{P}_{\alpha+1}$ adds a Cohen real $\dot{c}_\alpha \in X$ over $V^{\mathbb{P}_\alpha}$ for any $\alpha < \nu$, then $\mathbb{P}_\nu$ forces that $\{\dot{c}_\alpha : \alpha < \nu\}$ is a strongly $\nu$-**R**-unbounded family of size $\nu$.*

**Theorem 3.10.** *Let $\theta$ be an uncountable regular cardinal, $\mathbf{R} = \langle X, Y, \sqsubset\rangle$ a Prs, $\delta \geq \theta$ an ordinal and let $\langle \mathbb{P}_\alpha, \dot{\mathbb{Q}}_\alpha\rangle_{\alpha<\delta}$ be a FS iteration of non-trivial $\theta$-**R**-good $\theta$-cc posets. Then, $\mathbb{P}_\delta$ forces $\mathfrak{b}(\mathbf{R}) \leq \theta$ and $\mathfrak{d}(\mathbf{R}) \geq |\delta|$.*

**Proof.** See e.g. Thm. 4.15 of Ref. 17 or Cor. 3.6 of Ref. 9. □

Fix transitive models $M \subseteq N$ of ZFC and a Polish relational system $\mathbf{R} = \langle X, Y, \sqsubset\rangle$ coded in $M$. The following results are related to preservation of **R**-unbounded reals along coherent pairs of FS iterations.

**Lemma 3.11 (Thm. 7 in Ref. 5).** *Let $\mathbb{S}$ be a Suslin ccc poset coded in $M$. If $M \models$ "$\mathbb{S}$ is **R**-good" then, in $N$, $\mathbb{S}^N$ forces that every real in $X \cap N$ that is **R**-unbounded over $M$ is still **R**-unbounded over $M^{\mathbb{S}^M}$.*

**Corollary 3.12.** *Let $\Gamma \in M$ be a non-empty set. If $M \models$ "$\mathbb{B}_\Gamma$ is **R**-good" then $\mathbb{B}_\Gamma^N$, in $N$, forces that every real in $X \cap N$ that is **R**-unbounded over $M$ is still **R**-unbounded over $M^{\mathbb{B}_\Gamma^M}$.*

**Lemma 3.13 (Ref. 2, see also Lemma 5.13 of Ref. 21).** *Assume $\mathbb{P}$ is a poset in $M$. Then, in $N$, $\mathbb{P}$ forces that every real in $X \cap N$ that is **R**-unbounded over $M$ is still **R**-unbounded over $M^{\mathbb{P}}$.*

**Lemma 3.14 (Blass and Shelah[4], see also Ref. 2).** *Let $\mathbf{s}$ be a coherent pair of FS iterations (wlog $I^{\mathbf{s}} = \{0, 1\}$). Then, $\mathbb{P}_{0,\xi} \lessdot \mathbb{P}_{1,\xi}$ for all $\xi \leq \pi$.*

*Moreover, if $\dot{c}$ is a $\mathbb{P}_{1,1}$-name of a real in $X$, $\pi$ is limit and $\mathbb{P}_{1,\xi}$ forces that $\dot{c}$ is **R**-unbounded over $V_{0,\xi}$ for all $0 < \xi < \pi$, then $\mathbb{P}_{1,\pi}$ forces that $\dot{c}$ is **R**-unbounded over $V_{0,\pi}$.*

We finish this part with the main result of this subsection.

**Theorem 3.15.** *Let* $\mathbf{R} = \langle X, Y, \sqsubset\rangle$ *be a Prs,* $\theta$ *a cardinal of uncountable cofinality and let* $\mathbf{s}$ *be a standard coherent system of FS iterations of length* $\pi > 0$ *that satisfies the hypothesis of Lemma 2.7. Further assume that*

*(i)* $\Gamma \subseteq \Omega$ *has size* $\geq \theta$,
*(ii)* $D \in I$ *and* $\Gamma \subseteq D$,
*(iii) for each* $l \in \Gamma$, $\mathbb{P}_{D,1}$ *adds a real* $\dot{c}_l$ *in* $X$ *such that, for any* $A \subseteq D$ *in* $I \cap [\Omega]^{<\theta}$, *if* $l \in D \smallsetminus A$ *then* $\mathbb{P}_{D,1}$ *forces that* $\dot{c}_l$ *is* $\mathbf{R}$*-unbounded over* $V_{A,1}$, *and*
*(iv) for every* $\xi \in S^{\mathbf{s}}$ *and* $B \in I \cap [\Omega]^{<\theta}$, $\mathbb{P}_{B,\xi}$ *forces that* $\dot{\mathbb{Q}}_{B,\xi}$ *is* $\mathbf{R}$*-good.*

*Then* $\mathbb{P}_{D,\pi}$ *forces that the family* $\dot{F} := \{\dot{c}_l : l \in \Gamma\}$ *is strongly* $\theta$-$\mathbf{R}$*-unbounded. In particular,* $\mathbb{P}_{D,\pi}$ *forces* $\mathfrak{b}(\mathbf{R}) \leq |\dot{F}|$ *and, when* $\theta$ *is regular, this poset forces* $|\dot{F}| \leq \mathfrak{d}(\mathbf{R})$.

**Proof.** Let $\dot{y}$ be a $\mathbb{P}_{D,\pi}$-name of a member of $Y$. By Lemma 2.7, there is some $A \in I \cap [D]^{<\theta}$ such that $\dot{y}$ is a $\mathbb{P}_{A,\pi}$-name. Fix $l \in \Gamma \smallsetminus A$. By Lemmas 3.11, 3.13, 3.14 and Corollary 3.12 applied to the coherent pair $\mathbf{s}|\{A, D\}$, $\mathbb{P}_{D,\pi}$ forces that $\dot{c}_l$ is $\mathbf{R}$-unbounded over $V_{A,\pi}$, which implies that $\dot{c}_l \not\sqsubset \dot{y}$. Therefore, $\mathbb{P}_{D,\pi}$ forces that $\{x \in \dot{F} : x \sqsubset \dot{y}\} \subseteq \{\dot{c}_l : l \in \Gamma \cap A\}$, which has size $\leq |A| < \theta$.

The second statement is a consequence of Remark 3.2. □

### 3.2. *Preservation of mad families*

**Definition 3.16.** Fix $A \subseteq [\omega]^{\aleph_0}$.

(1) Let $P \subseteq [[\omega]^{\aleph_0}]^{<\aleph_0}$. For $x \subseteq \omega$ and $h : \omega \times P \to \omega$, define $x \sqsubset^* h$ by

$$\forall^\infty n < \omega \forall F \in P([n, h(n, F)) \smallsetminus \bigcup F \not\subseteq x).$$

(2) Define the relational system $\mathbf{Md}(A) := \langle [\omega]^{\aleph_0}, \omega^{\omega \times [A]^{<\aleph_0}}, \sqsubset^* \rangle$.

(3) Say that a poset $\mathbb{P}$ is *uniformly* $\mathbf{Md}(A)$*-good* if, for any $\mathbb{P}$-name $\dot{h}$ of a member of $\omega^{\omega\times[A]^{<\aleph_0}}$, there is a non-empty countable $H \subseteq \omega^{\omega\times[A]^{<\aleph_0}}$ (in the ground model) such that, for any countable $C \subseteq A$ and any $x \in [\omega]^{\aleph_0}$, if $x \not\sqsubset^* h' \restriction (\omega \times [C]^{<\aleph_0})$ for all $h' \in H$ then $\Vdash x \not\sqsubset^* \dot{h}\restriction(\omega \times [C]^{<\aleph_0})$.

Throughout this subsection, fix transitive models $M \subseteq N$ of ZFC and $A \in M$ such that $A \subseteq [\omega]^{\aleph_0} \cap M$. The relational system $\mathbf{Md}(A)$ helps to abbreviate the main notion presented in Ref. 2 for the preservation of mad families. What is defined in Def. 2 of Ref. 2 as $(\star^{M,N}_{A,a})$ for $a \in [\omega]^{\aleph_0}$, which is the same as "$a$ diagonalizes $M$ outside $A$" in Def. 4.2 of Ref. 1, actually means in our notation that

$a$ is $\mathbf{Md}(A)$-unbounded over $M$. Note that, for any countable $C \subseteq [\omega]^{\aleph_0}$, $\mathbf{Md}(C)$ is a Prs.

The following results from Ref. 2 indicate that the a.d. family added by $\mathbb{H}_\Omega$ is composed of unbounded reals in the sense of relational systems like in Definition 3.16(2), which in turn becomes a mad family when $\Omega$ is uncountable.

**Lemma 3.17 (Lemma 3 in Ref. 2).** *If $a^* \in [\omega]^{\aleph_0}$ is $\mathbf{Md}(A)$-unbounded over $M$ then $|a^* \cap x| = \aleph_0$ for any $x \in M \setminus \mathcal{I}(A)$ where $\mathcal{I}(A) := \{x \subseteq \omega : \exists F \in [A]^{<\aleph_0}(x \subseteq^* \bigcup F)\}$.*

**Lemma 3.18 (Lemma 4 in Ref. 2).** *Let $\Omega$ be a set, $z^* \in \Omega$ and $\dot{A} := \langle \dot{a}_z : z \in \Omega \rangle$ the a.d. family added by $\mathbb{H}_\Omega$. Then, $\mathbb{H}_\Omega$ forces that $\dot{a}_{z^*}$ is $\mathbf{Md}(\dot{A} \restriction (\Omega \setminus \{z^*\}))$-unbounded over $V^{\mathbb{H}_{\Omega \setminus \{z^*\}}}$.*

The known results about the preservation of $\mathbf{Md}(A)$-unbounded reals along coherent pairs of FS iterations are referred below. This is similar to the previous discussion about preservation of $\mathbf{R}$-unbounded reals for a Prs $\mathbf{R}$.

**Lemma 3.19 (Lemma 12 in Ref. 2).** *Let $\mathbf{s}$ be a coherent pair of FS iterations (wlog $I^{\mathbf{s}} = \{0, 1\}$) with $\pi = \pi^{\mathbf{s}}$ limit, $\dot{A}$ a $\mathbb{P}_{0,1}$-name of a family of infinite subsets of $\omega$ and $\dot{a}^*$ a $\mathbb{P}_{1,1}$-name for an infinite subset of $\omega$ such that*

$$\Vdash_{\mathbb{P}_{1,\xi}} \text{“}\dot{a}^* \text{ is } \mathbf{Md}(A)\text{-unbounded over } V_{0,\xi}\text{"}$$

*for all $0 < \xi < \pi$. Then, $\mathbb{P}_{0,\pi} \lessdot \mathbb{P}_{1,\pi}$ and $\Vdash_{\mathbb{P}_{1,\pi}}$ “$\dot{a}^*$ is $\mathbf{Md}(A)$-unbounded over $V_{0,\pi}$".*

**Lemma 3.20 (Lemma 11 in Ref. 2).** *Let $\mathbb{P} \in M$ be a poset. If $N \models$ “$a^*$ is $\mathbf{Md}(A)$-unbounded over $M$" then*

$$N^{\mathbb{P}} \models \text{“}a^* \text{ is } \mathbf{Md}(A)\text{-unbounded over } M^{\mathbb{P}}\text{”.}$$

**Corollary 3.21.** *If $\Omega \in M$ and $N \models$ “$a^*$ is $\mathbf{Md}(A)$-unbounded over $M$" then*

$$N^{\mathbb{C}_\Omega} \models \text{“}a^* \text{ is } \mathbf{Md}(A)\text{-unbounded over } M^{\mathbb{C}_\Omega}\text{”.}$$

*Likewise, $\mathbb{H}_\Omega$ satisfies a similar statement.*

**Lemma 3.22 (Lemma 4.8 and Cor. 4.11 in Ref. 1).** *Let $\mathbb{S}$ be either $\mathbb{E}$ or a random algebra. If $N \models$ “$a^*$ is $\mathbf{Md}(A)$-unbounded over $M$" then*

$$N^{\mathbb{S}^N} \models \text{“}a^* \text{ is } \mathbf{Md}(A)\text{-unbounded over } M^{\mathbb{S}^M}\text{”.}$$

The previous result indicates that $\mathbb{E}$ and random forcing, when used as iterands in a coherent pair of FS iterations, help to preserve $\mathbf{Md}(A)$-unbounded reals. To generalize this fact, we use the notion of ‘uniformly good’ introduced in Definition 3.16(3).

**Theorem 3.23.** *Let $\mathbb{S}$ be a Suslin ccc poset coded in $M$ and $A \in M$, $A \subseteq [\omega]^{\aleph_0}$. Assume*

($\star$) $[A]^{\aleph_0} \cap M$ *is cofinal in* $[A]^{\aleph_0} \cap N$.

*If $M \models$ "$\mathbb{S}$ is uniformly* **Md**$(A)$*-good" then, in $N$, $\mathbb{S}^N$ forces that every real in $[\omega]^{\aleph_0} \cap N$ that is* **Md**$(A)$*-unbounded over $M$ is still* **Md**$(A)$*-unbounded over $M^{\mathbb{S}^M}$.*

Note that, when $N$ is a generic extension of $M$ by a proper poset, ($\star$) holds.

**Proof.** Let $a \in [\omega]^{\aleph_0} \cap N$ be **Md**$(A)$-unbounded over $M$. Assume that $\dot{h} \in M$ is a $\mathbb{S}^M$-name of a function in $\omega^{\omega \times [A]^{<\omega}}$. As $\mathbb{S}$ is uniformly **Md**$(A)$-good in $M$, there is a family $\{h_n : n < \omega\} \subseteq \omega^{\omega \times [A]^{<\aleph_0}}$ (in $M$) that witnesses goodness for $\dot{h}$. Thus $a \not\subseteq^* h_n$ for every $n < \omega$, so we can find a countable $C \subseteq A$ such that $a \not\subseteq^* h_n{\restriction}(\omega \times [C]^{<\aleph_0})$ for every $n < \omega$. By ($\star$), wlog we can find such $C$ in $M$.

In $M$ the statement "for every $x \in [\omega]^{\aleph_0}$, if $x \not\subseteq^* h_n{\restriction}(\omega \times [C]^{<\aleph_0})$ for all $n < \omega$, then $\Vdash_{\mathbb{S}} x \not\subseteq^* \dot{h}{\restriction}(\omega \times [C]^{<\aleph_0})$" is true. Furthermore, as this statement is a conjunction of a $\mathbf{\Sigma}^1_1$-statement with a $\mathbf{\Pi}^1_1$-statement of the reals (see e.g. Claim 4.27 of Ref. 17), it is also true in $N$. In particular, since $a \not\subseteq^* h_n{\restriction}(\omega \times [C]^{<\aleph_0})$ for every $n < \omega$, $\Vdash^N_{\mathbb{S}^N} a \not\subseteq^* \dot{h}{\restriction}(\omega \times [C]^{<\aleph_0})$. □

Though $\mathbb{E}$ and $\mathbb{B}$ are indeed uniformly **Md**$(A)$-good (by Theorem 3.27 and Lemma 3.29), the application of Theorem 3.23 yields a version of Lemma 3.22 restricted to the condition ($\star$). To avoid this restriction, we consider an alternative generalization based on the following notion.

**Definition 3.24.** Let $\mathbb{P}$ be a poset.

(1) Say that a set $Q \subseteq \mathbb{P}$ is *Frechet-linked (in $\mathbb{P}$)*, abbreviated Fr-*linked*, if, for any sequence $\bar{p} = \langle p_n : n < \omega \rangle$ in $Q$, there is some $q \in \mathbb{P}$ that forces $\exists^\infty n < \omega (p_n \in \dot{G})$.

(2) Let $\mu$ be an infinite cardinal. Say that a poset $\mathbb{P}$ is *$\mu$-Frechet-linked* (often abbreviated $\mu$-Fr-*linked*) if there is a sequence $\langle Q_\alpha : \alpha < \mu \rangle$ of Fr-linked subsets of $\mathbb{P}$ such that $\bigcup_{\alpha<\mu} Q_\alpha$ is dense in $\mathbb{P}$.
By $\sigma$-Fr-linked we mean $\aleph_0$-Fr-linked.

(3) A poset $\mathbb{S}$ is *Suslin $\sigma$-Frechet-linked* if $\mathbb{S}$ is a subset of some Polish space, the relations $\leq$ and $\perp$ are $\mathbf{\Sigma}^1_1$ (in that Polish space) and $\mathbb{S} = \bigcup_{n<\omega} Q_n$ where each $Q_n$ is a Fr-linked $\mathbf{\Sigma}^1_1$ set.

Here, Fr denotes the Frechet filter on $\omega$. The reason of the terminology 'Frechet-linked' is that this notion corresponds to a particular case on Fr of a more general notion of linkedness with filters that we provide in Example 5.4.

**Remark 3.25.**

(1) The notion '$\mu$-Fr-linked' is a forcing property, i.e., if $\mathbb{P}$ and $\mathbb{Q}$ are posets, $\mathbb{P} \lessdot \mathbb{Q}$ (in the sense that the Boolean completion of $\mathbb{P}$ is completely embedded into the completion of $\mathbb{Q}$) and $\mathbb{Q}$ is $\mu$-Fr-linked then so is $\mathbb{P}$ (see more on this in Section 5).

(2) No Fr-linked subset of a poset can contain infinite antichains. In addition, if $\mathbb{P}$ is a poset and $Q \subseteq \mathbb{P}$, the statement "$Q$ does not contain infinite antichains" is absolute for transitive models of ZFC. This is because that statement is equivalent to say that "$T$ is a well-founded tree" where $T := \{s \in Q^{<\omega} : \operatorname{ran} s \text{ is an antichain}\}$.

(3) As a consequence of (2), $\mu$-Fr-linked posets are $\mu^+$-cc. Even more, by Thm. 2.4 of Ref. 22, they are $\mu^+$-Knaster (see more in Section 5).

(4) Any poset of size $\leq \mu$ is $\mu$-Fr-linked (witnessed by its singletons). In particular, Cohen forcing is $\sigma$-Fr-linked.

(5) By (3), any Suslin $\sigma$-Fr-linked poset is Suslin ccc. Moreover, if $\langle Q_n : n < \omega \rangle$ witnesses that a poset $\mathbb{S}$ is Suslin $\sigma$-Fr-linked then the statement "$Q_n$ is Fr-linked" is $\mathbf{\Pi}^1_2$ (by (6) below, its negation is equivalent to $\exists f \in Q_n^\omega \exists g \in \mathbb{S}^\omega(\{g(n) : n < \omega\}$ is a maximal antichain and $\forall n < \omega \exists m < \omega \forall k \geq m(g(n) \perp f(k)))$, which is $\mathbf{\Sigma}^1_2$). Therefore, if $M \models$"$\mathbb{S}$ is Suslin $\sigma$-Fr-linked" and $\omega_1^N \subseteq M$ then $N \models$"$\mathbb{S}$ is Suslin $\sigma$-Fr-linked".

(6) Let $\mathbb{P}$ be a poset and $Q \subseteq \mathbb{P}$. Note that a sequence $\langle p_n : n < \omega \rangle$ in $Q$ witnesses that $Q$ is <u>not</u> Fr-linked iff the set

$$\{q \in \mathbb{P} : \forall^\infty n < \omega(q \perp p_n)\}$$

is dense.

**Lemma 3.26.** *Let $\mathbb{P}$ be a poset and $Q \subseteq \mathbb{P}$ Fr-linked. If $\dot{n}$ is a $\mathbb{P}$-name of a natural number then there is some $m < \omega$ such that $\forall p \in Q(p \not\Vdash m < \dot{n})$.*

**Proof.** Towards a contradiction, assume that for each $m < \omega$ there is some $p_m \in Q$ that forces $m < \dot{n}$. Hence, as $Q$ is Fr-linked, there is some $q \in Q$ that forces $\exists^\infty m < \omega(p_m \in \dot{G})$, which implies $\exists^\infty m < \omega(m < \dot{n})$, a contradiction. □

**Theorem 3.27.** *Any $\sigma$-Fr-linked poset is uniformly* $\mathbf{Md}(B)$*-good for any $B \subseteq [\omega]^{\aleph_0}$.*

**Proof.** Let $\mathbb{P}$ be a $\sigma$-Fr-linked poset witnessed by $\langle Q_n : n < \omega \rangle$. Assume that $\dot{h}$ is a $\mathbb{P}$-name of a function in $\omega^{\omega \times [B]^{<\omega}}$. Fix $n < \omega$. For each $k < \omega$ and $F \in [B]^{<\aleph_0}$, by Lemma 3.26 there is some $h_n(k, F) < \omega$ such that $\forall p \in Q_n(p \not\Vdash h_n(k, F) < \dot{h}(k, F))$. This allows to define $h_n \in \omega^{\omega \times [B]^{<\omega}}$.

Assume that $C \subseteq B$ is infinite and that $x \in [\omega]^{\aleph_0}$ is $\mathbf{Md}(C)$-unbounded over $\{h_n \restriction (\omega \times [C]^{<\aleph_0}) : n < \omega\}$. We show that $\Vdash x \not\subseteq^* \dot{h} \restriction (\omega \times [C]^{<\aleph_0})$. Assume that $p \in \mathbb{P}$ and $m < \omega$. Wlog, we may assume that $p \in Q_n$ for some $n < \omega$. As $x \not\subseteq^* h_n \restriction (\omega \times [C]^{<\aleph_0})$, there are some $k > m$ and $F \in [C]^{<\aleph_0}$ such that $[k, h_n(k, F)) \setminus \bigcup F \subseteq x$. On the other hand, by the definition of $h_n$, there is some $q \leq p$ in $\mathbb{P}$ that forces $\dot{h}(k, F) \leq h_n(k, F)$. Hence $q \Vdash [k, \dot{h}(k, F)) \setminus \bigcup F \subseteq x$. □

As a consequence, Theorem 3.23 can be applied to Suslin $\sigma$-Fr-linked posets. However, it can be proved directly that Suslin $\sigma$-Fr-linked posets preserves $\mathbf{Md}(A)$-unbounded reals without the condition $(\star)$.

**Theorem 3.28.** *Let $\mathbb{S}$ be a Suslin $\sigma$-*Fr*-linked poset coded in $M$. Then, in $N$, $\mathbb{S}^N$ forces that every real in $[\omega]^{\aleph_0} \cap N$ that is $\mathbf{Md}(A)$-unbounded over $M$ is still $\mathbf{Md}(A)$-unbounded over $M^{\mathbb{S}^M}$.*

**Proof.** Let $\langle Q_n : n < \omega \rangle \in M$ be a sequence that witnesses that $\mathbb{S}$ is Suslin $\sigma$-Fr-linked. Let $a \in [\omega]^{\aleph_0} \cap N$ be $\mathbf{Md}(A)$-unbounded over $M$. Assume that $\dot{h} \in M$ is a $\mathbb{S}^M$-name of a function in $\omega^{\omega \times [A]^{<\omega}}$. Fix $p \in \mathbb{S}^N$ and $m < \omega$, so there is some $n < \omega$ such that $p \in Q_n^N$. Down in $M$, find $h_n \in \omega^{\omega \times [A]^{<\omega}} \cap M$ as in the proof of Theorem 3.27. As $a \not\subseteq^* h_n$, there are some $k \geq m$ and $F \in [A]^{<\aleph_0}$ such that $[k, h_n(k, F)) \setminus \bigcup F \subseteq a$. On the other hand, the statement "no $q \in Q_n$ forces that $h_n(k, F) < \dot{h}(k, F)$" is $\mathbf{\Pi}^1_1$, so it is absolute and, as true in $M$, it also holds in $N$. Therefore, in $N$, $p$ does not force $h_n(k, F) < \dot{h}(k, F)$, which implies that some $q \leq p$ in $\mathbb{S}^N$ forces the contrary. This clearly implies that $q$ forces $[k, \dot{h}(k, F)) \setminus \bigcup F \subseteq a$. □

Therefore, in conjunction with the following result, the previous theorem is a suitable generalization of Lemma 3.22.

**Lemma 3.29.** *The posets $\mathbb{E}$ and $\mathbb{B}$ are Suslin $\sigma$-*Fr*-linked. Moreover, any complete Boolean algebra that admits a strictly positive $\sigma$-additive measure (e.g. any random algebra) is $\sigma$-*Fr*-linked.*

**Proof.** For each $s \in \omega^{<\omega}$ and $m < \omega$ define $E_{s,m} := \{(t, \varphi) \in \mathbb{E} : t = s \text{ and } \forall i < \omega(|\varphi(i)| \leq m)\}$. This set is actually Borel in $\omega^{<\omega} \times \mathcal{P}(\omega)^{\omega}$ (the Polish space where $\mathbb{E}$ is defined). A compactness argument similar to the one in Ref. 19 shows that $E_{s,n}$ is Fr-linked in $\mathbb{E}$. Fix a non-principal ultrafilter $U$ on $\omega$ and let $\langle p_n : n < \omega \rangle$ be a sequence in $E_{s,m}$. Write $p_n = (s, \varphi_n)$. For each $i < \omega$ define $\varphi(i) \subseteq \omega$ such that $l \in \varphi(i)$ iff $\{n < \omega : l \in \varphi_n(i)\} \in U$. It can be proved that $|\varphi(i)| \leq m$, (so $q := (s, \varphi) \in \mathbb{E}$) and that $q$ forces $\exists^{\infty} n < \omega(p_n \in \dot{G})$ (that is, for any $q' \leq q$ and $n < \omega$, there is some $k \geq n$ such that $q'$ is compatible with $p_k$).

Now, consider random forcing as $\mathbb{B} = \bigcup_{m<\omega} B_m$ where[a]

$$B_m := \left\{T \subseteq 2^{<\omega} : T \text{ is a well-pruned tree and } \mathrm{Lb}([T]) \geq \frac{1}{m+1}\right\}.$$

Note that $B_m$ is Borel in $2^{2^{<\omega}}$. It is enough to show that $B_m$ is Fr-linked. Assume the contrary, so by Remark 3.25(6) there are a sequence $\langle T_n : n < \omega\rangle$ in $B_m$ and a partition $\langle A_n : n < \omega\rangle$ of $2^\omega$ into Borel sets of positive measure such that, for each $n < \omega$, $A_n \cap [T_k]$ has measure zero for all but finitely many $k < \omega$. Construct an increasing function $g : \omega \to \omega$ such that $A_n \cap [T_k]$ has measure zero for all $k \geq g(n)$. As $2^\omega = \bigcup_{n<\omega} A_n$, we can find some $n^* < \omega$ such that the measure of $A^* := \bigcup_{n<n^*} A_n$ is strictly larger than $1 - \frac{1}{m+1}$. Hence $A^* \cap [T]$ has positive measure for any $T \in B_m$, but this contradicts that $A^* \cap [T_k]$ has measure zero for all $k \geq g(n^* \quad 1)$.

A similar proof works for any complete Boolean algebra that admits a strictly positive $\sigma$-additive measure. □

Indeed, the notion of $\mu$-Fr-linked behaves well for preservation of **D**-unbounded families as shown in the following result. Even more, this generalizes the facts of Example 3.7(2).

**Theorem 3.30.** *If $\mu < \theta$ are infinite cardinals then any $\mu$-*Fr*-linked poset is $\theta$-***D***-good.*

**Proof.** This argument is very similar to the proof of Theorem 3.27. Let $\mathbb{P}$ be a $\mu$-Fr-linked poset witnessed by $\langle Q_\alpha : \alpha < \mu\rangle$, and let $\dot{y}$ be a $\mathbb{P}$-name of a member of $\omega^\omega$. Using Lemma 3.26, for each $\alpha < \mu$ find a $y_\alpha \in \omega^\omega$ such that, for every $i < \omega$, no member of $Q_\alpha$ forces that $y_\alpha(i) < \dot{y}(i)$. It can be proved that $\{y_\alpha : \alpha < \mu\}$ witnesses goodness for $\dot{y}$. □

**Remark 3.31.** There is a Suslin $\sigma$-Fr-linked poset that is not **Lc**-good. For $b, h \in \omega^\omega$ such that $\forall i < \omega(b(i) > 0)$ and $h$ goes to infinity, consider the poset

$$\mathbb{LOC}_{b,h} := \left\{p \in \prod_{i<\omega}[b(i)]^{\leq h(i)} : \exists m < \omega \forall^\infty i < \omega(|p(i)| \leq m)\right\}$$

ordered by $q \leq p$ iff $\forall i < \omega(p(i) \subseteq q(i))$. This poset adds a slalom $\dot{\varphi}_* \in \prod_{i<\omega}[b(i)]^{\leq h(i)}$, defined by $\dot{\varphi}_*(i) := \bigcup_{p\in\dot{G}} p(i)$, such that $x \in^* \varphi_*$ for every $x \in \prod_{i<\omega} b(i)$ in the ground model. Note that $\dot{\varphi}_*(i)$ is forced to either have size $h(i)$ (whenever $h(i) \leq b(i)$) or to be equal to $b(i)$ (whenever $b(i) \leq h(i)$).

---

[a]A *well-pruned tree* is a non-empty tree such that every node has a successor.

This poset is Suslin $\sigma$-Fr-linked. In fact, for any $s \in \bigcup_{n<\omega} \prod_{i<n} b(i)$ and $m < \omega$, the set

$$L_{b,h}(s,m) := \left\{p \in \prod_{i<\omega} [b(i)]^{\leq h(i)} : s \subseteq p, \text{ and } \forall i \geq |s| (|p(i)| \leq m)\right\}$$

is Borel and Fr-linked, and $\mathbb{LOC}_{b,h} = \bigcup_{s,m} L_{b,h}(s,m)$. To see that $L_{b,h}(s,m)$ is Fr-linked, assume that $\langle p_n : n < \omega \rangle$ is a sequence in $L_{b,h}(s,m)$ and fix a non-principal ultrafilter $U$ on $\omega$. For each $i \geq |s|$, as $[b(i)]^{\leq m}$ is finite, we can find $q(i) \in [b(i)]^{\leq m}$ and $a_i \in U$ such that $p_n(i) = q(i)$ for any $n \in a_i$. Hence $q \in L_{b,h}(s,m)$ where $q(i) := s(i)$ for all $i < |s|$. It remains to show that $q$ forces $|\{n < \omega : p_n \in \dot{G}\}| = \aleph_0$. If $r \leq q$ and $n_0 < \omega$, then we can find $k, m_0 < \omega$ such that $k \geq |s|$ and, for any $i \geq k$, $|r(i)| \leq m_0$ and $m_0 + m \leq h(i)$. Choose some $n \in \bigcap_{i<k} a_i \setminus n_0$ (put $a_i := \omega$ for $i < |s|$). Hence $q{\restriction}k = p_n{\restriction}k$ and $q'$ forces that $p_n \in \dot{G}$ where $q'(i) := r(i) \cup p_n(i)$ (for $i < k$, $q'(i) = r(i)$; for $i \geq k$, $|q'(i)| \leq m_0 + m \leq h(i)$).

Now, if $h = \text{id}$ and $b(i) = i^i$ for every $i < \omega$ then $\mathbb{LOC}_{b,h}$ is <u>not</u> **Lc**-good. Consider the name $\dot{\varphi}_*$ of the slalom that this poset adds, which is clearly a name of a member of $\mathcal{S}(\omega,\mathcal{H})$ (see Example 3.7(4)). If $H$ is a countable subset of $\mathcal{S}(\omega,\mathcal{H})$, then there exists an $x \in \prod_{i<\omega} b(i)$ such that $x$ is not localized by any member of $H$. On the other hand, $\mathbb{LOC}_{b,h}$ already forces that $x \in^* \dot{\varphi}_*$.

We finish this section with a general result about preservation of mad families through standard coherent systems.

**Theorem 3.32.** *Let $\nu$ be a cardinal of uncountable cofinality and let* **s** *be a standard coherent system that satisfies the hypothesis of Lemma 2.7 for $\theta = \nu$. Further assume that*

*(i) $\Gamma \subseteq \Omega$ has size $\geq \nu$,*

*(ii) $D \in I$ and $\Gamma \subseteq D$,*

*(iii) for each $l \in \Gamma$, $\mathbb{P}_{D,1}$ adds a real $\dot{a}_l$ in $[\omega]^{\aleph_0}$ such that, for any $Z \subseteq D$ in $I \cap [\Omega]^{<\nu}$, $\dot{A}{\restriction}Z := \langle \dot{a}_l : l \in Z \cap \Gamma \rangle$ is a $\mathbb{P}_{Z,1}$-name and, whenever $l \in D \setminus Z$, $\mathbb{P}_{D,1}$ forces that $\dot{a}_l$ is* $\mathbf{Md}(\dot{A}{\restriction}Z)$*-unbounded over $V_{Z,1}$, and*

*(iv) for every $\xi \in S^{\mathbf{s}}$ and $B \in I \cap [D]^{<\nu}$, $\mathbb{P}_{B,\xi}$ forces that $\dot{\mathbb{Q}}_{B,\xi}$ is either uniformly* $\mathbf{Md}(\dot{A}{\restriction}B)$*-good or a random algebra.*

*Then $\mathbb{P}_{D,\pi}$ forces that any infinite subset of $\omega$ intersects some member of $\dot{A} := \dot{A}{\restriction}\Gamma$. In particular, $\mathbb{P}_{D,\pi}$ forces $\mathfrak{a} \leq |\Gamma|$ whenever $\dot{A}$ is an a.d. family.*

**Proof.** Let $\dot{x}$ be a $\mathbb{P}_{D,\pi}$-name of an infinite subset of $\omega$. By Lemma 2.7, there is some $Z \subseteq D$ in $I \cap [\Omega]^{<\nu}$ such that $\dot{x}$ is a $\mathbb{P}_{Z,\pi}$-name. Thus, by Lemmas 3.19, 3.20 and Theorem 3.23, for any $l \in \Gamma \setminus Z$, $\mathbb{P}_{D,\pi}$ forces that $\dot{a}_l$ is $\mathbf{Md}(\dot{A}{\restriction}Z)$-unbounded over $V_{Z,\pi}$, so $\dot{x} \notin \mathcal{I}(\dot{A}{\restriction}Z)$ implies that $\dot{x} \cap \dot{a}_l$ is infinite by Lemma 3.17. As $\dot{x} \in \mathcal{I}(\dot{A}{\restriction}Z)$ implies that $\dot{x} \cap \dot{a}_j$ is infinite for some $j \in Z \cap \Gamma$, we are done. □

**Remark 3.33.** By Theorem 3.27, it is clear that in condition (iv) of Theorem 3.32 we can use Suslin $\sigma$-Fr-linked posets.

## 4. Applications

The following result generalizes Thm. 4.17 of Ref. 1 in the sense that it allows to preserve mad families of singular cardinality along a more general type of FS iterations.

**Theorem 4.1.** *Let $\theta$ be an uncountable regular cardinal and let $\Omega$ be a set of size $\geq \theta$. After forcing with $\mathbb{H}_\Omega$, any further FS iteration where each iterand is one of the following types preserves the mad family added by $\mathbb{H}_\Omega$.*

*(0) Suslin $\sigma$-Fr-linked.*
*(1) Random algebra.*
*(2) Hechler poset (for adding a mad family).*
*(3) Poset with ccc of size $< \theta$.*

**Proof.** Consider a FS iteration $\mathbb{P}_\pi = \langle \mathbb{P}_\xi, \dot{\mathbb{Q}}_\xi : \xi < \pi \rangle$ with $\pi > 0$ such that $\dot{\mathbb{Q}}_0 = \mathbb{P}_1 = \mathbb{H}_\Omega$ and, for $0 < \xi < \pi$, $\dot{\mathbb{Q}}_\xi$ is of one of the types above. To be more precise, let $\langle C_j : j < 4 \rangle$ be a partition of $[1, \pi)$ such that, for each $j < 4$ and $\xi \in C_j$, $\dot{\mathbb{Q}}_\xi$ is a $\mathbb{P}_\alpha$-name of a poset of type $(j)$. Note that this iteration can be defined as the standard coherent system $\mathbf{m}$ on $I^{\mathbf{m}} := \langle \mathcal{P}(\Omega), \subseteq \rangle$ such that

(o) $\mathbb{P}_{X,1} = \mathbb{H}_X$ for any $X \subseteq \Omega$;
(i) $S^{\mathbf{m}} = C_0 \cup C_1$, $C^{\mathbf{m}} = C_2 \cup C_3$;
(ii) $\Delta : [1, \pi) \to [\Omega]^{<\theta}$ such that $\Delta(\xi) = \emptyset$ for each $\xi \in C_1 \cup C_2$;
(iii) for $\xi \in C_0$, when $\dot{\mathbb{Q}}_\xi$ is coded in $V_{\Delta(\xi),\xi}$, $\dot{\mathbb{S}}_\xi = \dot{\mathbb{Q}}_\xi$ (or trivial otherwise, though this latter case will not happen) and, for $\xi \in C_1$, $\dot{\mathbb{S}}_\xi$ is the random algebra $\dot{\mathbb{Q}}_\xi$ (wlog its support is in the ground model);
(iv) for $\xi \in C_2$, $\dot{\mathbb{Q}}^{\mathbf{m}}_\xi = \dot{\mathbb{Q}}_\xi$ (wlog, the support of this Hechler poset is in the ground model) and, for $\xi \in C_3$, when $\dot{\mathbb{Q}}_\xi$ is forced to be in $V_{\Delta(\xi),\xi}$, $\dot{\mathbb{Q}}^{\mathbf{m}}_\xi = \dot{\mathbb{Q}}_\xi$ (or trivial otherwise, though this will not happen).

By recursion on $\xi \leq \pi$, $\mathbf{m}{\restriction}\xi$ and $\Delta{\restriction}\xi$ should be constructed and it should be guaranteed that $\mathbb{P}_{\Omega,\xi} = \mathbb{P}_\xi$. This is fine in the steps $\xi = 0, 1$ and the limit steps. Consider the successor step, i.e., assume we have the construction up to $\xi$. As $\xi \in C_j$ for some $j < 4$, we consider cases for each $j$. If $j = 0$ then, as a Suslin ccc poset is coded by reals and $\mathbb{P}_{\Omega,\xi} = \mathbb{P}_\xi$, by Lemma 2.7 there is some $\Delta(\xi) \in [\Omega]^{<\theta}$ such that $\dot{\mathbb{Q}}_\xi$ is (coded by) a $\mathbb{P}_{\Delta(\xi),\xi}$-name; if $j = 1, 2$ then, as the support of $\dot{\mathbb{Q}}_\xi$ can be assumed to be in the ground model, we can put $\Delta(\xi) = 0$; if $j = 3$ then, by Lemma 2.7 and the regularity of $\theta$, there is some $\Delta(\xi) \in [\Omega]^{<\theta}$ such that $\dot{\mathbb{Q}}_\xi$ is a

$\mathbb{P}_{\Delta(\xi),\xi}$-name. Therefore, in any case, we can define $\mathbf{m}{\restriction}(\xi+1)$ as required and it is clear that $\mathbb{P}_{\Omega,\xi}=\mathbb{P}_\xi$.

Now let $\dot{A}=\langle \dot{a}_l : l\in\Omega\rangle$ be the $\mathbb{P}_{\Omega,1}$-name of the generic a.d. family it adds. Note that $\dot{a}_l$ is a $\mathbb{P}_{\{l\},1}$-name and, by Lemma 3.18, for any $B\subseteq\Omega$ with $l\notin B$, $\mathbb{P}_{B\cup\{l\},1}$ forces that $\dot{a}_l$ is $\mathbf{Md}(\dot{A}{\restriction}B)$-unbounded over $V_{B,1}$. Hence, by Theorem 3.32, $\mathbb{P}_{\Gamma,\pi}$ forces that $\dot{A}$ is a mad family. □

**Remark 4.2.** The previous theorem remains true if we add the type

(0') Suslin ccc poset coded in the ground model such that, for any $X\subseteq\Omega$ in the ground model, it is uniformly $\mathbf{Md}(\dot{A}{\restriction}X)$-good in any ccc generic extension of $V^{\mathbb{H}_X}$.

By Theorem 3.27 Suslin $\sigma$-Fr-linked posets coded in the ground model satisfy (0').

The remaining results in this section are improvements of the consistency results of Sec. 5 of Ref. 1 about separating cardinals in Cichoń's diagram. Not only can we force an additional singular value, but the constructions are uniform in the sense that there is no need to distinguish between 2D or 3D constructions anymore since all the coherent systems can be constructed on a partial order of the form $\langle\mathcal{P}(\Omega),\subseteq\rangle$. In the following proofs, sum and product denote the corresponding operations in the ordinals, even when they are applied to cardinal numbers.

The following result improves Thm. 5.2 of Ref. 1 about separating Cichoń's diagram into 7 different values.

**Theorem 4.3.** *Assume that $\theta_0\leq\theta_1\leq\kappa\leq\mu$ are uncountable regular cardinals, $\nu\leq\lambda$ are cardinals such that $\mu\leq\nu$, $\nu^{<\kappa}=\nu$ and $\lambda^{<\theta_1}=\lambda$. Then there is a ccc poset that forces $\mathrm{MA}_{<\theta_0}$, $add(\mathcal{N})=\theta_0$, $cov(\mathcal{N})=\theta_1$, $\mathfrak{b}=\mathfrak{a}=\kappa$, $non(\mathcal{M})=cov(\mathcal{M})=\mu$, $\mathfrak{d}=\nu$ and $non(\mathcal{N})=\mathfrak{c}=\lambda$.*

**Proof.** Let $\Omega_0$ and $\Omega_1$ be disjoint sets of size $\kappa$ and $\nu$ respectively. Put $\Omega:=\Omega_0\cup\Omega_1$. As $\nu^{<\kappa}=\nu$, we can enumerate $[\Omega]^{<\kappa}:=\{W_\zeta:\zeta<\nu\}$. Fix a bijection $g=(g_0,g_1,g_2):\lambda\to 2\times\nu\times\lambda$ and a function $t:\nu\mu\to\nu$ such that, for any $\zeta<\nu$, $t^{-1}[\{\zeta\}]$ is cofinal in $\nu\mu$.[b] Put $\pi:=\lambda\nu\mu$, $S:=\{\lambda\rho:\rho<\nu\mu\}$ and define $\Delta:[1,\pi)\to[\Omega]^{<\kappa}$ such that $\Delta(\lambda\rho):=\emptyset$, $\Delta(\lambda\rho+1):=W_{t(\rho)}$ and $\Delta(\lambda\rho+2+\varepsilon)=W_{g_1(\varepsilon)}$ for each $\rho<\nu\mu$ and $\varepsilon<\lambda$.

Define the standard coherent system $\mathbf{m}$ of FS iterations of length $\pi$ on $\langle\mathcal{P}(\Omega),\subseteq\rangle$ such that $S^{\mathbf{m}}:=S$, $C^{\mathbf{m}}:=[1,\pi)\smallsetminus S$, $\Delta^{\mathbf{m}}:=\Delta$, $\dot{\mathbb{Q}}^{\mathbf{m}}_{X,0}:=\mathbb{H}_{X\cap\Omega_0}\times\mathbb{C}_{X\cap\Omega_1}$ and where the FS iterations at each interval of the form $[\lambda\rho,\lambda(\rho+1))$ for $\rho<\mu\nu$ is defined as follows. Assume that $\mathbf{m}{\restriction}\lambda\rho$ has already been defined. For each $\zeta<\nu$ choose

[b]For example, define $t(\nu\delta+\alpha)=\alpha$ for each $\delta<\mu$ and $\alpha<\nu$.

(0) an enumeration $\{\dot{\mathbb{Q}}_{0,\zeta,\gamma} : \gamma < \lambda\}$ of all the (nice) $\mathbb{P}_{W_\zeta,\lambda\rho}$-names for posets of size $< \theta_0$, with underlying set contained in $\theta_0$, that are forced by $\mathbb{P}_{\Omega,\lambda\rho}$ to have ccc; and

(1) an enumeration $\{\dot{\mathbb{Q}}_{1,\zeta,\gamma} : \gamma < \lambda\}$ of all the (nice) $\mathbb{P}_{W_\zeta,\lambda\rho}$-names for subalgebras of random forcing of size $< \theta_1$.

Put $\mathbb{S}^{\mathbf{m}}_{\lambda\rho} := \mathbb{E}$ (only when $\rho > 0$) and, for each $\xi \in (\lambda\rho, \lambda(\rho+1))$, put

(i) $\dot{\mathbb{Q}}^{\mathbf{m}}_\xi := \mathbb{D}^{V_{\Delta(\xi),\xi}}$ when $\xi = \lambda\rho + 1$, and

(ii) $\dot{\mathbb{Q}}^{\mathbf{m}}_\xi := \dot{\mathbb{Q}}_{g(\varepsilon)}$ when $\xi = \lambda\rho + 2 + \varepsilon$ for some $\varepsilon < \lambda$.

This construction is possible because, as $\lambda^{<\theta_1} = \lambda$, each $\mathbb{P}_{\Omega,\xi}$ has size $\leq \lambda$.

It remains to show that $\mathbb{P} := \mathbb{P}_{\Omega,\pi}$ forces what we want. First note that this poset can be obtained by the FS iteration $\langle \mathbb{P}_{\Omega,\xi}, \dot{\mathbb{Q}}_{\Omega,\xi} : \xi < \pi \rangle$, and observe that all these iterands are $\theta_0$-**Lc**-good and $\theta_1$-**Cn**-good. Hence, by Theorem 3.10, $\mathbb{P}$ forces $\mathrm{add}(\mathcal{N}) \leq \theta_0$, $\mathrm{cov}(\mathcal{N}) \leq \theta_1$ and $\lambda \leq \mathrm{non}(\mathcal{N})$. Actually, those are equalities, even more, $\mathbb{P}$ forces $\mathrm{MA}_{<\theta_0}$ (which implies $\mathrm{add}(\mathcal{N}) \geq \theta_0$). To see this, let $\dot{\mathbb{R}}$ is a $\mathbb{P}$-name of a ccc poset of size $< \theta_0$ and $\dot{\mathcal{D}}$ a family of size $< \theta_0$ of dense subsets of $\dot{\mathbb{R}}$. By Lemma 2.7 there is some $\zeta < \nu$ such that both $\dot{\mathbb{R}}$ and $\dot{\mathcal{D}}$ are $\mathbb{P}_{W_\zeta,\pi}$-names. Moreover, as $\pi$ has cofinality $\mu$, there is some $\rho < \nu\mu$ such both are $\mathbb{P}_{W_\zeta,\lambda\rho}$-names. Therefore, there is some $\gamma < \lambda$ such that $\dot{\mathbb{R}} = \dot{\mathbb{Q}}_{0,\zeta,\gamma}$, so the generic set added by $\dot{\mathbb{Q}}_{g(\varepsilon)} = \dot{\mathbb{Q}}^{\mathbf{m}}_{W_\zeta,\xi}$ intersects all the dense sets in $\dot{\mathcal{D}}$ where $\varepsilon := g^{-1}(0,\zeta,\gamma)$ and $\xi := \lambda\rho + 2 + \varepsilon$. In a similar way, it can be proved that $\mathbb{P}$ forces $\mathrm{cov}(\mathcal{N}) \geq \theta_1$. On the other hand, since $\Vdash_{\mathbb{P}} \mathfrak{c} \leq \lambda$ follows from $|\mathbb{P}| \leq \lambda$, together with $\mathrm{non}(\mathcal{N}) \geq \lambda$ (see above) it is forced that $\mathrm{non}(\mathcal{N}) = \mathfrak{c} = \lambda$.

As the FS iteration that determines $\mathbb{P}$ has cofinality $\mu$ and $\mu$-cofinally many full eventually different reals are added by $\mathbb{E}$, $\mathbb{P}$ forces $\mathrm{cov}(\mathcal{M}) \leq \mu \leq \mathrm{non}(\mathcal{M})$. Actually, $\mathrm{non}(\mathcal{M}) \leq \mu \leq \mathrm{cov}(\mathcal{M})$ is forced by Theorem 3.10 applied to the Prs **Ed**, so $\mathrm{non}(\mathcal{M}) = \mathrm{cov}(\mathcal{M}) = \mu$. Now we show that $\mathbb{P}$ forces $\mathfrak{a} \leq \kappa$ and $\nu \leq \mathfrak{d}$. Let $\dot{A} := \{\dot{a}_l : l \in \Omega_0\}$ be the $\mathbb{H}_{\Omega_0}$-name of the mad family added by $\mathbb{H}_{\Omega_0}$ and let $\{\dot{c}_l : l \in \Omega_1\} \subseteq \omega^\omega$ be the Cohen reals added by $\mathbb{C}_{\Omega_1}$. For any $X \subseteq \Omega$, $l \in \Omega_0$ and $l' \in \Omega_1$, it is clear that $\dot{a}_l$ is a $\mathbb{P}_{X,1}$-name whenever $l \in X$, and $\dot{c}_{l'}$ is a $\mathbb{P}_{X,1}$-name whenever $l' \in X$. On the other hand, if $l' \notin X$ then $\mathbb{P}_{X\cup\{l'\},1}$ forces that $\dot{c}_{l'}$ is Cohen over $V_{X,1}$, hence it is **D**-unbounded over it; and if $l \notin X$ then $\mathbb{P}_{X\cup\{l\},1}$ forces that $\dot{a}_l$ is $\mathbf{Md}(\dot{A}{\restriction}(X \cap \Omega_0))$-unbounded over $V_{X,1}$. The latter is a consequence of Lemma 3.20 applied to $\mathbb{C}_{X\cap\Omega_1}$. Therefore, by Theorems 3.15 and 3.32 applied to $\{\dot{c}_{l'} : l' \in \Omega_1\}$ and $\{\dot{a}_l : l \in \Omega_0\}$ respectively, $\mathbb{P}$ forces $\nu \leq \mathfrak{d}$ (because $\{\dot{c}_{l'} : l' \in \Omega_1\}$ is strongly $\kappa$-**D**-unbounded) and $\mathfrak{a} \leq \kappa$.

It remains to show that $\mathbb{P}$ forces $\kappa \leq \mathfrak{b}$ and $\mathfrak{d} \leq \nu$. For each $\rho < \nu\mu$ denote by $\dot{d}_\rho$ the (restricted) dominating real over $V_{W_{t(\rho)},\lambda\rho+1}$ added by $\dot{\mathbb{Q}}_{W_{t(\rho)},\lambda\rho+1}$. It is enough to show that $\mathbb{P}$ forces that any subset of $\omega^\omega$ of size $< \kappa$ is dominated by some $\dot{d}_\rho$

(hence $\{\dot{d}_\rho : \rho < \nu\mu\}$ is a dominating family of size $\nu$). Let $\dot{F}$ be a $\mathbb{P}$-name of such a subset of $\omega^\omega$. By Lemma 2.7 and because $\mathrm{cf}(\pi) = \mu$, there are $\zeta < \nu$ and $\rho_0 < \nu\mu$ such that $\dot{F}$ is a $\mathbb{P}_{W_\zeta,\lambda\rho_0}$-name. Thus, there is some $\rho \in [\rho_0, \nu\mu)$ such that $t(\rho) = \zeta$, so $\mathbb{P}_{W_\zeta,\lambda\rho+2}$ forces that $\dot{d}_\rho$ dominates $\dot{F}$. □

We summarize in the rest of this section the results from Ref. 5 and Sec. 5 of Ref. 1 that can be improved by the method of the previous proof. Note that, in the forcing constructions for Theorems 4.5(b) and 4.6(c),(d), we cannot preserve a mad family added by a poset of the form $\mathbb{H}_\Omega$ because their constructions require that full generic dominating reals are added. For these items, it is enough to base their constructions on $\langle \mathcal{P}(\nu), \subseteq \rangle$ and start with $\dot{\mathbb{Q}}_{X,0} := \mathbb{C}_X$ for any $X \subseteq \nu$. In addition, by an argument similar to Rem. 5.9 of Ref. 1, it can be additionally forced within these items that $\mathfrak{a} = \mu$.

**Theorem 4.4.** *Let $\theta_0 \leq \theta_1 \leq \kappa \leq \nu \leq \lambda$ be as in the statement of Theorem 4.3. Then there is a ccc poset forcing $\mathrm{MA}_{<\theta_0}$, $add(\mathcal{N}) = \theta_0$, $cov(\mathcal{N}) = \theta_1$, $\mathfrak{b} = \mathfrak{a} = non(\mathcal{M}) = \kappa$, $cov(\mathcal{M}) = \mathfrak{d} = \nu$ and $non(\mathcal{N}) = \mathfrak{c} = \lambda$.*

**Proof.** The construction of the standard coherent system that forces the above is very similar to the one in the proof of Theorem 4.3. The only changes are that $S^{\mathbf{m}} := \emptyset$ and, for each $\xi \in [\lambda\rho, \lambda(\rho+1))$, $\dot{\mathbb{Q}}^{\mathbf{m}}_\xi := \mathbb{D}^{V_{\Delta(\xi),\xi}}$ when $\xi = \lambda\rho$, and $\dot{\mathbb{Q}}^{\mathbf{m}}_\xi := \dot{\mathbb{Q}}_{g(\varepsilon)}$ when $\xi = \lambda\rho + 1 + \varepsilon$ for some $\varepsilon < \lambda$. □

**Theorem 4.5.** *Assume that $\theta_0 \leq \kappa \leq \mu$ are uncountable regular cardinals, $\nu \leq \lambda$ are cardinals such that $\mu \leq \nu$, $\nu^{<\kappa} = \nu$ and $\lambda^{<\theta_0} = \lambda$. Then, for each of the statements below, there is a ccc poset forcing it.*

*(a)* $\mathrm{MA}_{<\theta_0}$, *$add(\mathcal{N}) = \theta_0$, $\mathfrak{b} = \mathfrak{a} = \kappa$, $cov(\mathcal{I}) = non(\mathcal{I}) = \mu$ for $\mathcal{I} \in \{\mathcal{M}, \mathcal{N}\}$, $\mathfrak{d} = \nu$ and $cof(\mathcal{N}) = \mathfrak{c} = \lambda$.*

*(b)* $\mathrm{MA}_{<\theta_0}$, *$add(\mathcal{N}) = \theta_0$, $cov(\mathcal{N}) = \kappa$, $add(\mathcal{M}) = cof(\mathcal{M}) = \mu$, $non(\mathcal{N}) = \nu$ and $cof(\mathcal{N}) = \mathfrak{c} = \lambda$.*

*(c)* $\mathrm{MA}_{<\theta_0}$, *$add(\mathcal{N}) = \theta_0$, $cov(\mathcal{N}) = \mathfrak{b} = \mathfrak{a} = \kappa$, $non(\mathcal{M}) = cov(\mathcal{M}) = \mu$, $\mathfrak{d} = non(\mathcal{N}) = \nu$ and $cof(\mathcal{N}) = \mathfrak{c} = \lambda$.*

*(d)* $\mathrm{MA}_{<\theta_0}$, *$add(\mathcal{N}) = \theta_0$, $cov(\mathcal{N}) = \mathfrak{b} = \mathfrak{a} = non(\mathcal{M}) = \kappa$, $cov(\mathcal{M}) = \mathfrak{d} = non(\mathcal{N}) = \nu$ and $cof(\mathcal{N}) = \mathfrak{c} = \lambda$.*

**Theorem 4.6.** *Assume that $\kappa \leq \mu$ are uncountable regular cardinals, $\nu \leq \lambda$ are cardinals such that $\mu \leq \nu$, $\nu^{<\kappa} = \nu$ and $\lambda^{\aleph_0} = \lambda$. Then, for each of the statements below, there is a ccc poset forcing it.*

*(a)* *$add(\mathcal{N}) = cov(\mathcal{N}) = \mathfrak{b} = \mathfrak{a} = \kappa$, $non(\mathcal{M}) = cov(\mathcal{M}) = \mu$, $\mathfrak{d} = non(\mathcal{N}) = cof(\mathcal{N}) = \nu$ and $\mathfrak{c} = \lambda$.*

(b) $add(\mathcal{N}) = \mathfrak{b} = \mathfrak{a} = \kappa$, $cov(\mathcal{I}) = non(\mathcal{I}) = \mu$ *for* $\mathcal{I} \in \{\mathcal{M}, \mathcal{N}\}$, $\mathfrak{d} = cof(\mathcal{N}) = \nu$ *and* $\mathfrak{c} = \lambda$.

(c) $add(\mathcal{N}) = cov(\mathcal{N}) = \kappa$, $add(\mathcal{M}) = cof(\mathcal{M}) = \mu$, $non(\mathcal{N}) = cof(\mathcal{N}) = \nu$ *and* $\mathfrak{c} = \lambda$.

(d) $add(\mathcal{N}) = \kappa$, $cov(\mathcal{N}) = add(\mathcal{M}) = cof(\mathcal{M}) = non(\mathcal{N}) = \mu$, $cof(\mathcal{N}) = \nu$ *and* $\mathfrak{c} = \lambda$.

(e) $add(\mathcal{N}) = non(\mathcal{M}) = \mathfrak{a} = \kappa$, $cov(\mathcal{M}) = cof(\mathcal{N}) = \nu$ *and* $\mathfrak{c} = \lambda$.

*Moreover, if* $\lambda^{<\kappa} = \lambda$, $\mathrm{MA}_{<\kappa}$ *can be forced additionally at each of the items above.*

**Remark 4.7.** This method can be used to force values (even singular) to other cardinal invariants different than those from Cichoń's diagram. For instance, the results in Sec. 3 of Ref. 23 can be adapted to the present approach.

## 5. Bonus track: Linkedness properties

The notions of $\sigma$-linked, $\sigma$-centered, $\sigma$-Fr-linked, etc., can be put into the following general framework.

**Definition 5.1.** Say that $\Gamma$ is a *linkedness property (for subsets of posets)* if $\Gamma$ is a class-function with domain the class of posets such that, for any poset $\mathbb{P}$, $\Gamma(\mathbb{P}) \subseteq \mathcal{P}(\mathbb{P})$.[c] We define the following notions for a linkedness property $\Gamma$.

(1) $\Gamma$ is *basic* if $[\mathbb{P}]^{\leq 1} \subseteq \Gamma(\mathbb{P})$ for any poset $\mathbb{P}$.
(2) $\Gamma$ is *conic* if, for any poset $\mathbb{P}$, $P \subseteq \mathbb{P}$ and $Q \in \Gamma(\mathbb{P})$, if $P \subseteq \{p \in \mathbb{P} : \exists q \in Q(q \leq p)\}$ and $Q = \{q \in Q : \exists p \in P(q \leq p)\}$ then $P \in \Gamma(\mathbb{P})$.
(3) $\Gamma$ is a *downwards forcing linkedness property* if, for any complete embedding $\iota : \mathbb{P} \to \mathbb{Q}$ between posets, if $P \subseteq \mathbb{P}$ and $\iota[P] \in \Gamma(\mathbb{Q})$ then $P \in \Gamma(\mathbb{P})$.
(4) $\Gamma$ is an *upwards forcing linkedness property* if, for any complete embedding $\iota : \mathbb{P} \to \mathbb{Q}$ between posets, if $P \in \Gamma(\mathbb{P})$ and $\iota{\restriction}P$ is 1-1 then $\iota[P] \in \Gamma(\mathbb{Q})$.
(5) A *forcing linkedness property* is a downwards and upwards linkedness forcing property.
(6) $\Gamma$ is *appropriate* if it is a basic conic forcing linkedness property.
(7) $\Gamma$ is *closed* if, for any poset $\mathbb{P}$ and $Q \subseteq Q' \subseteq \mathbb{P}$, if $Q' \in \Gamma(\mathbb{P})$ then $Q \in \Gamma(\mathbb{P})$.
(8) Let $\mu$ be an infinite cardinal. A poset $\mathbb{P}$ is $\mu$-$\Gamma$-*covered* if it can be covered by $\leq \mu$-many sets from $\Gamma(\mathbb{P})$. When $\mu = \aleph_0$ we just say $\sigma$-$\Gamma$-*covered.*
(9) Let $\theta$ be an infinite cardinal. A poset $\mathbb{P}$ is $\theta$-$\Gamma$-*Knaster* if, for any $P \subseteq \mathbb{P}$ of size $\theta$, there is some $Q \subseteq P$ of size $\theta$ such that $Q \in \Gamma(\mathbb{P})$. For $\theta = \aleph_1$ we just say $\Gamma$-*Knaster.*

[c]Concretely, $\Gamma$ is a formula $\varphi(x, y)$ (with fixed parameters) in the language of ZF and $\Gamma(\mathbb{P}) := \{Q \subseteq \mathbb{P} : \varphi(Q, \mathbb{P})\}$.

(10) If $\Lambda$ is another linkedness property, say that $\Lambda$ *is stronger than* $\Gamma$ (or $\Gamma$ *is weaker than* $\Lambda$), denoted by $\Lambda \Rightarrow \Gamma$, if $\Lambda(\mathbb{P}) \subseteq \Gamma(\mathbb{P})$ for any poset $\mathbb{P}$. We say that both properties are *equivalent*, denote by $\Lambda \Leftrightarrow \Gamma$, when one is weaker and stronger than the other.

**Remark 5.2.** Let $\Gamma$ be a linkedness property and $\mu$ an infinite cardinal.

(1) If $\Gamma$ is closed, then any $\mu$-$\Gamma$-covered poset is $\mu^+$-$\Gamma$-Knaster.
(2) If $\Gamma$ is a closed conic forcing linkedness property then '$\mu$-$\Gamma$-covered' is a property of forcing notions.
(3) If $\Gamma$ is appropriate and $\mu$ is regular, then '$\mu$-$\Gamma$-Knaster' is a property of forcing notions.
(4) If a property $\Lambda$ is stronger than $\Gamma$ then any $\mu$-$\Lambda$-covered (Knaster) poset is $\mu$-$\Gamma$-covered (Knaster).

To see that closed is necessary in (1) and (2), consider the property $\Gamma_0$ defined by $Q \in \Gamma_0(\mathbb{P})$ iff $Q$ is not an antichain in $\mathbb{P}$ of size $\geq 2$. Note that $\Gamma_0$ is an appropriate linkedness property, but it is not closed. A poset $\mathbb{P}$ is $\mu$-$\Gamma_0$-covered iff either it is an antichain in itself of size $\leq \mu$, or it is not an antichain in itself. Hence, though $\omega_1^{\leq 1}$ and $\omega_1^1$ (as posets of sequences in $\omega_1$ ordered by $\supseteq$) are forcing equivalent, the first is $\sigma$-$\Gamma_0$-covered while the second is not. On the other hand, any poset is $\mu$-$\Gamma_0$-Knaster iff it is $\mu$-cc.

**Example 5.3.** The following are appropriate closed linkedness properties.

- $\Gamma_{\eta\text{-cc}}$ (when $2 \leq \eta \leq \omega$): $\eta$-cc, that is, $Q \in \Gamma_{\eta\text{-cc}}(\mathbb{P})$ iff $Q$ does not contain antichains in $\mathbb{P}$ of size $\eta$.
- $\Gamma_{\text{bd-cc}}$: $n$-cc for some $2 \leq n < \omega$.
- $\Lambda_n$ (when $2 \leq n < \omega$): $n$-linked.
- $\Lambda_\omega$: centered.
- $\Gamma_{\text{cone}}$: Say that $Q \in \Gamma_{\text{cone}}(\mathbb{P})$ if there is some $q \in \mathbb{P}$ such that $\forall p \in Q(q \leq^* p)$.[d]
- $\Lambda_{\text{Fr}}$: Frechet-linked.

It is clear that

$$\Gamma_{\text{cone}} \Rightarrow \Lambda_\omega \Rightarrow \Lambda_{n+1} \Rightarrow \Lambda_n \Rightarrow \Lambda_2 \Rightarrow \Gamma_{n\text{-cc}} \Rightarrow \Gamma_{n+1\text{-cc}} \Rightarrow \Gamma_{\text{bd-cc}} \Rightarrow \Gamma_{\omega\text{-cc}}$$

for $2 \leq n < \omega$ (actually $\Gamma_{2\text{-cc}} \Leftrightarrow \Lambda_2$). Also $\Gamma_{\text{cone}} \Rightarrow \Lambda_{\text{Fr}} \Rightarrow \Gamma_{\omega\text{-cc}}$ and $\Lambda_2 \Rightarrow \Gamma_0$. These properties determine some well-known forcing properties, for example, '$\mu$-$\Lambda_\omega$-covered' means '$\mu$-centered', '$\mu$-$\Lambda_\omega$-Knaster' means 'precaliber $\mu$', '$\mu$-$\Lambda_2$-covered' means '$\mu$-linked', '$\mu$-$\Lambda_2$-Kanster' is the typical $\mu$-Knaster property, and '$\mu$-$\Lambda_{\text{Fr}}$-covered' is what we defined as $\mu$-Fr-linked in Definition 3.24.

---

[d] Here, $\leq^*$ denotes the separable order of $\mathbb{P}$, that is, $q \leq^* p$ iff any condition compatible with $q$ in $\mathbb{P}$ is compatible with $p$.

By an argument similar to Thm. 2.4 of Ref. 22 it can be proved that, if $\theta$ is regular, then any $Q \in \Gamma_{\omega\text{-cc}}(\mathbb{P})$ of size $\theta$ contains a 2-linked subset of the same size, thus $\theta$-$\Lambda_2$-Knaster is equivalent to $\theta$-$\Gamma_{\omega\text{-cc}}$-Knaster. On the other hand, Todorčević[24,25] constructed a $\Lambda_\omega$-Knaster (i.e. $\aleph_1$-precaliber) poset that is not $\sigma$-$\Gamma_{\omega\text{-cc}}$-covered (i.e. $\sigma$-finite-cc) and, under $\mathfrak{b} = \aleph_1$, a $\sigma$-$\Lambda_n$-covered poset that is not $\Lambda_{n+1}$-Knaster. Todorčević[26] and Thümmel[27] constructed $\sigma$-finite-cc posets that are not $\sigma$-$\Gamma_{\text{bd-cc}}$-covered (i.e. $\sigma$-bounded-cc).

It is clear that any Boolean algebra that admits a strictly positive fam (finitely additive measure) is $\sigma$-bounded-cc and, by Lemma 3.29, any complete Boolean algebra that admits a strictly positive $\sigma$-additive measure is $\sigma$-$\Lambda_{\text{Fr}}$-covered. Note that $\mathbb{D}$ is a $\sigma$-centered poset that admits a strictly positive fam but it is not $\sigma$-Frechet-linked (otherwise it would contradict Theorem 3.30), and $\mathbb{B}_{\mathfrak{c}^+}$ is a complete Boolean algebra that admits a strictly positive $\sigma$-additive measure but it is not $\sigma$-linked (see Dow and Steprans[28]).

Note that any poset is $\mu$-$\Gamma_{\text{cone}}$-covered iff it is forcing equivalent to a poset of size $\leq \mu$, and the notion $\theta$-$\Gamma_{\text{cone}}$-Knaster is equivalent to $(\theta,\theta)$-caliber. The property $\Gamma_{\text{cone}}$ is the strongest of all the appropriate linkedness properties with respect to the class of separative posets, so it is morally the strongest appropriate linkedness property.

With the exception of $\Lambda_{\text{Fr}}$, all the other properties (including $\Gamma_0$) are absolute for transitive models of ZFC. Recall from Ref. 29 that there is a $\omega^\omega$-bounding proper poset that forces that $\mathbb{B}^V$ (random forcing from the ground model) adds a dominating real, so this poset, though $\sigma$-Frechet-linked in the ground model, is not forced to be so.

**Example 5.4.** In Ref. 30 (work in progress) we discuss properties stronger than $\Lambda_{\text{Fr}}$. Given a free filter $F$ of subsets of $\omega$, define $\Lambda_F$ such that, for any poset $\mathbb{P}$, $Q \in \Lambda_F(\mathbb{P})$ iff for any sequence $\langle p_n : n < \omega\rangle$ in $Q$ there is some $q \in \mathbb{P}$ that forces $\{n < \omega : p_n \in \dot{G}\} \in F^+$ (that is, it intersects every member of $F$), which is an appropriate closed linkedness property. Also define $\Lambda_{\text{uf}}(\mathbb{P}) := \bigcap\{\Lambda_F(\mathbb{P}) : F \text{ free filter}\}$. It is clear that $\Lambda_{\text{uf}} \Rightarrow \Lambda_{F'} \Rightarrow \Lambda_F \Rightarrow \Lambda_{\text{Fr}}$ whenever $F \subseteq F'$. Even more, we have the following equivalences.

**Lemma 5.5.**

*(a) For any free filter $F$ in $\omega$ generated by $< \mathfrak{p}$-many sets, $\Lambda_F \Leftrightarrow \Lambda_{\text{Fr}}$.*
*(b) For any $\mathfrak{p}$-cc poset $\mathbb{P}$, $\Lambda_{\text{uf}}(\mathbb{P}) = \Lambda_{\text{Fr}}(\mathbb{P})$.*

**Proof.** Both items can be proved simultaneously. Let $\mathbb{P}$ a poset, $F$ a free filter on $\omega$ and assume that either $F$ is generated by $< \mathfrak{p}$-many sets or $\mathbb{P}$ is $\mathfrak{p}$-cc. It is enough to show that $\Lambda_{\text{Fr}}(\mathbb{P}) \subseteq \Lambda_F(\mathbb{P})$. Assume that $Q \subseteq \mathbb{P}$ is Fr-linked but not in

$\Lambda_F(\mathbb{P})$, so there are a countable sequence $\langle p_n : n < \omega\rangle$ in $Q$, a maximal antichain $A \subseteq \mathbb{P}$ and a sequence $\langle a_r : r \in A\rangle$ in $F$ such that each $r \in A$ is incompatible with $p_n$ for every $n \in a_r$. In any of the two cases of the hypothesis, it can be concluded that there is some pseudo-intersection $a \in [\omega]^{\aleph_0}$ of $\langle a_r : r \in A\rangle$. Hence each $r \in A$ forces $p_n \in \dot{G}$ for only finitely many $n \in a$, which means that $\mathbb{P}$ forces the same. However, as $Q$ is Fr-linked, there is some $q \in \mathbb{P}$ that forces $\exists^\infty n \in a(p_n \in \dot{G})$, a contradiction. □

As a consequence of the previous result and Lemma 3.29, $\mathbb{E}$ and any complete Boolean algebra that admits a strictly positive $\sigma$-additive measure are $\sigma$-$\Lambda_{\mathrm{uf}}$-covered. In Ref. 30 we show that $\Lambda_{\mathrm{Fr}}$-Knaster posets do not add dominating reals. Hence, $\mathbb{D}$ becomes an example of a $\sigma$-centered poset that is not $\Lambda_{\mathrm{Fr}}$-Knaster. Also note that these properties associated with filters are not absolute.

To finish, in the general context of Definition 5.1, we provide simple conditions to understand when the FS iteration of $\theta$-$\Gamma$-Knaster (or covered) posets is $\theta$-$\Gamma$-Knaster (or covered), likewise for FS products. These conditions are summarized in the following definition, and they just represent facts extracted from the typical proofs of the iteration results for $\sigma$-linked and $\Lambda_2$-Knaster. At the end of this section, we relate the linkedness properties presented so far with the notions below.

**Definition 5.6.** Let $\Gamma$ be a linkedness property.

(1) $\Gamma$ is *productive* if, for any posets $\mathbb{P}$ and $\mathbb{Q}$, and for any $Q \subseteq \mathbb{P} \times \mathbb{Q}$, if $\mathrm{dom} Q \in \Gamma(\mathbb{P})$ and $\mathrm{ran} Q \in \Gamma(\mathbb{Q})$ then $Q \in \Gamma(\mathbb{P} \times \mathbb{Q})$.
(2) $\Gamma$ is *FS-productive* if for any sequence $\langle \mathbb{P}_i : i \in I\rangle$ of posets, $n < \omega$ and any $Q \subseteq \{p \in \prod_{i\in I}^{<\omega} \mathbb{P}_i : |\mathrm{dom} p| \leq n\}$ (FS product), if $\{\mathrm{dom} p : p \in Q\}$ forms a $\Delta$-system with root $s$ and $\{p{\restriction}s : p \in Q\} \in \Gamma(\prod_{i\in s} \mathbb{P}_i)$ then $Q \in \Gamma(\prod_{i\in I}^{<\omega} \mathbb{P}_i)$.
(3) $\Gamma$ is *strongly productive* if, for any sequence $\langle \mathbb{P}_i : i \in I\rangle$ of posets, $Q \in \Gamma(\prod_{i\in I}^{<\omega} \mathbb{P}_i)$ whenever
   (i) there is some $n < \omega$ such that $Q \subseteq \{p \in \prod_{i\in I}^{<\omega} \mathbb{P}_i : |\mathrm{dom} p| \leq n\}$ and
   (ii) $\{p(i) : p \in Q\} \in \Gamma(\mathbb{P}_i)$ for any $i \in I$.
(4) $\Gamma$ is *two-step iterative* if, for any poset $\mathbb{P}$, any $\mathbb{P}$-name $\dot{\mathbb{Q}}$ of a poset and any $Q \subseteq \mathbb{P} * \dot{\mathbb{Q}}$, if $\mathrm{dom} Q \in \Gamma(\mathbb{P})$ and $\mathbb{P}$ forces that $\{\dot{q} : \exists p \in \dot{G}((p, \dot{q}) \in Q)\} \in \Gamma(\dot{\mathbb{Q}})$, then $Q \in \Gamma(\mathbb{P} * \dot{\mathbb{Q}})$.
(5) $\Gamma$ is *direct-limit iterative* if whenever
   (i) $\theta$ is an uncountable regular cardinal,
   (ii) $\langle \mathbb{P}_\alpha : \alpha \leq \delta\rangle$ is an increasing $\lessdot$-sequence of posets such that $\mathrm{cf}(\delta) = \theta$ and $\mathbb{P}_\gamma = \mathrm{limdir}_{\alpha<\gamma} \mathbb{P}_\alpha$ for any limit $\gamma \leq \delta$,
   (iii) $f : \theta \to \delta$ is increasing,

(iv) $Q = \{p_\xi : \xi < \theta\} \subseteq \mathbb{P}_\delta$ such that each $p_\xi \in \mathbb{P}_{f(\xi+1)}$, and
(v) for each $\xi < \theta$, $r_\xi \in \mathbb{P}_{f(\xi)}$ is a reduction of $p_\xi$,

if there is some $\gamma < \delta$ such that $r_\xi \in \mathbb{P}_\gamma$ for every $\xi < \theta$ and $\{r_\xi : \xi < \theta\} \in \Gamma(\mathbb{P}_\gamma)$, then $Q \in \Gamma(\mathbb{P}_\delta)$.

(6) $\Gamma$ is *strongly iterative* if, for any FS iteration $\mathbb{P}_\delta = \langle \mathbb{P}_\alpha, \dot{\mathbb{Q}}_\alpha : \alpha < \delta \rangle$, $Q \in \Gamma(\mathbb{P}_\delta)$ whenever

(i) there is some $n < \omega$ such that $Q \subseteq \{p \in \mathbb{P}_\delta : |\mathrm{dom} p| \leq n\}$ and
(ii) for any $\alpha < \delta$, if $Q{\restriction}\alpha \in \Gamma(\mathbb{P}_\alpha)$ then

$$\Vdash_{\mathbb{P}_\alpha} \{p(\alpha) : p{\restriction}(\alpha+1) \in Q{\restriction}(\alpha+1),\ p{\restriction}\alpha \in \dot{G}_\alpha\} \in \Gamma(\dot{\mathbb{Q}}_\alpha).$$

Note that any strongly productive linkedness property is both productive and FS-productive. On the other hand, if $\Gamma$ is strongly iterative, $\mathbb{P}_\delta = \langle \mathbb{P}_\alpha, \dot{\mathbb{Q}}_\alpha : \alpha < \delta \rangle$ is a FS iteration and $Q \subseteq \{p \in \mathbb{P}_\delta : |\mathrm{dom} p| \leq n\}$ satisfies (6)(i),(ii) then $Q{\restriction}\alpha \in \Gamma(\mathbb{P}_\alpha)$ for any $\alpha \leq \delta$. It is clear that any strongly iterative property is two-step iterative and satisfies a weak form of direct-limit iterative (which we leave implicit in the proof of Corollary 5.10).

The following is a general result about FS products.

**Theorem 5.7.** *Let $\mu$ be an infinite cardinal, $\theta$ an uncountable regular cardinal, and let $\Gamma$ be an appropriate linkedness property.*

*(a) If $\Gamma$ is productive then any finite product of $\mu$-$\Gamma$-covered sets is $\mu$-$\Gamma$-covered.*
*(b) If $\Gamma$ is closed and productive then and any finite product of $\theta$-$\Gamma$-Knaster posets is $\theta$-$\Gamma$-Knaster.*
*(c) If $\Gamma$ is FS-productive and $\langle \mathbb{P}_i : i \in I \rangle$ is a sequence of $\theta$-$\Gamma$-Knaster posets, then $\prod^{<\omega}_{i\in I} \mathbb{P}_i$ is $\theta$-$\Gamma$-Knaster iff $\prod_{i\in s} \mathbb{P}_i$ is $\theta$-$\Gamma$-Knaster for every $s \in [I]^{<\aleph_0}$.*
*(d) If $\Gamma$ is strongly productive, $\langle \mathbb{P}_i : i \in I \rangle$ is a sequence of $\mu$-$\Gamma$-covered posets and $|I| \leq 2^\mu$, then $\prod^{<\omega}_{i\in I} \mathbb{P}_i$ is $\mu$-$\Gamma$-covered.*

**Proof.** Items (a),(b) are easy and (c) follows by a classical $\Delta$-system argument. Item (d) uses the following result.

**Lemma 5.8 (Engelking and Karłowicz[31]).** *If $\mu$ is an infinite cardinal and $I$ is a set of size $\leq 2^\mu$ then there exists a set $H \subseteq \mu^I$ of size $\leq \mu$ such that any finite partial function from $I$ to $\mu$ is extended by some member of $H$.*

For each $i \in I$ choose a sequence $\langle Q_{i,\zeta} : \zeta < \mu \rangle$ of non-empty sets in $\Gamma(\mathbb{P}_i)$ that covers $\mathbb{P}_i$. Let $H$ be as in Lemma 5.8. By Definition 5.6(3), the set $Q^*_{h,n} := \{p \in \prod^{<\omega}_{i\in I} Q_{i,h(i)} : |\mathrm{dom} p| = n\}$ is in $\Gamma(\prod^{<\omega}_{i\in I} \mathbb{P}_i)$ and it is clear that $\langle Q^*_{h,n} : h \in H,\ n < \omega \rangle$ covers $\prod^{<\omega}_{i\in I} \mathbb{P}_i$. □

Now we turn to a general result about FS iterations.

**Theorem 5.9.** *Let $\mu$ be an infinite cardinal, $\theta$ an uncountable regular cardinal, and let $\Gamma$ be an appropriate linkedness property.*

*(a) If $\Gamma$ is two-step iterative, $\mathbb{P}$ is $\mu$-$\Gamma$-covered, and $\mathbb{P}$ forces that $\Gamma$ is basic and that $\dot{\mathbb{Q}}$ is a $|\mu|$-$\Gamma$-covered poset, then $\mathbb{P} * \dot{\mathbb{Q}}$ is $\mu$-$\Gamma$-covered.*

*(b) Let $\mathbb{P}$ be a $\theta$-$\Gamma$-Knaster poset and let $\dot{\mathbb{Q}}$ be a $\mathbb{P}$-name of a $\theta$-$\Gamma$-Knaster poset. Assume in addition that either $\Gamma = \Gamma_0$, or $\Gamma$ is closed, two-step iterative, $\mathbb{P}$ is $\theta$-cc and $\mathbb{P}$ forces that $\Gamma$ is closed and basic. Then $\mathbb{P} * \dot{\mathbb{Q}}$ is $\theta$-$\Gamma$-Knaster.*

*(c) If $\Gamma$ is direct-limit iterative, $\delta$ is a limit ordinal and $\langle \mathbb{P}_\alpha : \alpha \leq \delta \rangle$ is an increasing $\lessdot$-sequence of $\theta$-$\Gamma$-Knaster posets such that $\mathbb{P}_\gamma = \mathrm{limdir}_{\alpha<\gamma}\mathbb{P}_\alpha$ for any limit $\gamma \leq \delta$, then $\mathbb{P}_\delta$ is $\theta$-$\Gamma$-Knaster.*

*(d) If $\Gamma$ is strongly iterative and any poset forces that $\Gamma$ is still basic, then any FS iteration of length $< (2^\mu)^+$ of $\mu$-$\Gamma$-covered posets is $\mu$-$\Gamma$-covered.*

**Proof.** To see (a), it is enough to show that, for any poset $\mathbb{P}$ that forces $\Gamma$ to be still basic and any $\mathbb{P}$-name $\dot{\mathbb{Q}}$ for a poset, if $P \in \Gamma(\mathbb{P})$ and $\dot{Q}$ is a $\mathbb{P}$-name of a non-empty set in $\Gamma(\dot{\mathbb{Q}})$, then $P * \dot{Q} := \{(p, \dot{q}) \in \mathbb{P} * \dot{\mathbb{Q}} : p \in P \text{ and } p \Vdash \dot{q} \in \dot{Q}\}$ is in $\Gamma(\mathbb{P} * \dot{\mathbb{Q}})$. Clearly $\mathrm{dom}(P * \dot{Q}) = P$. On the other hand, any $p_0 \in P$ forces that $\dot{R} := \{\dot{q} : \exists p \in \dot{G}((p, \dot{q}) \in P * \dot{Q})\} = \dot{Q}$, so $\mathbb{P}$ forces that $\dot{R}$ is either $\dot{Q}$ or the empty set, so $\dot{R} \in \Gamma(\dot{\mathbb{Q}})$. By Definition 5.6(4), it follows that $P * \dot{Q} \in \Gamma(\mathbb{P} * \dot{\mathbb{Q}})$.

Item (b) is well-known when $\Gamma = \Gamma_0$, so assume that $\Gamma$ is closed, two-step iterative, $\mathbb{P}$ is $\theta$-cc and $\mathbb{P}$ forces that $\Gamma$ is closed and basic. Let $\{(p_\alpha, \dot{q}_\alpha) : \alpha < \theta\} \subseteq \mathbb{P} * \dot{\mathbb{Q}}$. As $\mathbb{P}$ forces $\dot{\mathbb{Q}}$ to be $\theta$-$\Gamma$-Knaster, there is some $\mathbb{P}$-name $\dot{K}$ for a subset of $\theta$ such that $\mathbb{P}$ forces that "whenever $|\{\alpha < \theta : p_\alpha \in \dot{G}\}| = \theta$, $\dot{K} \subseteq \{\alpha < \theta : p_\alpha \in \dot{G}\}$ has size $\theta$ and $\{\dot{q}_\alpha : \alpha \in \dot{K}\} \in \Gamma(\dot{\mathbb{Q}})$, otherwise $\dot{K} = \emptyset$". Set $K_0 := \{\alpha < \theta : \nVdash \alpha \notin \dot{K}\}$, which has size $\theta$ (otherwise $\mathbb{P}$ would force that $|\{\alpha < \theta : p_\alpha \in \dot{G}\}| < \theta$, which contradicts that $\mathbb{P}$ is $\theta$-cc). For each $\alpha \in K_0$ choose an $r_\alpha \leq p_\alpha$ that forces $\alpha \in \dot{K}$. Hence, there is some $K_1 \subseteq K_0$ of size $\theta$ such that $\{r_\alpha : \alpha \in K_1\} \in \Gamma(\mathbb{P})$.

As $\Gamma$ is conic and $r_\alpha \leq p_\alpha$ for any $\alpha \in K_1$, it is enough to show that $Q := \{(r_\alpha, \dot{q}_\alpha) : \alpha \in K_1\} \in \Gamma(\mathbb{P} * \dot{\mathbb{Q}})$. It is clear that $\mathrm{dom} Q \in \Gamma(\mathbb{P})$. On the other hand, $\mathbb{P}$ forces that $\dot{R} := \{\dot{q}_\alpha : r_\alpha \in \dot{G},\ \alpha \in K_1\} \subseteq \{\dot{q}_\alpha : \alpha \in \dot{K}\}$, so $\dot{R} \in \Gamma(\dot{\mathbb{Q}})$ because $\Gamma$ is closed. As $\Gamma$ is two-step iterative, we are done.

Now we show (c). Let $\{p_\xi : \xi < \theta\} \subseteq \mathbb{P}_\delta$. If $\mathrm{cf}(\delta) \neq \theta$ then there are some $\alpha < \delta$ and a $K \subseteq \theta$ of size $\theta$ such that $\{p_\xi : \xi \in K\} \subseteq \mathbb{P}_\alpha$, so there is some $K' \subseteq K$ of size $\theta$ such that $\{p_\xi : \xi \in K\} \in \Gamma(\mathbb{P}_\alpha)$ (note that $\Gamma(\mathbb{P}_\alpha) \subseteq \Gamma(\mathbb{P}_\delta)$ because $\Gamma$ is appropriate). Assume that $\mathrm{cf}(\delta) = \theta$ and choose an increasing continuous cofinal function $g : \theta \to \delta$ such that each $g(\xi)$ is a limit ordinal. For each $\xi < \theta$ choose a reduction $r_\xi \in \mathbb{P}_{g(\xi)}$ of $p_\xi$. As $g(\xi)$ is limit, there is some $h(\xi) < \xi$ such that $r_\xi \in \mathbb{P}_{g(h(\xi))}$. Hence, by Fodor's Lemma, there is some stationary set $S \subseteq \theta$

such that $h[S] = \{\eta\}$ for some $\eta < \theta$, that is, $r_\xi \in \mathbb{P}_{g(\eta)}$ for every $\xi \in S$. By recursion define $j : \theta \to S$ increasing such that $j(0) > \eta$ and, for any $\zeta < \theta$, $p_{j(\zeta)} \in \mathbb{P}_{g(j(\zeta+1))}$. As $\mathbb{P}_{g(\eta)}$ is $\theta$-$\Gamma$-Knaster, there is some $K \subseteq \theta$ of size $\theta$ such that $\{r_{j(\zeta)} : \zeta \in K\} \in \Gamma(\mathbb{P}_{g(\eta)})$. Let $i : \theta \to K$ be the increasing enumeration of $K$.

Put $f := g \circ j \circ i$ and $\gamma := g(\eta)$. Note that $\{p_{j(i(\beta))} : \beta < \theta\}$, $\{r_{j(i(\beta))} : \beta < \theta\}$, $f$ and $\gamma$ satisfy the conditions of Definition 5.6(5) so, as $\Gamma$ is direct-limit iterative, $\{p_{j(i(\beta))} : \beta < \theta\} \in \Gamma(\mathbb{P}_\delta)$.

To finish, we show (d). Let $\delta < (2^\mu)^+$ and let $\mathbb{P}_\delta = \langle \mathbb{P}_\alpha, \dot{\mathbb{Q}}_\alpha : \alpha < \delta\rangle$ be a FS iteration of $\mu$-$\Gamma$-covered sets. For each $\alpha < \delta$ choose a sequence $\langle \dot{Q}_{\alpha,\zeta} : \zeta < \mu\rangle$ of $\mathbb{P}_\alpha$-names of sets in $\Gamma(\dot{\mathbb{Q}}_\alpha)$ that is forced to cover $\dot{\mathbb{Q}}_\alpha$. For $\alpha \leq \delta$ define $\mathbb{P}^*_\alpha \subseteq \mathbb{P}_\alpha$ such that $p \in \mathbb{P}^*_\alpha$ iff $p \in \mathbb{P}_\alpha$ and, for any $\xi \in \mathrm{dom} p$, there is some $\zeta < \mu$ such that $p{\restriction}\xi \Vdash_{\mathbb{P}_\xi} p(\xi) \in \dot{Q}_{\xi,\zeta}$. By induction it can be proved that $\mathbb{P}^*_\alpha$ is dense in $\mathbb{P}_\alpha$.

Now choose $H$ as in Lemma 5.8 and, for each $h \in H$ and $n < \omega$, define $Q_{h,n}$ as the set of $p \in \mathbb{P}^*_\delta$ such that $|\mathrm{dom} p| \leq n$ and, for any $\alpha \in \mathrm{dom} p$, $p{\restriction}\alpha \Vdash_{\mathbb{P}_\alpha} p(\alpha) \in \dot{Q}_{\alpha,h(\alpha)}$. It is clear that $\langle Q_{h,n} : h \in H,\ n < \omega\rangle$ covers $\mathbb{P}^*_\delta$, so it remains to show that $Q_{h,n} \in \Gamma(\mathbb{P}_\delta)$. If $\alpha < \delta$ and $Q_{h,n}{\restriction}\alpha \in \Gamma(\mathbb{P}_\alpha)$ then a similar argument as in (a) shows that $\mathbb{P}_\alpha$ forces $\{p(\alpha) : p{\restriction}(\alpha+1) \in Q_{h,n}{\restriction}(\alpha+1),\ p{\restriction}\alpha \in \dot{G}\} \in \Gamma(\dot{\mathbb{Q}}_\alpha)$. Therefore, as $\Gamma$ is strongly iterative, $Q_{h,n} \in \Gamma(\mathbb{P}_\delta)$. □

**Corollary 5.10.** *Let $\theta$ an uncountable regular cardinal and assume that $\Gamma$ is either*

*(i) $\Gamma_0$ or*

*(ii) a closed appropriate linkedness property that is closed and basic in any generic extension, and that it is either strongly iterative, or two-step and direct-limit iterative.*

*Then any FS iteration of $\theta$-$\Gamma$-Knaster $\theta$-cc posets is $\theta$-$\Gamma$-Knaster.*

**Proof.** Case (i) and case (ii) when $\Gamma$ is two-step and direct-limit iterative follow directly from Theorem 5.9. Case (ii) when $\Gamma$ is strongly iterative is a bit similar but requires a bit more work. If $\langle \mathbb{P}_\alpha, \dot{\mathbb{Q}}_\alpha : \alpha < \delta\rangle$ is a FS iteration of $\theta$-$\Gamma$-Knaster $\theta$-cc posets, it is enough to show by induction on $\alpha \leq \delta$ that, for any sequence $\langle p_\beta : \beta < \theta\rangle$ in $\mathbb{P}_\alpha$ there are some $K \subseteq \theta$ of size $\theta$ and some sequence $\langle r_\beta : \beta \in K\rangle$ in $\mathbb{P}_\alpha$ that satisfies (i) and (ii) of Definition 5.6(6) (with respect to $\mathbb{P}_\alpha$) and such that $r_\beta \leq p_\beta$ for any $\beta \in K$. The successor step is exactly like the proof of Theorem 5.9(b) and the limit step is very similar to Theorem 5.9(c). We just look at the case $\mathrm{cf}(\alpha) = \theta$. Let $\langle p_\beta : \beta < \theta\rangle$ be a sequence in $\mathbb{P}_\alpha$. Exactly like in the proof of Theorem 5.9(c), we can find a $\gamma < \alpha$, a $K_0 \subseteq \theta$ of size $\theta$ and an increasing function $f : K_0 \to \alpha \smallsetminus \gamma$ such that, for each $\beta \in K_0$, $p_\beta{\restriction}f(\beta) \in \mathbb{P}_\gamma$ and $p_\beta \in \mathbb{P}_{f(\beta+1)}$. Even more, we may assume that there is some $n_1 < \omega$ such that $|\mathrm{dom} p_\beta \smallsetminus f(\beta)| = n_1$ for all $\beta \in K$. By the inductive hypothesis, there are $K \subseteq K_0$ of size $\theta$ and a sequence

$\langle r^0_\beta : \beta \in K\rangle$ of conditions in $\mathbb{P}_\gamma$ that satisfies (i) (for some $n_0 < \omega$) and (ii) of Definition 5.6(6) (with respect to $\mathbb{P}_\gamma$) and such that $r^0_\beta \leq p_\beta \restriction f(\beta)$ for any $\beta \in K$. The set $\{r^0_\beta \cup p_\beta \restriction (f(\beta+1) \setminus f(\beta)) : \beta \in K\}$ is as required. □

**Remark 5.11.** Table 1 illustrates which productive or iterative notions are satisfied by the linkedness properties discussed so far.

Table 1. A circle means that the linkedness property satisfies the corresponding productive or iterative notion (see Definition 5.6) on the left, an × means that such notion is not satisfied, and a question mark means unclear.

| | $\Gamma_0$ | $\Gamma_{n\text{-cc}}$ $(3 \leq n)$ | $\Gamma_{\omega\text{-cc}}$ | $\Gamma_{\text{bd-cc}}$ | $\Lambda_n$ $(2 \leq n)$ | $\Lambda_\omega$ | $\Gamma_{\text{cone}}$ | $\Lambda_{\text{Fr}}$ | $\Lambda_F$ | $\Lambda_{\text{uf}}$ |
|---|---|---|---|---|---|---|---|---|---|---|
| Prod. | × | × | ○ | ○ | ○ | ○ | ○ | ? | ? | ○ |
| FS Prod. | ○ | ○ | ○ | ○ | ○ | ○ | × | ○ | ○ | ○ |
| Str. Prod. | × | × | ○ | × | ○ | ○ | × | ? | ? | ? |
| Two-step it. | ○ | × | × | × | ○ | ○ | ○ | ○ | ? | ○ |
| Dir.-lim. it. | ○ | ○ | ○ | ○ | ○ | ○ | × | ○ | ○ | ○ |
| Str. it | ○ | × | × | × | ○ | ○ | × | ○ | ? | ? |

We explain some of the facts indicated in the table. First, we show that $\Gamma_{\omega\text{-cc}}$ is strongly productive. Let $Q \subseteq \{p \in \prod^{<\omega}_{i\in I} \mathbb{P}_i : |\text{dom}p| \leq n\}$ and assume that $\{p(i) : p \in Q\} \in \Gamma_{\omega\text{-cc}}(\mathbb{P}_i)$ for every $i \in I$. Fix a countable sequence $\langle p_k : k < \omega\rangle$ in $Q$. As the size of the domains of the members of the sequence are bounded by $n$, we can find a $W \in [\omega]^{\aleph_0}$ such that $\langle p_k : k \in W\rangle$ forms a $\Delta$-system with root $R$. By Ramsey's Theorem it can be proved that $\{p_k \restriction R : k \in W\}$ is not an antichain, so neither is $\langle p_k : k < \omega\rangle$.

Ramsey's Theorem also implies that $\Gamma_{\text{bd-cc}}$ is productive. However, it is not FS-productive (consider the set of conditions with domain of size 1 of the FS product $\prod^{<\omega}_{i\in\omega} \mathbb{P}_i$ where each $\mathbb{P}_i$ is an antichain of size $i+1$). The following example indicates that both $\Gamma_{\omega\text{-cc}}$ and $\Gamma_{\text{bd-cc}}$ are not two-step iterative. Consider $P := \{(p_n, \dot{q}_n) : n < \omega\} \subseteq \mathbb{C} * \dot{\mathbb{C}}$ such that each $p_n$ is a sequence of zeros of length $n+1$, $p_n \Vdash \dot{q}_n = \langle n\rangle$ but $r \Vdash \dot{q}_n = \langle\,\rangle$ for any $r \in \mathbb{C}$ incompatible with $p_n$. Though $P$ is an antichain in $\mathbb{C} * \dot{\mathbb{C}}$, dom$P$ is centered and $\mathbb{C}$ forces that $\{\dot{q}_n : p_n \in \dot{G},\ n < \omega\}$ is a finite antichain.

As $\mathbb{C}_{\omega_1}$ is uncountable and it has not $(\aleph_1, \aleph_0)$-precaliber, Theorems 5.7(c),(d) and 5.9(c),(d) cannot be applied to $\Gamma_{\text{cone}}$. Hence, $\Gamma_{\text{cone}}$ does not satisfy the properties indicated with × in the table.

It is unclear whether $\Lambda_F$ is productive in general, but it is proved in Ref. 30 that it is when $F$ is an ultrafilter. Therefore, $\Lambda_{\text{uf}}$ is productive. By a $\Delta$-system argument, $\Lambda_{\text{Fr}}$ is strongly iterative. To see this, assume that $\mathbb{P}_\delta$, $Q$ and $n < \omega$ satisfy the conditions in Definition 5.6(6). It is enough to show, by induction on $\alpha \leq \delta$, that $Q{\restriction}\alpha \in \Lambda_{\mathrm{Fr}}(\mathbb{P}_\alpha)$. Since $\Lambda_{\mathrm{Fr}}$ is two-step iterative, we only need to prove the limit step. Let $\langle p_k : k < \omega \rangle$ be a sequence in $Q{\restriction}\alpha$. As $|\mathrm{dom} p_k| \leq n$ for any $k < \omega$, there is some infinite $W \subseteq \omega$ such that $\{\mathrm{dom} p_k : k \in W\}$ forms a $\Delta$-system with root $R$, and there is some $\xi < \alpha$ such that $R \subseteq \xi$. By the inductive hypothesis, $Q{\restriction}\xi \in \Lambda_{\mathrm{Fr}}(\mathbb{P}_\xi)$, so there is some $q \in \mathbb{P}_\xi$ that forces $\exists^\infty k \in W(p_k{\restriction}\xi \in \dot{G}_\xi)$. Therefore, it can be proved that $q$ forces (in $\mathbb{P}_\alpha$) that $\exists^\infty k \in W(p_k \in \dot{G}_\alpha)$.

By a similar argument, if $\Lambda_{\mathrm{Fr}}$ were productive then it would be strongly productive. In particular, by Lemma 5.5, $\Lambda_{\mathrm{Fr}}$ restricted to the class of Knaster posets is strongly productive, so Theorem 5.7 is valid for $\Lambda_{\mathrm{Fr}}$ for FS products of Knaster posets (or just FS products that have ccc).

## Acknowledgments

This work was supported by grant no. IN201711, Dirección Operativa de Investigación, Institución Universitaria Pascual Bravo, and by the Grant-in-Aid for Early Career Scientists 18K13448, Japan Society for the Promotion of Science.

The author would like to thank Miguel Cardona for the very useful discussions that helped this work to take its final form. He is also very grateful with the anonymous referee for his/her very useful comments, specially for asking whether any $\sigma$-Fr-linked poset is **Lc**-good (which is answered negative with a counterexample in Remark 3.31).

## References

1. V. Fischer, S. D. Friedman, D. A. Mejía and D. C. Montoya, Coherent systems of finite support iterations, *J. Symbolic Logic* **83**, 208 (2018).
2. J. Brendle and V. Fischer, Mad families, splitting families and large continuum, *J. Symbolic Logic* **76**, 198 (2011).
3. J. Brendle, Larger cardinals in Cichoń's diagram, *J. Symbolic Logic* **56**, 795 (1991).
4. A. Blass and S. Shelah, Ultrafilters with small generating sets, *Israel J. Math.* **65**, 259 (1989).
5. D. A. Mejía, Matrix iterations and Cichon's diagram, *Arch. Math. Logic* **52**, 261 (2013).
6. D. A. Mejía, Models of some cardinal invariants with large continuum, *Kyōto Daigaku Sūrikaiseki Kenkyūsho Kōkyūroku* **1851**, 36 (2013).
7. A. Dow and S. Shelah, On the cofinality of the splitting number, *Indag. Math.* **29**, 382 (2018).
8. M. Goldstern, J. Kellner and S. Shelah, Cichoń's maximum. arXiv:1708.03691.
9. M. Goldstern, D. A. Mejía and S. Shelah, The left side of Cichoń's diagram, *Proc. Amer. Math. Soc.* **144**, 4025 (2016).
10. J. Kellner, S. Shelah and A. Tănasie, Another ordering of the ten cardinal characteristics in Cichoń's diagram. arXiv:1712.00778.

11. J. Kellner and S. Shelah, Decisive creatures and large continuum, *J. Symbolic Logic* **74**, 73 (2009).
12. J. Kellner and S. Shelah, Creature forcing and large continuum: the joy of halving, *Arch. Math. Logic* **51**, 49 (2012).
13. A. Fischer, M. Goldstern, J. Kellner and S. Shelah, Creature forcing and five cardinal characteristics in Cichoń's diagram, *Arch. Math. Logic* **56**, 1045 (2017).
14. H. Judah and S. Shelah, The Kunen-Miller chart (Lebesgue measure, the Baire property, Laver reals and preservation theorems for forcing), *J. Symbolic Logic* **55**, 909 (1990).
15. S. H. Hechler, Short complete nested sequences in $\beta N \backslash N$ and small maximal almost-disjoint families, *General Topology and Appl.* **2**, 139 (1972).
16. H. Judah and S. Shelah, Souslin forcing, *J. Symbolic Logic* **53**, 1188 (1988).
17. M. A. Cardona and D. A. Mejía, On cardinal characteristics of Yorioka ideals. arXiv:1703.08634.
18. T. Bartoszyński and H. Judah, *Set Theory: On the Structure of the Real Line* (A K Peters, Wellesley, Massachusetts, 1995).
19. A. W. Miller, Some properties of measure and category, *Trans. Amer. Math. Soc.* **266**, 93 (1981).
20. A. Kamburelis, Iterations of Boolean algebras with measure, *Arch. Math. Logic* **29**, 21 (1989).
21. D. A. Mejía, Template iterations with non-definable ccc forcing notions, *Ann. Pure Appl. Logic* **166**, 1071 (2015).
22. A. Horn and A. Tarski, Measures in Boolean algebras, *Trans. Amer. Math. Soc.* **64**, 467 (1948).
23. D. A. Mejía, Some models produced by 3D-iterations, *Kyōto Daigaku Sūrikaiseki Kenkyūsho Kōkyūroku* **2042**, 75 (2017).
24. S. Todorčević, Two examples of Borel partially ordered sets with the countable chain condition, *Proc. Amer. Math. Soc.* **112**, 1125 (1991).
25. S. Todorčević, Remarks on cellularity in products, *Compositio Math.* **57**, 357 (1986).
26. S. Todorčcević, A Borel solution to the Horn-Tarski problem, *Acta Math. Hungar.* **142**, 526 (2014).
27. E. Thümmel, The problem of Horn and Tarski, *Proc. Amer. Math. Soc.* **142**, 1997 (2014).
28. A. Dow and J. Steprāns, The $\sigma$-linkedness of the measure algebra, *Canad. Math. Bull.* **37**, 42 (1994).
29. J. Pawlikowski, Adding dominating reals with $\omega^\omega$ bounding posets, *J. Symbolic Logic* **57**, 540 (1992).
30. J. Brendle, M. A. Cardona and D. A. Mejía, Filter-linkedness and its effect on the preservation of cardinal characteristics. In preparation.
31. R. Engelking and M. Karłowicz, Some theorems of set theory and their topological consequences, *Fund. Math.* **57**, 275 (1965).

# Moral Dilemmas and the Contrary-to-Duty Scenarios in Dynamic Logic of Acts of Commanding — The Significance of Moral Considerations Behind Moral Judgments —

Tomoyuki Yamada

*Department of Philosophy, Hokkaido University, Sapporo, Hokkaido 060-0810, Japan*
*E-mail: yamada@let.hokudai.ac.jp*

In the literature on deontic logic, there have been many proposals to weaken or abandon some of the principles of standard deontic logic so that moral dilemmas can be accommodated without triggering deontic explosion. Since the so-called principle (D) precludes the possibility of moral dilemmas in standard deontic logic, such proposals usually include the rejection of (D). If we abandon (D), however, it is possible to avoid deontic explosion in most real cases without abandoning or weakening any other principles of standard deontic logic.

What is important to note here is the fact that the moral consideration that leads to a moral dilemma can be divided into two or more moral sub-considerations in such a way that each of them leads to one of the conflicting moral judgments that jointly constitute the very dilemma in question. Then by *personifying* the sub-considerations as different command giving authorities, we can represent such a moral dilemma as an example of a deontic dilemma generated by conflicting commands given to one and the same agent by different authorities without triggering deontic explosion in a dynamic logic of acts of commanding that extends multi-agent variant of standard deontic logic without (D). The basic idea is to relativize deontic operators to commandees and commanders. This enables us to "minimize" the possibility of deontic explosion, so to speak. Deontic explosion occurs only when a set of conflicting commands are given to one and the same commandee by one and the same commander or a self-contradictory command is given. But then, in the case of acts of commanding, a rational commander has every reason to avoid such a situation, and in the case of moral dilemmas, we show that it is always possible to divide the moral consideration that leads to a dilemma into two or more moral sub-considerations.

We then show how contrary-to-duty scenarios can be formalized in the same dynamic logic. We apply the same personification of moral sub-considerations to capture the intuition that the relevant fact and obligations included in a scenario are logically independent from each other. The crucial independence will be shown to be the independence of each of the underlying moral sub-considerations from the other moral sub-considerations and the fact mentioned in the scenario.

*Keywords*: Moral dilemma; deontic explosion; contrary-to-duty paradox; dynamified deontic logic; act of commanding; moral sub-consideration.

## 1. Introduction

As Kant's categorical imperative exemplifies, moral principles and moral judgments can be expressed in the form of imperatives. Since moral considerations tell us what we should (or should not) do through the moral judgments they support, we can think of moral considerations as a kind of moral command issuing authorities (moral commanders, for short) somewhat similar to the "daimonion" whose voice Socrates claimed to hear.[a] This suggests the possibility of applying the methods and techniques of the recently developed dynamic logic of acts of commanding to the analysis of various problems in moral philosophy and deontic logic. The purpose of this paper is to show how two old problems of deontic logic, namely the problem of deontic explosion generated by moral dilemmas and that of the paradox generated by contrary-to-duty scenarios can be solved in an intuitively satisfactory way in a version of such dynamic logic, $\mathsf{DMDL}^{+}\mathsf{II}$ developed in Yamada (2007b, 2008b)[1,2].[b] It enables us to do justice to the difference of moral considerations behind the moral judgments involved in moral dilemmas and contrary-to-duty scenarios by distinguishing them as different moral commanders, and thus opens the possibility of taking their difference into account in representing moral dilemmas and contrary-to-duty scenarios.

With respect to the problem of deontic explosion, there have been many proposals to abandon or weaken some of the principles of standard deontic logic (SDL, for short) so that moral dilemmas can be accommodated without triggering deontic explosion as the study by Goble (2015)[7] shows. Since the so-called principle (D) precludes the possibility of moral dilemmas in SDL, such proposals usually include the rejection of (D). If we abandon (D), however, it is possible to avoid deontic explosion in real life cases without abandoning or weakening any other principles of SDL.[c]

---

[a] According to Socrates, the "daimonion" only stopped him when he was about to do something wrong but never told him what to do, though (Plato, *Apology*, 31d and 40a).

[b] $\mathsf{DMDL}^{+}\mathsf{II}$ is a refinement of $\mathsf{DMDL}^{+}$ (Dynamified Multi-agent Deontic Logic with Alethic Modality) developed in Yamada (2007a)[3]. There is a further refinement $\mathsf{DMDL}^{+}\mathsf{III}$ that deals with acts of promising along with acts of commanding (Yamada 2008a)[4] and its extension DMEDL that deals with acts of requesting and asserting as well as acts of commanding and promising (Yamada 2011, 2016)[5,6]. But for the present purpose we only need $\mathsf{DMDL}^{+}\mathsf{II}$. $\mathsf{DMDL}^{+}$ and $\mathsf{DMDL}^{+}\mathsf{II}$ were originally called ECL (Eliminative Command Logic) and ECLII respectively in Yamada (2007a, b, etc.), but after the subsequent development of dynamified logics of speech acts, the names are changed in order to make the naming more systematic.

[c] There are also proposals that keep (D). For example, Prakken (1996)[8] has argued for embedding a deontic logic validating (D) in a version of non-monotonic logic. We will not examine these proposals in this paper, however, as it is not clear whether any substantial role is left for (D) to play in deontic reasoning once we admit the possibility of moral or deontic dilemmas.

What is important to note here is the fact that when a moral consideration leads to a moral dilemma, it can be divided into two (or more) sub-considerations in such a way that each of them leads to one of the conflicting moral judgments that jointly constitute the very dilemma in question. Then by *personifying* the sub-considerations as different moral command issuing authorities, we can represent such a moral dilemma as an example of a deontic dilemma generated by conflicting commands given to one and the same agent by different authorities without triggering deontic explosion in $\mathsf{DMDL}^{+}\mathsf{II}$, which is a dynamic extension of multi-agent variant of SDL without (D). The crucial idea is to relativize deontic operators to commandees and commanders. This enables us to "minimize" the possibility of deontic explosion, so to speak. Deontic explosion occurs only when a set of conflicting commands are given to one and the same commandee by one and the same commander, or a self-contradictory command is given. But then, in the case of acts of commanding, a rational commander has every reason to avoid such a situation, and in the case of moral dilemmas, we will see that it is always possible to divide the moral consideration that lead to a dilemma into two or more sub-considerations.

With respect to the contrary-to-duty paradox, there are many proposals that introduce a dyadic deontic operator.[d] We can apply, however, the above idea of representing the different moral considerations as different moral commanders to this paradox as well. It enables us to capture the intuition that the relevant fact and obligations included in a contrary-to-duty scenario are independent from each other even with monadic deontic operators. The crucial independence will be shown to be the logical independence of each of the underlying moral sub-considerations from the other moral sub-considerations and the fact mentioned in the scenario.

Considering moral judgments as commands given to agents by moral considerations opens the possibility of dealing with the problem of deontic explosion and the paradox of the contrary to duty scenarios. It does so by enabling us to appreciate the importance of the fact that moral judgments are supported by moral considerations.

This paper is structured as follows. In Section 2, we review how moral dilemmas give rise to deontic explosion in SDL without (D). Then in Section 3, we review how a deontic dilemma where one and the same commandee receives a pair of conflicting commands can be represented without triggering deontic explosion in $\mathsf{DMDL}^{+}\mathsf{II}$. In Section 4, we show how the same method can be applied to moral

[d]For overviews of earlier proposals, see Prakken & Sergot (1997) and Carmo and Jones (2001)[9,10]. A more recent proposal that includes the analysis of information dynamics and deontic dynamics can be found in van Benthem, Grossi & Liu (2014)[11].

dilemmas. Then in Section 5, we show how contrary-to-duty scenarios can be represented without leading to a paradox in DMDL$^+$II. And finally, in Section 6, we conclude with a short summary of the results so far and brief discussions of remaining tasks left for further study.

## 2. Moral dilemmas and deontic explosion in SDL$^-$

In the language of SDL the formula of the form $OA$ means that it ought to be the case that $A$, and thus a moral dilemma can be considered as a situation in which we have $OA \wedge O\neg A$.[e] Since the following principle (D) precludes the possibility of such a situation, we have to abandon it if we wish to accommodate moral dilemmas.

$$OA \to \neg O \neg A \ . \tag{D}$$

Even if we abandon the principle (D), however, we still have to face the problem of deontic explosion. Let SDL$^-$ be the logic obtained from SDL by deleting (D) from the proof system. As the following principle, which is often called "ex falso quodlibet", is a tautology, it is included in any modal logic that includes classical propositional logic, and SDL$^-$ is such a logic.

$$(A \wedge \neg A) \to B \ . \tag{EFQ}$$

Thus, we can derive the following formula from (EFQ) by applying the necessitation rule for the deontic operator $O$ in SDL$^-$.

$$O((A \wedge \neg A) \to B) \ . \tag{1}$$

Now the following formula is an instance of the axiom K for the operator $O$.

$$O((A \wedge \neg A) \to B) \to (O(A \wedge \neg A) \to OB) \ . \tag{2}$$

So, we can derive the following principle from (1) and (2) by applying modus ponens.

$$O(A \wedge \neg A) \to OB \ . \tag{3}$$

Moreover, SDL$^-$ inherits the following principle from SDL.

$$(OA \wedge OB) \to O(A \wedge B) \ . \tag{AND}$$

Now the following formula is an instance of (AND).

$$(OA \wedge O\neg A) \to O(A \wedge \neg A) \ . \tag{4}$$

[e]Or more broadly, we will be in a moral dilemma if we have $(OA \wedge OB)$ but $A$ and $B$ are logically, metaphysically, physically, or in some other way mutually incompatible.

Thus, we can derive the following deontic explosion principle from (3) and (4) by propositional logic.

$$(OA \wedge O\neg A) \rightarrow OB\ . \tag{DEX}$$

Since B here is arbitrary, (DEX) means that if there is a moral dilemma, everything ought to be the case, hence "deontic explosion".

Naturally, there have been various attempts to weaken $\mathsf{SDL}^-$ so that moral dilemmas can be accommodated without triggering deontic explosion. As Goble (2015, pp. 465–473)[7] has shown, however, none of the resulting systems earlier than the one proposed in Goble (2015) are satisfactory. Some of them are too weak to support the full range of normative inferences that seem valid while the others are not weak enough to avoid the derivation of untoward consequences from the existence of a moral dilemma. The system proposed in Goble (2015, pp. 473–476, 480–482.) can be said to be better in these respects than earlier ones, but still suffers from "a conflict of intuitions" (pp. 476–477). I will not go into details of these systems in this paper, as it is possible to avoid deontic explosion in most cases without weakening $\mathsf{SDL}^-$. In the next section I will show how deontic dilemma generated by a pair of conflicting commands can be represented without triggering deontic explosion in a dynamic logic of acts of commanding which extends a multi-agent variant of $\mathsf{SDL}^-$.

## 3. Conflicting commands in $\mathsf{DMDL^+II}$

In this section, we review how deontic dilemmas generated by conflicting commands can be represented without triggering deontic explosion in $\mathsf{DMDL^+II}$. $\mathsf{DMDL^+II}$ is a dynamic extension of $\mathsf{MDL^+II}$ (Multi-agent Deontic Logic Plus an alethic modality with Double indexing), which is a combination of multi-agent variant of $\mathsf{SDL}^-$ and a simplified version of alethic modal logic. Thus, in the language $\mathcal{L}_{\mathsf{MDL^+II}}$ of $\mathsf{MDL^+II}$, we have formulas of the form $O_{(i,j)}\varphi$ for each pair of agents $i$ and $j$ taken from a given finite set of agent $I$.

Intuitively $O_{(i,j)}\varphi$ means that it is obligatory upon an agent $i$ to see to it that $\varphi$ by the name of an authority $j$. We use the first index in order to distinguish a particular agent to whom a command is given from the other agents. When an agent $a$, but not $b$, is commanded to see to it that $p$ by an authority $c$, we wish to be able to say that it is obligatory upon $a$, but not upon $b$, to see to it that $p$ by the name of $c$, and this is expressed by $O_{(a,c)}p \wedge \neg O_{(b,c)}p$.[f]

[f]Although commands can be given to a group $G \subseteq I$ of agents, we will only consider commands given to individual agents for the sake of simplicity.

We also use an alethic modal operator $\Box$ in order to say things like $\neg\Box\neg\varphi \land O_{(a,b)}\neg\varphi$. It intuitively means that $\varphi$ is possible but $a$ is forbidden by the name of $b$ to see to it that $\varphi$.

The truth definition for this language is given in a completely standard way with reference to relational models called $\mathcal{L}_{\mathsf{MDL^+II}}$-models. Given an $\mathcal{L}_{\mathsf{MDL^+II}}$-model $\mathcal{M}$ with a non-empty set $W^{\mathcal{M}}$ of "possible worlds" or "states", we have an arbitrary binary relation $A^{\mathcal{M}}$ on $W^{\mathcal{M}}$ to be used in interpreting the alethic modality $\Box$ and a separate binary relation $D^{\mathcal{M}}_{(i,j)}$ on $W^{\mathcal{M}}$ for each pair $\langle i, j\rangle \in I \times I$ to be used in interpreting each deontic operator $O_{(i,j)}$. The clause for the formula of the form $O_{(i,j)}\varphi$, for example, reads as follows:

$$\mathcal{M}, w \models_{\mathsf{MDL^+}} O_{(i,j)}\varphi \quad \text{iff} \quad \text{for every } v \text{ such that } \langle w, v\rangle \in D^{\mathcal{M}}_{(i,j)}, \ \mathcal{M}, v \models_{\mathsf{MDL^+}} \varphi\,.$$

Each of $D^{\mathcal{M}}_{(i,j)}$ is required to be a subset of $A^{\mathcal{M}}$.[g] Apart from this condition, no other conditions are imposed on each $D^{\mathcal{M}}_{(i,j)}$.

A Hilbert style proof system for $\mathsf{MDL^+}$ consists of (a) all instantiations of propositional tautologies over $\mathcal{L}_{\mathsf{MDL^+}}$, (b) the axioms K for $\Box$ and $O_{(i,j)}$ for each $\langle i, j\rangle \in I \times I$, (c) the necessitation rules for $\Box$ and $O_{(i,j)}$ for each $\langle i, j\rangle \in I \times I$, (d) modus ponens, and (e) the following axiom:[h]

$$P_{(i,j)}\varphi \to \Diamond\varphi\,. \qquad \text{(MIX)}$$

As it is easy and safe to add restrictions on the alethic accessibility relation $A^{\mathcal{M}}$, we do not bother to do so.[i] Thus, apart from (MIX), $\mathsf{MDL^+II}$ is just a multi-K system. The above proof system is sound and complete with respect to $\mathcal{L}_{\mathsf{MDL^+II}}$-models (Yamada 2007b[1]).

$\mathsf{MDL^+II}$ is developed by extending $\mathsf{MDL^+II}$ just as PAL (Public Announcement Logic) is developed by extending a multi-agent variant of the standard epistemic logic.[j] Thus, the language $\mathcal{L}_{\mathsf{DMDL^+II}}$ of $\mathsf{DMDL^+II}$ (Dynamified $\mathsf{MDL^+II}$) is given by adding the formula of the form $[!_{(i,j)}\varphi]\psi$ to the language $\mathcal{L}_{\mathsf{MDL^+II}}$ of the static base logic. The formula of this form means that whenever an act of type $!_{(i,j)}\varphi$ is performed, $\psi$ holds in the resulting situation. The expression $!_{(i,j)}\varphi$ stands for the type of every act in which an agent $i$ is commanded by an authority $j$ to see to it that $\varphi$. Note that $i$ stands for the commandee and $j$ stands for the commander.

[g]This condition guarantees the validity of the axiom called (MIX), which will be shown shortly.

[h]In addition to standard abbreviations, we abbreviate $O_{(i,j)}\neg\varphi$ as $F_{(i,j)}\varphi$ and $\neg O_{(i,j)}\neg\varphi$ as $P_{(i,j)}\varphi$.

[i]For more on this, see Footnote l.

[j]PAL is the earliest system of DEL (Dynamic Epistemic Logic). A detailed textbook exposition of PAL and DEL is given in van Ditmarsch, van der Hoek & Kooi (2007)[12]. It includes a succinct description of the development of these systems. On more recent development, see Baltag & Rene (2016)[13].

Then a truth definition for this language is given with reference to $\mathcal{L}_{\mathsf{MDL^+II}}$-models by adding the clause for the dynamic formula, which reads as follows:[k]

$$\mathcal{M}, w \models_{\mathsf{DMDL^+II}} [!_{(i,j)}\varphi]\psi \quad \text{iff} \quad \mathcal{M}_{!_{(i,j)}\varphi}, w \models_{\mathsf{DMDL^+II}} \psi\ ,$$

where $\mathcal{M}_{!_{(i,j)}\varphi}$ is the $\mathcal{L}_{\mathsf{MDL^+II}}$-model obtained from $\mathcal{M}$ by replacing the deontic accessibility relation $D^{\mathcal{M}}_{(i,j)}$ for the commandee $i$ and the commander $j$ with its subset

$$D^{\mathcal{M}_{!_{(i,j)}\chi}}_{(i,j)} = \{\langle x, y\rangle \in D^{\mathcal{M}}_{(i,j)} \mid \mathcal{M}, y \models_{\mathsf{DMDL^+II}} \varphi\}\,.$$

If $\langle x, y\rangle \in D^{\mathcal{M}}_{(i,j)}$, let us say that $\langle x, y\rangle$ is a $D^{\mathcal{M}}_{(i,j)}$-arrow. Then, intuitively, this model updating operation cuts all and only the $D^{\mathcal{M}}_{(i,j)}$-arrows that arrive in non-$\varphi$ worlds in the original model $\mathcal{M}$. Thus, it is called link cutting. Note that it cuts no $D^{\mathcal{M}}_{(k,l)}$-arrows if $\langle k, l\rangle \neq \langle i, j\rangle$.

Note also that the truth of the formula $[!_{(i,j)}\varphi]\psi$ at $w$ in $\mathcal{M}$ is defined in terms of the truth of its sub-formula $\psi$ at $w$ in the updated model $\mathcal{M}_{!_{(i,j)}\varphi}$. So, the updated model needs to be an $\mathcal{L}_{\mathsf{MDL^+II}}$-model. Since $D^{\mathcal{M}_{!_{i}\varphi}}_{(i,j)} \subseteq D^{\mathcal{M}}_{(i,j)} \subseteq A^{\mathcal{M}}$, $\mathcal{M}_{!_{(i,j)}\varphi}$ satisfies the condition that every deontic accessibility relation is a subset of the alethic accessibility relation. Thus $\mathcal{M}_{!_{(i,j)}\varphi}$ is guaranteed to be an $\mathcal{L}_{\mathsf{MDL^+II}}$-model.[l]

The proof system of $\mathsf{DMDL^+II}$ is obtained by adding the necessitation rule for the dynamic operator $[!_{(i,j)}\varphi]$ for each $\langle i, j\rangle \in I \times I$ and a set of recursion axioms to the proof system of $\mathsf{MDL^+II}$ with tautologies extended to $\mathcal{L}_{\mathsf{DMDL^+II}}$.[m] It is shown to be complete with respect to the class of $\mathcal{L}_{\mathsf{MDL^+II}}$-models. For the list of the recursion axioms, see Yamada (2007b)[1] or Yamada (2008b)[2], and for the details of the completeness proof, see Yamada (2018)[14].[n]

Now let us say that a formula $\varphi$ is $(i, j)$-free if $\varphi$ is a formula of the static base logic $\mathsf{MDL^+II}$ and the deontic operator $O_{(i,j)}$ does not occur in $\varphi$. For example, if $p$ is an atomic proposition, $p$, $\neg p$, $\Box p$, and $O_{(k,l)}p$ (if $\langle k, l\rangle \neq \langle j, j\rangle$) are $(i, j)$-free but $O_{(i,j)}p$ is not. Then the above truth definition gives us the following results.

---

[k]The clauses in the definition of $\models_{\mathsf{MDL^+II}}$ relation should of course be reproduced mutatis mutandis as the clauses for defining $\models_{\mathsf{DMDL^+II}}$ relation.

[l]This is the reason I wrote earlier that it is easy and safe to add restrictions on the alethic accessibility relation. Whatever condition we impose on $A^{\mathcal{M}}$, $D^{\mathcal{M}_{!_{(i,j)}\varphi}}_{(i,j)} \subseteq A^{\mathcal{M}}$ since $D^{\mathcal{M}_{!_{(i,j)}\varphi}}_{(i,j)} \subseteq D^{\mathcal{M}}_{(i,j)}$ by the definition of $D^{\mathcal{M}_{!_{(i,j)}\varphi}}_{(i,j)}$ and $D^{\mathcal{M}}_{(i,j)} \subseteq A^{\mathcal{M}}$ by the definition of $\mathcal{L}_{\mathsf{MDL^+II}}$-models.

[m]Recursion axioms are also known as reduction axioms in the literature.

[n]Exactly speaking, Yamada (2018) presents the details of the completeness proof of $\mathsf{ECL}$, which dynamifies $\mathsf{MDL^+}$. Although $\mathsf{MDL^+}$ utilizes indexing by a given finite set $I$ of agents where $\mathsf{MDL^+II}$ uses indexing by $I \times I$, indexing by $I \times I$ is just another instance of indexing by a finite set. Thus $\mathsf{MDL^+II}$ is just an instance of $\mathsf{MDL^+}$, and similarly, $\mathsf{DMDL^+II}$ is an instance of $\mathsf{ECL}$. Therefor the procedure for proving the completeness of $\mathsf{ECL}$ can be used mutatis mutandis for proving that of $\mathsf{DMDL^+II}$.

**Proposition 3.1 (Partial Inertia Principle).** *If $\varphi$ is $(i, j)$-free, the following equivalence holds.*

$$\mathcal{M}, w \models_{\mathsf{DMDL^{+}II}} \varphi \textit{ iff } \mathcal{M}_{[!_{(i,j)}\varphi]\psi}, w \models_{\mathsf{DMDL^{+}II}} \varphi \ .$$

**Proposition 3.2 (CUGO Principle).** *If $\varphi$ is $(i, j)$-free, the following formula is valid.*

$$[!_{(i,j)}\varphi]O_{(i,j)}\varphi \ .$$

This principle means that **C**ommands **U**sually **G**enerates **O**bligations; hence the acronym.

Since no deontic operator occur in an atomic proposition, if $p$ is an atomic proposition, CUGO Principle and Partial Inertia Principle jointly guarantee the validity of the following formula.

$$[!_{(a,b)}p][!_{(a,c)}\neg p](O_{(a,b)}p \wedge O_{(a,c)}\neg p) \ . \tag{5}$$

This implies the following.

$$(\mathcal{M}_{!_{(a,b)}p})_{!_{(a,c)}\neg p}, w \models_{\mathsf{DMDL^{+}II}} O_{(a,b)}p \wedge O_{(a,c)}\neg p \ . \tag{6}$$

The situation $((\mathcal{M}_{!_{(a,b)}p})_{!_{(a,c)}\neg p}, w)$ is brought about when $a$ is commanded by $c$ to see to it that $\neg p$ after $a$ is commanded by $b$ to see to it that $p$ in $(\mathcal{M}, w)$. In this situation, $a$ has to see to it that $p$ if $a$ wishes to obey $b$'s command, but if $a$ does so, $a$ will go against $c$'s command. In order to obey $c$'s command, $a$ has to see to it that $\neg p$, but if $a$ does so, $a$ will go against $b$'s command. Since $p \wedge \neg p$ is a contradiction, it is not possible for $a$ to obey both commands jointly. So, $a$ has to choose which command to obey, but whichever command $a$ chooses to obey, $a$ will go against one of the two commands. Thus, $a$ is caught in a deontic dilemma, but this dilemma is represented without triggering deontic explosion.

When the relevant impossibility is not the logical impossibility but some other kind of impossibility, we may have the following situation.

$$(\mathcal{M}_{!_{(a,b)}p})_{!_{(a,c)}q}, w \models_{\mathsf{DMDL^{+}II}} O_{(a,b)}p \wedge O_{(a,c)}q \wedge \neg\Diamond(p \wedge q) \ , \tag{7}$$

where $\Diamond$ stands for the relevant kind of possibility.[o]

In both cases, if $a$, $b$ and $c$ all belong to one and the same hierarchical organization, there may be a rule for deciding whose command override whose, but if $b$ and $c$ belong to different organizations and $a$ belong to both, it is unlikely that such a

[o]For a concrete real life example, see Yamada (2007b), pp. 303, 307–308.[1]

rule is available. In such a case, unless one of the two commanders withdraws her command, $a$ will not be able to get out of the deontic dilemma.

Now a pair of conflicting commands generate deontic explosion in DMDL$^+$II only if they are given to one and the same commandee by one and the same commander (for example, when $b = c$ in the above two situations). Here we need to note three points. First, the distinction between the commanders is crucial in avoiding deontic explosion as the following DMDL$^+$II-analogue of (AND) is valid.

$$(O_{(i,j)}\varphi \wedge O_{(i,j)}\neg\varphi) \rightarrow O_{(i,j)}(\varphi \wedge \neg\varphi). \tag{8}$$

Second, the sameness of the commandee is essential for the generation of a dilemma. Consider the following Example.

**Example 3.1.** The wife $a$ of the first son of the emperor secretly hired an assassin $b$ and commanded $b$ to assassinate the second son of the emperor. Soon after that, the wife $c$ of the second son of the emperor happened to hire a master $d$ of martial art and secretly commanded $d$ to protect her husband from the assassination.

Let $p$ be the proposition that $c$'s husband is assassinated. Then CUGO Principle and Partial Inertia Principle jointly guarantees the validity of the following formula.

$$[!_{(b,a)}p][!_{(d,c)}\neg p](O_{(b,a)}p \wedge O_{(d,c)}\neg p). \tag{9}$$

But this by itself does not generate a deontic dilemma if $b \neq d$.

Third, a situation in which one and the same commandee is given a pair of conflicting commands by one and the same commander is possible but not very likely. If the commander is rational, she has every reason to avoid issuing such a pair of commands, since she could not expect such commands to be jointly obeyed. This suggests that we should avoid weakening deontic logic if the purpose of doing so is just to avoid deontic explosion. The absurdity of the situation where deontic explosion occurs represents the absurdity of the decision of the commander who issues a pair of conflicting commands or a self-contradictory command. It is not the logic but the commander that is to blame for generating such an absurd situation.

It may be noted here, however, that the formula $(O_{(a,b)}p \wedge O_{(a,c)}\neg p)$ above is compatible with the following DMDL$^+$II analogue (DD) of the principle (D).

$$\neg(O_{(i,j)}\varphi \wedge O_{(i,j)}\neg\varphi) \; . \tag{DD}$$

Does this mean that the principle (D) should be recovered as (DD)? To do so, however, would preclude the possibility of an inadvertent commander. A rational commander who has inadvertently issued a pair of conflicting commands will see the need to withdraw at least one of her commands when she realizes the conflict. She needs to do so, not because one of her commands is void, but because otherwise both of her commands will remain in force.[p] Since the following formula is a contradiction, (DD) should not be validated in DMDL$^{+}$II.

$$(O_{(a,b)}p \wedge O_{(a,b)}\neg p) \wedge \neg(O_{(a,b)}p \wedge O_{(a,b)}\neg p) \ . \tag{10}$$

The situation is similar to the situation where one and the same agent, say $d$, has inadvertently asserted both $p$ and $\neg p$. Let $[\text{a-cmt}]_i\varphi$ mean that the agent $i$ has an assertoric commitment to $\varphi$.[q] Then we can say in the current situation we have

$$[\text{a-cmt}]_d p \wedge [\text{a-cmt}]_d \neg p \ . \tag{11}$$

Now the following principle (AD) is an assertoric commitment analogue of (DD).

$$\neg([\text{a-cmt}]_i\varphi \wedge [\text{a-cmt}]_i\neg\varphi) \ . \tag{AD}$$

Just as (DD) precludes the possibility of an inadvertent commander, (AD) precludes the possibility of an inadvertent asserter. If $d$ is rational, she will see the need to withdraw at least one of her assertions when she realizes their incompatibility. She needs to do so, not because one of her assertions is void, but because otherwise she will remain committed to both $p$ and $\neg p$. Although both such a commander and such an asserter are inadvertent, the inadvertence of this kind is possible and should not be precluded by logic. Validating (DD) in the logic that deals with the effects of acts of commanding is as misguided as validating (AD) in the logic that deals with the effects of acts of asserting.[r]

## 4. Moral dilemmas in DMDL$^{+}$II

In this section we show how DMDL$^{+}$II can be applied to moral dilemmas. Generally speaking, when an agent $a$ is caught in a moral dilemma in which two moral judgments are in conflict by a moral consideration $c$, it seems in principle possible to think of $c$ as a combination of two sub-considerations $c_1$ and $c_2$ such that

$$c_1 \text{ leads to the conclusion that } a \text{ should see to it that } \varphi \,, \tag{a}$$

$$c_2 \text{ leads to the conclusion that } a \text{ should see to it that } \psi \ \text{, and} \tag{b}$$

$$\varphi \text{ and } \psi \text{ are in some way mutually incompatible.} \tag{c}$$

Such a situation can be represented in the language of DMDL$^{+}$II by personifying $c_1$ and $c_2$ as two command giving authorities. If $\varphi$ is both $(a, c_1)$-free and $(a, c_2)$-free

[p]This seems to be closely related to the role of meta-norms and law-givers or norm-authorities in the situation where a set of norms is found to be inconsistent discussed by von Wright (1991, p. 277)[15].

[q]For more on assertoric commitments, see Yamada (2016)[6].

[r]This of course does not mean that it is illegitimate to say "it is usually, or normally, the case that (DD) and (AD) hold".

and $\psi$ is $(a, c_2)$-free, for some $\mathcal{M}$ and $w$ we have:

$$\mathcal{M}, w \models_{\mathsf{DMDL^{+}II}} [!_{(a,c_1)}\varphi][!_{(a,c_2)}\psi](O_{(a,c_1)}\varphi \wedge O_{(a,c_2)}\psi \wedge \neg\Diamond(\varphi \wedge \psi)) \,, \tag{12}$$

where $\Diamond$ stands for the relevant kind of possibility.

The two sub-considerations $c_1$ and $c_2$ here may include different moral principles. But it is also possible that one and the same moral principle, when combined with different facts, lead to a moral dilemma. Consider the following example.

**Example 4.1.** Sophie and her two children were caught and sent to Auschwitz during the World War II. When they arrived, the doctor there said to Sophie, "You may keep one of your children. The other one will have to go. Which one will you keep?" (Stylon, 1979, p. 594.[16])

Here "to go" means "to go to the gas chamber in Birkenau".

The following moral principle and facts are relevant to Sophie's quandary.

Parents ought to see to it that their children are safe. (i)

Eva is one of Sophie's children. (ii)

Jan is one of Sophie's children. (iii)

Let $c_1$ be the consideration that includes (i) and (ii), $c_2$ be the consideration that includes (i) and (iii), $s$ be Sophie, and $d$ be the doctor. In addition, let $E$ be the proposition that Eva goes to Birkenau, and $J$ be the proposition that Jan goes to Birkenau. Then in the situation after the doctor's words, CUGO Principle and Partial Inertia Principle gives us, for some $\mathcal{M}$ and $w$, the following.

$$\begin{aligned}((\mathcal{M}_{!_{(s,d)}(E\vee J)})_{!_{(s,c_1)}\neg E})_{!_{(s,c_2)}\neg J}, w \models_{\mathsf{DMDL^{+}II}} \\ O_{(s,d)}(E \vee J) \wedge O_{(s,c_1)}\neg E \wedge O_{(s,c_2)}\neg J \,.\end{aligned} \tag{13}$$

Obviously, this does not generate deontic explosion, yet it represents a moral dilemma. As the doctor had the power to enforce his words, it was not practically possible for Sophie to see to it that $\neg E \wedge \neg J$ in that situation. So, here we also have

$$\begin{aligned}((\mathcal{M}_{!_{(s,d)}(E\vee J)})_{!_{(s,c_1)}\neg E})_{!_{(s,c_2)}\neg J}, w \models_{\mathsf{DMDL^{+}II}} \\ O_{(s,c_1)}\neg E \wedge O_{(s,c_2)}\neg J \wedge \neg\Diamond(\neg E \wedge \neg J) \,,\end{aligned} \tag{14}$$

where $\Diamond$ stands for the relevant kind of possibility. She was forced to make a choice. Since $O_{(s,c_1)}\neg E \wedge O_{(s,c_2)}\neg J$ and $E \vee J$ are jointly satisfiable in $\mathsf{DMDL^{+}II}$, the logical consistency of moral principles is not enough to enable Sophie to get out of her quandary. What she needed was some real life means of preventing the doctor's words from being enforced as Marcus (1980)[17] would say.

Note that the distinction between $c_1$ and $c_2$ is crucial to avoiding deontic explosion. Although these two considerations can be of exactly the same form, they are still distinct in that (ii) refers to Eva but not to Jan while (iii) refers to Jan but not to Eva.[s]

In regard to the generation of a dilemma, the sameness of the agent who is obligated to see to it that $\neg E$ and the agent who is obligated to see to it that $\neg J$ is essential. Consider a slightly different case in which Jan is not Sophie's child but Wanda's, who was also caught and sent to Auschwitz with Sophie. Let $b$ be Wanda, and let $!_{(s\oplus b,d)}(E \vee J)$ be the doctor's command given to $s$ and $b$ jointly to see to it that $E \vee J$.[t] Then we will have something like the following.

$$((\mathcal{M}_{!_{(s\oplus b,d)}(E\vee J)})_{!_{(s,c_1)}\neg E})_{!_{(b,c_3)}\neg J}, w \models_{\mathsf{DMDL^+II}} O_{(s,c_1)}\neg E \wedge O_{(b,c_3)}\neg J \wedge \neg\Diamond(\neg E \wedge \neg J) \;, \tag{15}$$

where $c_3$ is the consideration that involves (i) and the fact that Jan is Wanda's child. Obviously, this does not by itself represent a dilemma. Nor does the following formula by itself do so.

$$((\mathcal{M}_{!_{(s\oplus b,d)}(E\vee J)})_{!_{(s,c_1)}\neg E})_{!_{(b,c_3)}\neg J}, w \models_{\mathsf{DMDL^+II}} O_{(s\oplus b,d)}(E \vee J) \wedge O_{(s,c_1)}\neg E \wedge O_{(b,c_3)}\neg J \;. \tag{16}$$

Although Sophie and Wanda are required to see to it that $E \vee J$ jointly, for Sophie seeing to it that $J$ is a way of seeing to it that $E \vee J$ and for Wanda seeing to it that $E$ is a way of seeing to it that $E \vee J$. Of course if some other moral consideration commands Sophie to see to it that $\neg J$ or commands Wanda to see to it that $\neg E$, a moral dilemma arises. But then the agent who is obligated to see to it that $\neg E$ and the agent who is obligated to see to it that $\neg J$ will be the same.

The example of Sophie's quandary shows that it is at least sometimes possible to represent even a moral dilemma where only one moral principle is involved without triggering deontic explosion in DMDL⁺II by treating moral judgments not in isolation but with reference to moral considerations that support them. Here we need to ask whether it is always possible to represent a moral dilemma without triggering deontic explosion in DMDL⁺II. The answer seems to be "yes". Since a

[s]The consideration $c_1$ can be considered as a practical reasoning that involves (i) and (ii) and leads to the conclusion expressed by the command $!_{(s,c_1)}\neg E$ and the consideration $c_2$ can be considered as another practical reasoning that involves (i) and (iii) and leads to the conclusion expressed by the command $!_{(s,c_2)}\neg J$. We will discuss the possibility of formalizing these considerations in deontic logic later in the concluding remarks.

[t]How joint decision is to be analyzed is an interesting question but since it goes beyond the scope of this paper, we will treat $s \oplus b$ just as an individual such that $s \neq (s \oplus b) \neq b$ formally, and only informally consider $O_{(s\oplus b,d)}\varphi$ as requiring both $s$ and $b$ to do something in order to see to it that $\varphi$.

moral dilemma is a situation in which two or more conflicting moral judgments are involved, it is always possible to divide the moral consideration that leads to a moral dilemma into as many sub-considerations as needed to support the moral judgments that jointly constitute the very dilemma. Moreover, Sophie's example shows that even a consideration that includes only one moral principle can be divided into two different sub-considerations. Note that the above moral principle (i) does not by itself imply either the judgment that Sophie ought to see to it that $\neg E$ or the judgment that Sophie ought to see to it that $\neg J$. It needs to be combined with (ii) and (iii).

It might be said here that the following judgment ($\alpha$) implies both ($\beta$) and ($\gamma$).

$$\text{Sophie ought to see to it that } \neg E \text{ and see to it that } \neg J. \qquad (\alpha)$$

$$\text{Sophie ought to see to it that } \neg E. \qquad (\beta)$$

$$\text{Sophie ought to see to it that } \neg J. \qquad (\gamma)$$

In DMDL$^+$II, however, ($\alpha$) must be formalized as

$$O_{(s,c)}\neg E \wedge O_{(s,c)}\neg J \qquad (17)$$

for some $c$. Then $c$ may be divided into two sub-considerations that lead to ($\beta$) and ($\gamma$) respectively. The sub-considerations $c_1$ and $c_2$ we considered above can be seen as just such sub-considerations.

It might be said here that no such considerations were necessary for Sophie to feel she ought to keep Eva and Jan. Note, however, that we are not dealing with the actual process of Sophie's conscious experience. We are dealing with the logic behind the generation of the moral dilemma. A moral dilemma arises only when two or more moral judgments supported by moral sub-considerations are in conflict. Otherwise, there would be no difficult moral choice.[u] What Sophie's example shows us is the fact that even one and the same moral principle can support two different moral judgments through two different moral sub-considerations when combined with different facts.[v]

[u]Even if Sophie felt she ought to keep Eva and Jan without any step by step considerations, if she had been asked why she felt so, she could have said something like this: "They are my children. I have to keep them safe." In this sense, something like the principle (i) and the facts (ii) and (iii) are relevant.

[v]I suggested a possibility of avoiding deontic explosion by personifying moral principles in Yamada (2008a, p. 311)[2]. Example 4.1 has shown us that it is not good enough. What we should personify as commanders who have jointly produced a moral dilemma are the moral sub-considerations that lead to the moral judgments which constitute the dilemma, and not just the moral principles involved.

## 5. Contrary-to-duty scenarios in DMDL$^{+}$II

There are many contrary-to-duty scenarios (CTD scenarios, for short) in the literature. The following example is taken from Chisholm (1963).[18]

**Example 5.1.** Chisholm (1963)

It ought to be that a certain man go to the assistance of his neighbors. (a)

It ought to be that if he does go he tell them he is coming. (b)

If he does not go then he ought not tell them he is coming. (c)

He does not go. (d)

The problem discussed in the literature using CTD scenarios of this form is that all the four possible formalizations of the scenarios in SDL fail to satisfy one or the other of the following two conditions.

These four statements are consistent. (i)

Each of them is logically independent from the others. (ii)

Let $g$ and $t$ be the proposition that a certain man goes to the assistance of his neighbors and the proposition that he tells them he is coming respectively. Then Table 1 shows the four formalizations in SDL and their consequences on the conditions (i) and (ii).

Table 1. Formalizations in SDL.

| (F1) | (F2) | (F3) | (F4) |
|---|---|---|---|
| $Og$ | $Og$ | $Og$ | $Og$ |
| $O(g \to t)$ | $O(g \to t)$ | $g \to Ot$ | $g \to Ot$ |
| $\neg g \to O\neg t$ | $O(\neg g \to \neg t)$ | $\neg g \to O\neg t$ | $O(\neg g \to \neg t)$ |
| $\neg g$ | $\neg g$ | $\neg g$ | $\neg g$ |
| (i) fails. | (i) holds. | (i) holds. | (i) holds. |
| (ii) holds. | (ii) fails. | (ii) fails. | (ii) fails. |

Note that in Chisholm's scenario the antecedent of the conditional is inside the scope of the modal operator in (b) while the antecedent of the conditional is outside the scope of the modal operator in (c), and (F1) faithfully reproduces this characteristic.

Chisholm uses his scenario to show that the systems of deontic logic that allow both formulas of the form '$O$( if $a$ then $b$)' and formulas of the form 'if $a$ then $Ob$'

give rise to contradiction (Chisholm, 1963, pp. 34–35). Thus, he derives $Ot$ from (a) and (b) with the help of a version of the axiom K for $O$, and $O\neg t$ from (c) and (d). Then he derives a contradiction by appealing to a version of (D).

Since the final step of Chisholm's derivation can be reproduced in SDL, but not in SDL$^-$, (F1) satisfies both of the conditions (i) and (ii) in SDL$^-$. Then the situation (F1) describes becomes an example of a moral dilemma, and we can represent it without triggering deontic explosion in DMDL$^+$II. Moreover, examining DMDL$^+$II analogues of the above four formalizations will shed some light on the condition (ii).

Now it is clear that the three moral judgments (a), (b) and (c) in Chisholm's scenario are supported by different moral sub-considerations. The crucial parts of them seem to be the following.

Everyone ought to be helpful to his neighbors. ($\alpha$)

If he does go to the assistance of his neighbors, by letting them know that he is coming, he can enable them to take his assistance into consideration when they plan what to do. ($\beta$)

If he does not go, telling them that he is coming amounts to telling a lie. ($\gamma$)

Let $c_\alpha$, $c_\beta$ and $c_\gamma$ be the sub-considerations that includes $\alpha$, $\beta$ and $\gamma$ respectively and $a$ be the relevant person. Then Table 2 gives the summary of the four possible formalizations of Chisholm's scenario in DMDL$^+$II and their consequences on the conditions (i) and (ii).

Table 2. Formalizations in DMDL$^+$II.

| (RF1) | (RF2) | (RF3) | (RF4) |
|---|---|---|---|
| $O_{(a,c_\alpha)}g$<br>$O_{(a,c_\beta)}(g \to t)$<br>$\neg g \to O_{(a,c_\gamma)}\neg t$<br>$\neg g$ | $O_{(a,c_\alpha)}g$<br>$O_{(a,c_\beta)}(g \to t)$<br>$O_{(a,c_\gamma)}(\neg g \to \neg t)$<br>$\neg g$ | $O_{(a,c_\alpha)}g$<br>$g \to O_{(a,c_\beta)}t$<br>$\neg g \to O_{(a,c_\gamma)}\neg t$<br>$\neg g$ | $O_{(a,c_\alpha)}g$<br>$g \to O_{(a,c_\beta)}t$<br>$O_{(a,c_\gamma)}(\neg g \to \neg t)$<br>$\neg g$ |
| (i) holds.<br>(ii) holds. | (i) holds.<br>(ii) holds. | (i) holds.<br>(ii) fails. | (i) holds.<br>(ii) fails. |

Note that two revised formalizations, (RF1) and (RF2), satisfy the conditions (i) and (ii). Thus, we might be expected to say which is better. As the moral judgment (c) and the statement (d) seems to jointly imply that $a$ ought not tell his neighbors he is coming, (RF1) may look better. But things are not so simple.

Suppose $(\mathcal{M}, w)$ is a situation where all the four formulas in (RF2) hold. Since we have $\mathcal{M}, w \models_{\mathsf{DMDL^{+}II}} \neg g$, if $\mathcal{M}, w \models_{\mathsf{DMDL^{+}II}} t$, we have $\langle w, w \rangle \notin D_{(a,c_\gamma)}$. This means that, given $\mathcal{M}, w \models_{\mathsf{DMDL^{+}II}} \neg g$, we must have $\mathcal{M}, w \models_{\mathsf{DMDL^{+}II}} \neg t$ in order for $w$ to be permissible according to $c_\gamma$.

So, it is not easy to find out which of (RF1) and (RF2) is better than the other. Moreover, here we also have yet another option of introducing dyadic deontic operators to deal with conditional obligations. Although we have seen that introduction of dyadic deontic operator is not necessary if all we need is just to satisfy the conditions (i) and (ii), it may be desirable for some other reasons. As the issues to be discussed in this connection are beyond the scope of the present paper, we leave them for further study.[w]

Before concluding, however, we would like to make one observation on (RF3) and (RF4). They fail to satisfy the condition (ii) because we have

$$\neg g \vdash g \rightarrow \varphi, \tag{18}$$

in propositional logic. But we could ask here whether it really shows that (RF3) and (RF4) are bad formalizations. Of course, we feel that the moral judgment (b) in Chisholm's Scenario is independent from the statement (d). The reason why we feel this way, however, seems to be that the judgment (b) is based on some consideration that include something like ($\beta$) and the statement like (d) is irrelevant to such a consideration.

We should not conclude from this, however, that this phenomenon is peculiar to moral or deontic reasoning as distinguished from factual reasoning. Compare this with the following pair of statements.

(1) If John is Bob's son, John will be bold.
(2) John is not Bob's son.

As Grice (1967)[19] has pointed out, it is odd to assert (1) on the grounds that (2) is true. Consider the following conversation.

**Example 5.2.**

**A:** If John is Bob's son, John will be bold.
**B:** Why?
**A:** Boldness is a hereditary trait, and Bob is bold.
**C:** But John is not Bob's son.
**A:** Oh, then John might not become bold.

---

[w]We also have still another option of introducing some kind of conditional other than the material conditional, but pursuing this possibility is also beyond the scope of this paper.

As this conversation shows, a factual conditional statement can be supported by some known or presumed regularity that connects its antecedent and its consequent, and when we do have such a support, the information to the effect that the antecedent is false will only make it practically pointless to assert the conditional. It will not provide any additional support to A's act of asserting the conditional in such a circumstance.

In a similar way, although $\neg g$ implies $g \rightarrow O_{(a,c_\beta)}t$ , there is no role to be played by this implication in deciding what the person $a$ ought to do. The command the consideration $c_\beta$ gives can be useful to $a$ if there is a possibility that he does go, but given $\neg g$, the consideration to which he should listen is not $c_\beta$ but $c_\gamma$. As the crucial independence of the judgment (b) from the statement (d) is the independence of the moral consideration $c_\beta$ from (d), even (RF3) and (RF4) can be considered not as bad as Table 2 might suggest.

## 6. Concluding remarks

We have seen that by personifying moral considerations that support moral judgments as moral command issuing authorities, it becomes possible to represent moral dilemmas without triggering deontic explosion in a dynamic logic of acts of commanding $\mathsf{DMDL}^+\mathsf{II}$. As $\mathsf{DMDL}^+\mathsf{II}$ is a dynamic extension based on a multi-agent variant of $\mathsf{SDL}^-$, this means that moral dilemmas can be accommodated without weakening $\mathsf{SDL}^-$ any further.

We have also seen that a CTD scenario can be formalized in $\mathsf{DMDL}^+\mathsf{II}$ in such a way that the three moral judgments and the one statement involved are consistent and mutually logically independent. Moreover, even in the formalizations where logical independence is lost, the intuition of independence can be captured by the logical independence of each of the three moral considerations that support three moral judgments from the other moral sub-considerations and the fact involved in the scenario. Here again, taking moral considerations behind moral judgments into account plays a pivotal role.

The fact that the paradox of a CTD imperative is avoided in $\mathsf{DMDL}^+\mathsf{II}$ shows that introduction of a dyadic deontic operator is not necessary to avoid it. As we have noted in the last section, however, introduction of a dyadic deontic operator may be desirable for some other reasons, and there seems to be no insurmountable difficulty in indexing dyadic deontic operators by the set $I \times I$. In addition, we also have still another option of introducing some conditional other than the material conditional. So, one of the remaining tasks for further research is that of examining how conditional obligations and conditional commands should be formalized.

Another issue we need to consider is the following. Moral considerations that lead to moral judgments themselves include moral or deontic reasoning. So, it must be possible to use deontic logic to formalize them. In order to apply our dynamic approach to such considerations, however, we need to extend the logic substantially. First note that such considerations usually refer to general moral principles such as (i) in the Example 4.1 or ($\alpha$) discussed in relation to Example 5.1. In order to formalize them without losing their generality, we need to introduce quantification. Then we will be able to say something like "For any $x$ and $y$, if $y$ is a child of $x$, $x$ ought by the name of the consideration $c$ to see to it that $y$ is safe", or "For any $x$, $x$ ought by the name of the consideration $c$ to see to it that for any $y$, if $y$ is a child of $x$, $y$ is safe" for some consideration $c$.[x] Second, such general moral principles themselves may be derived from more fundamental moral principles. The consideration $c$ mentioned in the above principles can be a consideration that derives them from such relatively more fundamental principles. Unless we wish to introduce infinite regress, however, we need to introduce special constant, say $M$, so that the most fundamental moral principles that are not derived from others can be expressed by saying that everyone ought by the name of $M$ to see to it that so and so. We may think of $M$ as the ultimate moral commander, the morality itself, so to speak.

Another worry here may be this. By the DMDL$^{+}$II analogues of the CUGO Principle and the Partial Inertia Principle, we have

$$[!_{(i,M)}(p \wedge \neg p)]O_{(i,M)}(p \wedge \neg p) \ . \tag{19}$$

Obviously, this generates deontic explosion. The only thing we need $M$ to do, however, is to give us fundamental moral principles, and $O_{(i,M)}(p \wedge \neg p)$, or its quantified version "For any $x$, $O_{(x,M)}(p \wedge \neg p)$", will not qualify as a fundamental moral principle. So, we should just avoid letting $M$ to give us a command of the form $!_{(i,M)}(\varphi \wedge \neg\varphi)$ when we construct a substantial moral theory. As introducing $M$ is in this sense safe, the task of introducing quantification into dynamified deontic logic is another remaining task for further study.

Obviously, there are still many other issues left for further study. We conclude by mentioning just two of them.

We need to note that the avoidance of deontic explosion by itself does not give us a solution to a moral dilemma. An agent who is caught in a dilemma still needs to decide what to do. Since it was not possible for Sophie to see to it that $\neg E \wedge \neg J$, she had no choice but to abandon that goal. This was inevitable. But then, what else she could do?

Intuitively speaking, it seems that Sophie had to prevent the worst result that $E \wedge J$. This means that she had to see to it that $\neg(E \wedge J)$, which is equivalent to

[x]Term-modal logics developed by Fitting *et al.* (2001)[20] may be of much help here.

$\neg E \vee \neg J$. To view things this way is to see the world where $\neg E \vee \neg J$ holds better than the world where $E \wedge J$ holds. In order to model this view, however, we need some ordering on the set of possible worlds. The "betterness" ordering based on the syntactic priority ordering applied to the discussion of CTD scenarios in van Benthem, Grossi & Liu (2014)[11], for example, is such an ordering. Examining the possibility of combining this ordering and the double indexing of this paper is an interesting task for further study.

Now, if the above intuitive remark is correct, Sophie had to see to it that $\neg E \vee \neg J$. This sounds like a form of disjunctivism propounded, for example, by Brink (1994)[21]. The main difference between his view and ours consists in the fact that he keeps the principle (D) in various forms while we reject it. Thus, Brink distinguishes prima facie oughts from all-things-considered oughts and only admits conflicts of prima facie oughts. According to his proposal, the situation Sophie was in can be represented by the following set of formulas(*op. cit.*, p. 238).[y]

$$o(\neg E), \quad \text{(a)} \qquad \neg\Diamond(\neg E \wedge \neg J), \quad \text{(e)}$$
$$o(\neg J), \quad \text{(b)} \qquad O(\neg E \vee \neg J), \quad \text{(f)}$$
$$\neg(o(\neg E) > o(\neg J)), \quad \text{(c)} \qquad \neg O(\neg E), \quad \text{(g)}$$
$$\neg(o(\neg J) > o(\neg E)), \quad \text{(d)} \qquad \neg O(\neg J), \quad \text{(h)}$$

where $o$ stands for prima facie obligation, $O$ for all-things-considered obligation, and $o(A) > o(B)$ means that $o(A)$ is weightier than $o(B)$.

The same situation can be represented in $\mathsf{DMDL^{+}II}$ as follows.

$$O_{(s,c_1)}(\neg E), \quad \text{(a*)} \qquad \neg\Diamond(\neg E \wedge \neg J), \quad \text{(e*)}$$
$$O_{(s,c_2)}(\neg J), \quad \text{(b*)} \qquad O_{(s,c_3)}(\neg E \vee \neg J), \quad \text{(f*)}$$

where $c_3$ stands for the informal consideration mentioned above, which can be considered as including $c_1$, $c_2$, and (e*).[z]

Brink denies the existence of moral dilemmas by (g) and (h) because they will lead to paradoxes when combined with the principle (D) (Brink, 1994, pp. 232–236). Since we reject the principle (D), however, we don't have to deny the existence of moral dilemmas in this way. But at least we have to admit that (a*)

[y]The actual symbol Brink used to represent the relevant kind of possibility was ♦. See *op. cit.*, p. 226. Although he doesn't mention (e) in his discussion on p. 238, he uses it in discussing the consequence of admitting a moral dilemma on pp. 228, 233, 235. I add (e) here to indicate the existence of the conflict of obligations.

[z]If we borrow $>$, we can also add the analogues of (c) and (d). Informally, they mean that neither of (a*) and (b*) is weightier than the other. Incidentally, we also have $O_{(s,c_1)}(\neg E \vee \neg J)$ and $O_{(s,c_2)}(\neg E \vee \neg J)$ in this situation. This may be of some interest when we study Horty's deontic logic to be mentioned shortly.

and (b*) cannot be said to be all-things-considered obligations as $c_1$ and $c_2$ are sub-considerations. Thus, the following questions are yet to be asked. Is (f*) an instance of all-things-considered obligations? Do we have to accept the following analogues of (g) and (h) even in the absence of the principle (D)?

$$\neg O_{(s,c_3)}(\neg E)\,, \tag{g*}$$

$$\neg O_{(s,c_3)}(\neg J)\,. \tag{h*}$$

In studying these questions, the theory of reasons developed in the form of default logic in Horty (2012)[22], I hope, may shed light on the nature of the reasoning in moral considerations for two reasons. First, as moral principles involved in moral considerations can be considered as defeasible default rules, moral considerations may be formalized in deontic logic based on default logic. Second, and more relevantly to the above questions, one of the two deontic logics developed and examined by Horty is meant to be a formally precise version of the disjunctivist approach while the other one allows a conflict of all-things-considered obligations (*op. cit.*, pp. 73-74). Examining them is another, and urgent, task left for further research.

**Acknowledgments**

This paper is based on the research supported by JSPS Grant-in-Aid for Scientific Research (B) (KAKENHI 17H02258). An earlier version of this paper was presented in ALC 2017 (The 15th Asian Logic Conference, July 10–14, 2017, National Institute for Mathematical Science, Daejeon, Korea). I am grateful to the Program Committee of ALC 2017 for inviting me to give a lecture and the participants for helpful comments and discussions. I would also like to thank two anonymous referees of this proceedings for their helpful and illuminating comments. They were of great help for making the discussions in this paper clearer. The remaining unclarity and mistakes are of course due to me.

**References**

1. T. Yamada, Logical dynamics of commands and obligations, in *New Frontiers in Artificial Intelligence, JSAI 2006 Conference and Workshops, Tokyo, Japan, June 2006, Revised Selected Papers*, eds. T. Washio, K. Satoh, H. Takeda and A. Inokuchi, Lecture Notes in Artificial Intelligence, Vol. 4384 (Springer-Verlag, Berlin / Heidelberg / New York, 2007).
2. T. Yamada, Logical dynamics of some speech acts that affect obligations and preferences, *Synthese* **165**, 295 (2008).
3. T. Yamada, Acts of commanding and changing obligations, in *Computational Logic in Multi-Agent Systems, 7th International Workshop, CLIMA VII, Hakodate, Japan, May*

*2006, Revised Selected and Invited Papers*, eds. K. Inoue, K. Sato and F. Toni, Lecture Notes in Artificial Intelligence, Vol. 4371 (Springer-Verlag, Berlin / Heidelberg / New York, 2007).
4. T. Yamada, Acts of promising in dynamified deontic logic, in *New Frontiers in Artificial Intelligence, JSAI 2007 Conference and Workshops, Miyazaki, Japan, June 18-22, 2007, Revised Selected Papers*, eds. K. Sato, A. Inokuchi, K. Nagao and T. Kawamura, Lecture Notes in Artificial Intelligence, Vol. 4914 (Springer-Verlag, Berlin / Heidelberg / New York, 2008).
5. T. Yamada, Acts of requesting in dynamic logic of knowledge and obligation, *European Journal of Analytic Philosophy* **7**, 59 (2011).
6. T. Yamada, Assertions and commitments, *The Philosophical Forum* **47**, 475 (2016).
7. L. Goble, A logic for deontic dilemmas, *Journal of Applied Logic* **3**, 461 (2015).
8. H. Prakken, Two approaches to the formalisation of defeasible deontic reasoning, *Studia Logica* **57**, 73 (1996).
9. H. Prakken and M. Sergot, Dyadic deontic logic and contrary-to-duty obligations, in *Defeasible Deontic Logic*, ed. D. Nute, Synthese Library, Vol. 263 (Kluwer Academic Publishers, Dordrecht / Boston / London, 1997) pp. 223–252.
10. J. Carmo and A. Jones, Deontic logic and contrary-to-duties, in *Handbook of Philosophical Logic*, eds. D. M. Gabbay and F. Guenthner (Kluwer Academic Publishers, Dordrecht / Boston / London, 2001) pp. 265–343, second edn.
11. J. van Benthem, D. Grossi and F. Liu, Priority structures in deontic logic, *Theoria* **80**, 116 (2014).
12. H. van Ditmarsch, W. van der Hoek and B. Kooi, *Dynamic Epistemic Logic*, Synthese Library, Vol. 337 (Springer, Dordrecht/Berlin, 2007).
13. A. Baltag and B. Renne, Dynamic epistemic logic The Stanford Encyclopedia of Philosophy (Winter 2016 Edition), https://plato.stanford.edu/entries/dynamic-epistemic/, (2016).
14. T. Yamada, The completeness of ECL: a revised proof, Unpublished draft, (2018).
15. G. H. von Wright, Is there a logic of norms?, *Ratio Juris* **4**, 265 (December 1991).
16. W. Stylon, *Sophie's Choice* (Vintage, London, 2004), First published in Great Britain by Jonathan Cape 1979. Page references are to the version published by Vintage 2004.
17. R. B. Marcus, Moral dilemmas and consistency, *Journal of Philosophy* **77**, 121 (1980), Reprinted in Ruth Barcan Marcus: *Modalities: Philosophical Essays*, Oxford University Press, New York / Oxford, 1993, 125-141.
18. R. M. Chisholm, Contrary-to-duty imperatives and deontic logic, *Analysis* **24**, 33 (1963).
19. P. Grice, Indicative conditionals (1989), pp. 58-85.
20. M. Fitting, L. Thalmann and A. Voronsky, Term-modal logics, *Studia Logica* **69**, 133 (2001).
21. D. O. Brink, Moral conflict and its structure, *The Philosophical Review* **103**, 215 (1994).
22. J. F. Horty, *Reasons as Defaults* (Oxford University Press, Oxford, 2012).

# Plural Arithmetic

Byeong-uk Yi

*Department of Philosophy, University of Toronto,*
*Toronto, ON M5R 2M8, Canada*
and
*Department of Philosophy, Kyung Hee University,*
*Seoul 130-701, South Korea*
*E-mail: b.yi@utoronto.ca*
*http://philosophy.utoronto.ca/directory/byeong-uk-yi/*

This paper presents an analysis of arithmetic based on the view that natural numbers are properties.

*Keywords*: Natural number; arithmetic; plural logic; plural predicate; plural property; one/many; potential infinity; Gottlob Frege; Peano axioms.

## 1. Introduction: The property view of natural numbers

Consider some statements with words for natural numbers (e.g., '2', 'two'):

(1) a. *2 < 3*.
b. *Two* is smaller than *three*.
c. The number of logicians who coauthored PM is *two*.[1]
d. Russell and Whitehead are *two* logicians.

(1a), it seems, has the same meaning or content as (1b), and we can take the Arabic numerals '2' and '3' to have the same meanings or functions as the English number words 'two' and 'three', respectively. If so, what is the meaning or content of (1b)? And what does the number word 'two', for example, mean or refer to? To answer these question, it is useful to consider a variety of truths involving numerals, such as (1b)–(1d). In these statements, 'two' seems to have different meanings or functions. In (1b) and (1c), it occurs as a singular term that seems comparable to proper names. In (1d), by contrast, it occurs like an adjective to modify the nominal 'logicians'. Does this mean that the numeral 'two' has the same kind of ambiguity as 'bank'?

[1] I use 'PM' in this paper as a name for the book *Principia Mathematica*.

I think not. It is a purely accidental feature of English that the language has a word with the same sound and spelling for both some financial establishments and some areas of land along rivers. Many languages have two unrelated words matching the two uses of 'bank'. By contrast, the two uses of 'two' are closely related. Consider (2):

(2) Russell and Whitehead are *two* logicians, and Hilbert and Bernays are *two* other logicians. So the former and the latter are *four* logicians, for *two* plus *two* is *four*.

This reasoning involves a step that connects an arithmetical truth (viz., two plus two is four) to an ordinary numerical truth: some logicians (viz., Russell and Whitehead) and some other logicians (viz., Hilbert and Bernays) are four logicians. This would be a non-sequitur involving the fallacy of equivocation, if the two uses of 'two' were as semantically unrelated as the two uses of 'bank'. But the step is clearly legitimate.[2]

I think this means that both arithmetical and ordinary numerical truths pertain to natural numbers. If so, can we give an account of numbers[3] that yields proper analyses of both kinds of truths that respect, and help to explain, connections between them?

It is usual to take numerals to occur primarily as singular terms comparable to proper names: 'Russell', 'Venus', 'London', etc. In defending logicism, the view that all arithmetical truths are reducible to logical (or analytic) truths,[4] Gottlob Frege forcefully presents this approach. He takes numerals to be proper names that refer to objects of a special kind, and holds that "Every individual number is a self-subsistent object" ( [ 2 ], 67). Specifically, he identifies numbers or the referents of numerals as certain classes or "extensions of concepts" ( [ 2 ], 79f),[5] which he considers logical objects. His identification of numbers with classes has a devastating problem, as Bertrand Russell has shown (the so-called Russell's paradox). Nevertheless, the view that numbers are objects of a special kind (some

[2]Yi ( [ 1 ], §§4.1–4.2) points out logical connections between substantive and predicative uses of numerals.

[3]In this paper, I use 'number' as a short for 'natural number'. Although some systems include 0 among natural numbers, I regard 1 as the smallest natural number and use 'number' for (positive) natural numbers. (See also note 32.)

[4]On logicism, all arithmetical truths are *analytic* truths, truths that depend on logic and definitions of arithmetical expressions. But I use 'logical truth' in a broad sense to include analytic truths that rest on logical definitions.

[5]He says "the Number which belongs to the concept $F$ is the extension . . . of the concept 'equal to the concept $F$'" ( [ 2 ], 79f). See also Russell [ 3 ], who says "a number is nothing but a class of similar classes" ( [ 3 ], 18; original italics). Russell ( [ 4 ], 18) makes essentially the same statement although he no longer accepts the existence of classes.

abstract objects) because numerals refer to objects in the likes of (2a)–(2c) remains the standard view in contemporary metaphysics and philosophy of mathematics.

This view, I think, has serious difficulties in accounting for the adjectival or predicative use of numerals, such as that of 'two' in (1d). The number word does not refer to any object in (1d). Moreover, (1d) logically follows from (3):

(3) Russell is a logician, Whitehead is a logician, and Russell is not Whitehead.

(3) does not imply the existence of *any* object other than Russell and Whitehead. So it is hard to see how any other object, be it a class or something of another kind, is relevant to the truth of (1d).[6]

Interestingly, Frege seems to have reached the same conclusion at a point in his later thought. In the diary entry dated 24.3.1924 published posthumously in Frege [ 5 ], he writes:

> Indeed, when one has been occupied with these questions [e.g., what numbers are] for a long time one comes to suspect that our way of using language is misleading, that number words are not proper names of objects at all . . . and that consequently a sentence like 'Four is a square number' . . . cannot be construed like the sentence 'Sirius is a fixed star.' But how then is it to be construed? (Frege [ 5 ], 263).[7]

In the subsequent entry, dated 25.3.1924, he explores a potential answer to the question how to construe number words:

> It is already a step forward when a number is seen not as a thing but as *something belonging to a thing*, where the view is that different things, in spite of their differences, can possess the same One, as different leaves can, say, all possess the color green. Now which things possess the number one? Does not the number one belong to each and every thing? (Frege [ 5 ], 264; my italics)

By "something belonging to a thing", Frege means a *property* that some objects have. And the view he explores in this entry holds that numbers (or referents of numerals) are properties of objects. Call it the *property view of numbers*.

While noting that the property view would be preferable to his earlier view that numbers are objects, Frege suggests that the view has a serious problem. But I do not think the problem arises from the property view *per se*. And I think the view

---

[6]For more on this, see Yi ([ 6 ], §1).

[7]This entry is discussed in Hodes ([ 7 ], 143f).

yields an adequate account of numerical statements (e.g., (1a)–(1d)) that explains connections between the substantival and predicative uses of number words.

It would be useful to compare this view about numbers with a plausible view about colors, the view that colors are not objects but properties of objects. Consider (4a)–(4b):

(4) a. *Red* is brighter than black.
b. *Red* is the color of maple leaf symbols in Canadian flags.
c. Maple leaf symbols in Canadian flags are *red*.

While the color word 'red' occurs predicatively in (4c), it occurs as an abstract singular noun in (4a)–(4b). This does not mean that the word signifies or refers to an object in these statements. The predicative 'red' in (4c) refers to a property of some material objects (e.g., strawberries), and (4c) is true because maple leaf symbols in Canadian flags have this color property. And we can take the abstract noun 'red' in (4a)–(4b) to refer to the same property. In this view, (4b) states that the property it refers to is identical with the color property shared by maple leaf symbols in Canadian flags, and (4a) states that this property holds a certain relation (viz., being brighter than) to the referent of 'black'.

Similarly, we can take number words to refer to properties to reach the property view of numbers. Consider a statement involving the predicative use of 'two':

(5) Russell and Whitehead are *two*.

One cannot take the numeral in (5) to refer to an object, concrete or abstract. It refers, it seems, to a property of some objects (e.g., Russell and Whitehead), as 'red' does. The numeral in (1d) can also be taken to refer to the number property, for (1d) can be analyzed as the conjunction of (5) and 'Russell and Whitehead are logicians.' If so, we can take the numeral to have the same reference in (1b)–(1c). In this view, (1c) identifies the number property of the logicians who coauthored PM (i.e., Russell and Whitehead) with the property the numeral refers to, and (1b) states that this property has a certain relation (viz., being smaller than) to the property the numeral 'three' refers to.[8]

And we can see that the property view of numbers yields a straightforward explanation of connections between arithmetical and ordinary numerical truths invoked in (2) and the like. Consider (6)–(8):

[8] In holding the property view of numbers, I assume realism about properties, the view that properties exist, and take properties to be universals. I argue for realism about properties in Yi [ 8,9 ], and argue against the trope account, a version of the realist view of properties that denies the existence of universals, in Yi [ 10 ].

(6) *Two* plus *two* is *four* (or $2 + 2 = 4$).

(7) If some things are *two* and some other things are *two*, then the former and the latter are *four*.

(8) If some things are *two* logicians and some other things are *two* logicians, then the former and the latter are *four* logicians.

The last step of (2) invokes (6) to assume (8). We can use an analysis of (6) based on the property view to show that this is a valid step. To do so, it is sufficient to show that (6) implies (7), for (7) implies (8).[9] Now, we can define a function for number properties, the *successor function* $S$, that satisfies (9):

(9) If some things are $N$ (e.g., two) and something is not one of them, then the former and the latter are $S(N)$ (e.g., $S$(two)).

And we can define the plus operation for numerical properties in terms of the successor function: two is the successor of one (or $S$(one)), three the successor of two (or $S$(two)), four the successor of three (or $S$(three)), etc. Then we can show:

(10) If some things are $N$ (e.g., two) and some other things are two, then the former and the latter are $S(S(N))$ (e.g., $S(S$(two))).

This implies (7), for four is the successor of the successor of two.

So I hold the property view of numbers. On this view, numerals refer to properties and natural numbers are numerical properties that they refer to. In this paper, I develop this view by presenting an analysis of arithmetic based on the view. In the next section, §2, I argue that numerical properties are properties of a special kind that I call *plural properties*. In §3, I give a sketch of an account of the logic and semantics of expressions signifying or referring to plural properties. The account develops regimented languages, (*regimented*) *plural languages*, that have symbolic counterparts of plural constructions of natural languages (e.g., (5)), and explains the logic of natural language plurals by characterizing the logic of regimented plural languages. In §4, I present an analysis of arithmetic based on the property view of numbers that can explain the logical connections between different uses of number words. To do so, I formulate the Peano-style axioms of arithmetic in plural languages and show that all those axioms except one (viz., PA4) are logical truths of plural languages. In §5, I conclude with remarks on the relation between logic and arithmetic.[10]

---

[9](8) follows from (7) and a logical truth: 'If some things are logicians and some (other) things are logicians, then the former and the latter are logicians.'

[10]§§2–3 prepare for the analysis of arithmetic in §4 by articulating the notion of *plural property* and

## 2. The plural property view

The property view identifies natural numbers as some number properties, which numerals can refer to. The number properties can be signified or referred to by predicates (e.g., 'be two') or other predicable expressions (e.g., the adjective 'two'). To refer to them, we can also use abstract nouns related to predicable expressions, such as the noun 'two', 'twoness', and 'being two'. Using the gerunds related to numerals (e.g., 'being two') as the canonical nouns for referring to number properties, we can state the property view as follows:

> *The Property View*: Natural numbers are properties that numerals refer to. Specifically, the numbers one, two, three, etc. are being one, being two, being three, etc., respectively.

It is useful to motivate this view, as we have seen, by comparing number words to color words. And the view, I think, helps to explain how we can have knowledge of numbers, as Maddy [ 11,12 ] and Kim [ 13 ] argue.[11] But there is a significant difference between number properties and color properties. Number properties differ from the usual properties recognized in standard accounts of attributes[12] (e.g., color properties) in that they in a sense pertain to *many* things taken together.[13] Unlike color properties, they are what I call *plural properties*. It would be useful to elaborate on this to meet usual objections to the property view.

### 2.1. *Numbers as plural properties*

While noting that the property view would be preferable to the standard view that numbers are objects, as we have seen, Frege suggests that the view has a serious problem by asking "Does not the number one belong to each and every thing?" ( [ 5 ], 264). In posing this question, he assumes that the answer is yes, and suggests that this gives rise to a problem for the property view because something that is many (e.g., two), in this view, must also be one if any number other than one (e.g., two) is to be a property that can be instantiated. I think he is right to assume the affirmative answer to the question, but I do not think this means that being two, for

presenting (regimented) plural languages and logical systems, systems of *plural logic*, that characterize the logics of first- and higher-order plural languages. Readers familiar with topics of the preparatory sections might skip them. (Note, however, that I formulate the analysis of arithmetic in higher-order plural languages, which go beyond the usual, first-order plural languages.)

[11] See also Yi [ 14 ].

[12] I use 'attribute' in a broad sense for both properties and relations (see §2.2).

[13] I use 'many' as a short for 'more than one' for convenience of exposition.

example, cannot instantiated at all or that something that is two must also be one. Unlike the usual properties (e.g., being red), being two is a property pertaining to many things taken together, and can be instantiated without overlapping with being one in instantiation. The number property is instantiated by, for example, the authors of PM (i.e., Russell and Whitehead) taken together, and these are not one but two.

Call properties that, like being two, pertain to many thing *taken together* (in short, *as such*) *plural properties*. They contrast with the usual properties that standard accounts of attributes recognize (e.g., being red). These are properties that pertain in a sense to one thing only, so to speak, at a time. Call them *singular properties*.[14] Most proponents of the property view assume standard accounts of attributes and identify natural numbers with singular properties. For example, J. S. Mill holds that a number is a property of an "aggregate", a whole composed of parts, such as an aggregate of pebbles ( [ 15 ], Bk. III, Ch. XXIV, §5), and Maddy [ 11,12 ] holds that numbers are cardinality properties of sets (e.g., being two-membered or having two members).[15] I think these views have a serious problem as Frege suggests, for any aggregate or set is one object and has the property of being one.[16]

But this problem does not arise because the views identify numbers as *properties* but because they identify them with *singular* properties. We can reject this identification and hold the natural view that number properties are plural properties. Combining this view with the property view of numbers yields the *plural property view of numbers*:

> *The Plural Property View*: Natural numbers (e.g., 1, 2) are plural properties (e.g., being one, being two).

To articulate this view, it is necessary to elaborate on the notion of plural property and explain the way in which plural properties pertain to the many.

---

[14]The above characterization of singular and plural properties is a preliminary meant to suggest the account given in §2.2, where I explain that being one is also a plural property.

[15]See also Kim [ 13 ], who identifies number properties with properties of sets or classes. Note that Maddy's recent view about numbers and arithmetic is considerably different from her earlier view presented in Maddy [ 11,12 ]. See, e.g., Maddy [ 16 ] for her recent view.

[16]So they must hold that some things (e.g., an aggregate with two parts, a set with two members) must be both one and two. See also Whitehead and Russell [ 17 ], who argue that classes are incomplete symbols because the existence of classes gives rise to "the ancient problem of the One and the Many" ( [ 17 ], 75). See Yi ( [ 18 ], §5; 2018b, §4.2) for discussions of this argument and their no-class theory. Some philosophers, who hold the thesis that composition is identity, argue that some things that are many must also be one because they compose some one thing. But this view is a logical contradiction, as I argue in Yi [ 19,20 ]. (See PL2 and PL5 in §4.)

### 2.2. *Plural attributes*

Consider (11a)–(11b):

(11) a. Serena *is a tennis player.*
  b. Serena *likes* Beyoncé.

We can analyze these statements as resulting from combining the italicized predicates with one or more terms. It is usual to take the singular terms 'Serena' and 'Beyoncé' to refer to objects (or non-predicable entities), and the predicates 'is a tennis player' and 'likes' to signify or refer to attributes (or predicable entities). As 'Serena' refers to Serena, 'is a tennis player' and 'likes' refer to being a tennis player and liking, respectively. Serena is not an attribute but an object, for no predicate can refer to her. By contrast, being a tennis player and liking, which predicates refer to, are attributes, which include properties and relations. Properties are one-place attributes, those with one argument place, and relations are multi-place attributes, those with more than one argument place. Thus being a tennis player is a property, for it has only one argument place (viz., the one matching the argument place of 'is a tennis player' that 'Serena' fills in (11a)); liking is a relation, for it has two argument places (viz., those matching the argument places of 'likes' that 'Serena' and 'Beyoncé' fill in (11b)). Now, we can invoke the semantics of terms and predicates figuring in (11a) and (11b) to characterize their truth conditions:

> (11a) is true, if the object that 'Serena' refers to instantiates the property that 'is a tennis player' refers to.
>
> (11b) is true, if the object that 'Serena' refers to holds the relation that 'likes' refers to with regard to the object that 'Beyoncé' refers to.

The usual semantic accounts apply the analysis sketched above only to singular constructions, which include (a) the singular terms 'Serena' and 'the author of *Hamlet*', and (b) the singular predicates 'is a tennis player' and 'likes'. But we can extend the analysis to plural constructions, which include (a*) the plural term 'Venus and Serena' and 'the authors of PM', and (b*) the plural predicates 'are siblings' and 'are two'.

Consider (12a)–(12c) and (13a)–(13d):

(12) a. *Venus and Serena* **are siblings**.
  b. *Venus and Serena* **are two**.
  c. *Venus and Serena* **collaborate**.

(13) a. *Venus and Serena* **won** *the 2000 Wimbledon women's doubles title.*
  b. *Russell and Whitehead* **wrote** PM.

c. Russell **is one of** *the authors of PM.*
d. *The Chinese millennial boys* **outnumber** *the Chinese millennial girls.*

The italicized phrases in these sentences are *plural terms*, and the boldface expressions are *plural predicates*, namely, predicates that can combine with plural terms. In (12a)–(12c), one-place predicates combine with plural terms. In (12b), for example, the predicate 'are two' combines with the plural term 'Venus and Serena', which fills its only argument place. In (13a)–(13d), two-place predicates admit plural terms in at least one of their argument places. In (13c), for example, the plural term 'the authors of PM' fills the second argument place of 'is one of'; in (13d), 'the Chinese millennial boys' and 'the Chinese millennial girls' fill the first and second argument places of 'outnumber', respectively.

We can now extend the notion of reference for plural terms and predicates. Like their singular cousins, plural predicates refer to attributes. For example, 'are siblings' and 'collaborate' refer to the property of being siblings and that of collaborating, respectively, and 'wrote' and 'outnumber' refer to the two-place relations of having written and outnumbering, respectively. How about plural terms? For example, the plural term 'Venus and Serena' refers to Venus and Serena, just as the singular term 'Serena' refers to Serena. But the plural term refers to two objects (taken together) while the singular term refers to one object (viz., Serena). For Venus and Serena are two humans, not one, while Serena is one human, not two. Similarly, the plural definite description 'the authors of PM' refers to the two logicians who wrote PM together: Russell and Whitehead. Unlike singular terms, then, a usual plural term refers to many objects (taken together). We can now invoke the semantics of plural terms and predicates to characterize truth conditions of plural predications, sentences resulting from combining plural predicates with plural terms. We can characterize the truth conditions of (12a) and (13b), for example, as follows:

> (12b) is true, if the objects that 'Venus and Serena' refers to instantiate the property that 'are two' refers to.

> (13c) is true, if the object that 'Russell' refers to holds the relation that 'is one of' refers to with regard to the objects that 'the authors of PM' refers to.

Now, attributes that plural predicates refer to have a special character. They differ from their singular cousins in having argument places that pertain to many things taken together.

Plural predicates can combine with plural terms because they have argument places that can admit plural terms. For example, the one-place predicate 'be two' (of which 'are two' is the plural form) has an argument place that can admit the plural term 'Venus and Serena', as in (12b), and the two-place predicate 'is one of' has an argument place (the second) that can admit 'the author of PM', as in (13c). Call such argument places *plural argument places*. Then a plural predicate is one that has at least one plural argument place. By contrast, singular predicates have no plural argument place. For example, 'is a tennis player' cannot combine with the plural term 'Venus and Serena' ('*Venus and Serena is a tennis player' is ill-formed), because its argument place can admit only singular terms. Call such argument places *singular argument places*. Now, call argument places of attributes matching singular argument places of predicates *singular*, and those matching plural argument places of predicates *plural*. And call attributes with only singular argument places *singular attributes*, and those with plural argument places *plural attributes*. Then attributes that singular predicates refer to (e.g., being a tennis player) are singular attributes. And those that plural predicates refer to are plural attributes. For example, being two is a plural property and being one of is a plural relation.

We can now see the way in which plural attributes pertain to the many while singular attributes do not. A plural attribute has a plural argument place and this can admit the objects that a plural term (e.g., 'Venus and Serena') refers to, for it matches an argument place of a predicate that can admit plural terms. For example, the argument place of being two can admit two tennis players (e.g., Venus and Serena) taken together. By contrast, every argument place of a singular attribute can admit only one thing, so to speak, at a time. For example, being a tennis player can admit Venus or Serena, but not the two taken together. The property can admit both, but only in the sense that it, so to speak, occurs twice to admit Venus in one occurrence and Serena in another. By contrast, the argument place of being two can admit both Venus and Serena, so to speak, in one occurrence. Similarly, the plural relation of having written has an argument place (the first) that can admit many things (e.g., Russell and Whitehead) taken together.

Now, a plural property might manifest its pertaining to the many by being *instantiated* by many things as such (i.e., taken together). For example, being two is instantiated by Venus and Serena (as such), which means that they, together, must fill the argument place of the property. But the two tennis players, for example, might fill the plural argument place of a property without instantiating the property. The property of being one, for example, is not instantiated by Venus and Serena (they are not one), but its argument place can still admit them (as such). For the complement of the property (viz., *not being one*) can admit them, who instantiate it. This means that the argument place of being one can also admit the two tennis players (as such), for it has the same character as the argument place of not being one. So a property can pertain to some things that do not instantiate it, just as predicates can legitimately combine with terms without yielding truths.

Call properties instantiated by many things (as such) *plurally instantiated properties*. Then all plurally instantiated properties are plural properties. But the converse does not hold. Being one is a plural property but is not plurally instantiated, as we have seen. Still all plural properties (plurally instantiated or not) pertain to the many in the sense explained above.

Note that the account of attributes sketched above departs radically from standard views of attributes. Surely, a property can be instantiated by each one of many things, and a two-place relation, for example, can relate two things. But it is usually taken for granted that there is no property instantiated by many things (as such), and this thesis presupposes that there is no plural attribute. I think the two related theses, which I call the *Principle of Singular Instantiation* and the *Principle of Singularity* (Yi [ 6 ], 167ff), are the main obstacles to reaching proper accounts of the logic and semantics of plural constructions. Those who assume the theses would argue that there can be no genuine plural predicates (and, thus, no genuine plural terms) because there are no attributes such predicates could refer to. So they would conclude that although natural languages might seem to have plural constructions, they must be considered devices for abbreviating purely singular constructions (i.e., constructions involving no plural predicates). But this stilted view of plurals conflict with robust logical relations pertaining to plurals, as I have argued in Yi [ 21,22 ]. So I reject the two theses about attributes underlying the bias against plurals and accept a liberal view that recognizes plural attributes as well as their singular cousins. I call this view the *plural view of attributes*. Embracing the view, we can accept the natural view of plural constructions: plural constructions (e.g., 'Venus and Serena') are devices for talking about the many. This view provides the basis for an account of the logic and meaning of plural constructions that underlies the plural property view of numbers.[17]

## 3. Plural languages and plural logic

In this section, I prepare for giving an analysis of arithmetic based on the plural property view. I formulate the analysis in a regimented or symbolic language that includes counterparts of plural constructions of natural languages. I call such regimented languages (*regimented*) *plural languages*. The basic ones among them are first-order plural languages, which result from adding symbolic counterparts of basic plural constructions of natural languages to elementary languages, the

[17] Yi [ 6 ] develops the plural view of attributes. See also Yi ( [ 23 ], 104–8; [ 22 ], 251ff).

standard first-order languages. Unlike elementary languages (and their higher-order extensions), first-order plural languages have counterparts of number predicates: 'be one', 'be two', 'be three', etc. We can obtain higher-order plural languages (i.e., higher-order extensions of first-order plural languages) by adding predicates, variables, and quantifiers that relate to predicates of first-order plural languages. And we can formulate an analysis of arithmetic based on the plural property view in higher-order plural languages. Let me give a sketch of plural languages and a logical system, *plural logic*, that captures the logic of plural languages.[18]

### 3.1. *First-order plural languages*

The usual regimented languages (e.g., elementary languages) can be taken to result from regimenting singular constructions of natural languages. We can develop extensions of those languages by adding refinements of plural constructions. We can call the extended languages *(regimented) plural languages* to contrast them with the usual regimented languages, which we can call *(regimented) singular languages*. The basic ones among regimented plural languages, first-order plural languages, are first-order extensions of elementary languages. In addition to expressions of elementary languages, they have symbolic counterparts of basic plural constructions of natural languages:

(i) (first-order) plural variables: '*xs*', '*ys*', etc.

(ii) (first-order) plural predicates: '**C** (**cooperate**)', '**H** (**is.one.of**)', etc.

(iii) (first-order) plural quantifiers: '**Π**' (the universal) and '**Σ**' (the existential).

Plural variables are plural cousins of the usual, singular variables: '*x*', '*y*', etc. They are refinements of anaphorically used plural pronouns (e.g., 'they'), as singular variables are refinements of anaphorically used singular pronouns (e.g., 'it').[19] The plural quantifiers, which bind plural variables, are plural cousins of the usual, singular quantifiers '∀' and '∃'. They result from refining the plural quantifier phrases 'Any things are such that . . .' and 'There are some things such that . . .', as the singular quantifiers '∀' and '∃' result from refining the singular quantifier phrases 'Anything is such that . . .' and 'There is something such that . . .'.[20] And

[18]This section gives a sketch of my account of the logic and semantics of plural constructions presented in, e.g., Yi [ 1,21,22 ]. The account departs radically from traditional approaches to their logic and semantics. For other accounts that take the same approach as mine, see, e.g., Linnebo [ 24 ], McKay [ 25 ], Oliver and Smiley [ 26 ], and Rayo [ 27 ].

[19]Pronouns are used anaphorically in, e.g., 'There are some cats and a dog over there, and *they* are surrounding *it*.'

[20]While singular variables relate to any one thing, plural variables relate to any *one or more* things. Thus plural quantifiers are semantically neutral about number and amount to 'Any one or more things

plural predicates are refinements of natural language predicates (e.g., 'cooperate') that can combine with plural pronouns or other plural terms (e.g., 'they', 'the authors of *PM*'). Elementary language predicates (e.g., '=', '$L$'), which are singular predicates resulting from refining singular forms of natural language predicates (e.g., 'is identical with', 'is a logician'), combine only with refinements of singular terms (e.g., the variable '$x$', the constant '$c$'). By contrast, plural predicates can combine with refinements of plural terms (e.g., '$xs$') and have *plural argument places*, namely, those that can admit plural terms (as well as singular terms).[21] For example, '**C**' (which amounts to 'cooperate') is a one-place plural predicate and '$\mathbf{C}(xs)$' is well-formed; and '**H**' (the plural language counterpart of 'is one of') is a two-place plural predicate and '$\mathbf{H}(x, ys)$' (in short, '$x\mathbf{H}ys$'), which amounts to 'It *is one of* them', is well-formed.

The predicate '**H** (**is.one.of**)' signifies a plural relation, one that holds between an object and any objects that include it. Like the elementary language predicate '=', it is a logical predicate. Its logical status helps to explain the logical equivalence between, e.g., 'Any one of Russell and Whitehead is a logician'[22] and 'Russell is a logician and Whitehead is a logician.' And one can use the predicate to define the plural cousin of the singular identity predicate, '≈ (**be.the.same.things.as**)':

**Definition 3.1 (Plural identity).** $xs \approx ys \equiv_{df} \forall z(z\mathbf{H}xs \leftrightarrow z\mathbf{H}ys)$.

And we can give a contextual definition of the term-connective '@ (**and**)':

**Definition 3.2 (Term-connective).**

$$\pi([xs@ys]) \equiv_{df} \Sigma zs[\forall x(x\mathbf{H}zs \leftrightarrow x\mathbf{H}xs \vee x\mathbf{H}ys) \wedge \pi(zs)],$$

where $\pi$ is a predicate and '$zs$' does not occur free in $\pi([xs@ys])$.

The term connective is a symbolic counterpart of the 'and' that occurs in, e.g., 'Russell and Whitehead are two' (=(5)).

are such that . . .' and 'There are some one or more things such that . . .', to which I think the simpler expressions 'Any things are such that . . .' and 'There are some things such that . . .', respectively, are equivalent.

[21] I call plural argument places that can also admit singular terms *neutral argument places* (and those that do not *exclusively plural argument places*). See Yi ([ 21 ], 479f). Note that most argument places of natural language predicates can admit both singular and plural terms. For example, 'be one' and 'be two' can admit singular terms (while taking the singular forms), as in 'Cicero is not two but one.'

[22] This sentence can be given as the analysis of the plural predication 'Russell and Whitehead are logicians.' To put it more precisely, the predicate 'be logicians' can be analyzed as 'be such that any one of them is a logician'. We can then explain that the plural predication is logically equivalent to the conjunction 'Russell is a logician and Whitehead is a logician.' See Yi ([ 21 ], 482ff).

### 3.2. *First-order plural logic*

We can give an account of the logic of plural constructions of natural languages by characterizing the logic of (regimented) plural languages. Call the logic of (first-order) plural languages *(first-order) plural logic*. To give a complete characterization of this logic, it is necessary to use the model-theoretic method because the logic is not axiomatizable. But we can formulate logical systems that capture enough of the logic for most purposes. Let me state some basic logical truths of plural languages that are axioms or theorems of one such system, $S_{\mathrm{PL}}$.[23]

Let $PL^1$ be a first-order plural language that extends an elementary language.[24] Then logical truths of $PL^1$ include all the logical truths of the elementary language. And it has other logical truths, which pertain to additional logical expressions of the plural language.[25]

Plural quantifiers are governed by essentially the same logic as their singular cousins, except that we can get instances of plural quantifications by replacing plural variables with singular as well as plural terms (see Axioms 5–7 in Appendix).[26] Thus '$\mathbf{\Pi} xs \exists y\, y\mathbf{H}xs$', for example, implies both '$\exists y\, y\mathbf{H}xs$' and '$\exists y\, y\mathbf{H}x$', which are its instances.[27] $PL^1$ has additional logical truths, such as P1–P4:

**P1**. $\mathbf{\Pi} xs \exists y\, y\mathbf{H}xs$. (Any things include something, i.e., given any things, there is something that is one of them.)

**P2**. $\forall x \forall y (x\mathbf{H}y \rightarrow x = y)$. (Anything that is one of something is identical with it.)

**P3**. *Substitutivity of Plural Identity* (*Schema*):
$[xs{\approx}ys \wedge \Phi] \rightarrow \Phi(xs/ys)$, where $\Phi(xs/ys)$ is the result of properly substituting '$ys$' for '$xs$' in $\Phi$. (Any things that are the same things as some things that are so-and-so must also be so-and-so.)

**P4**. *Plural Comprehension* (*Schema*):
$\exists y \Phi \rightarrow \mathbf{\Sigma} xs \forall y (y\mathbf{H}xs \leftrightarrow \Phi)$, where '$xs$' does not occur free in $\Phi$. (If there is

---

[23]See Yi ([ 6 ], 181; [ 22 ], 257 & 262) about the non-axiomatizability of first-order plural logic. For model-theoretic characterizations of the logic, see Yi ([ 1 ], §2.2.2; [ 22 ], §6). For partial axiomatizations thereof, see Yi ([ 6 ], 179ff; [ 1 ], 57f; [ 22 ], §7). See also Appendix, which gives the axioms of $S_{\mathrm{PL}}$.

[24]PL$^1$ contains plural variables, quantifiers, and the logical predicate '**H**'. For the present purpose, we may take the language to have no non-logical expressions.

[25]Plural logic is a conservative extension of elementary logic. See Yi ([ 1 ]; [ 22 ], 260f).

[26]For plural existential and universal quantifiers are semantically neutral about number (see note 20).

[27]In subsequent discussions, I invoke inferences invoking the logic of plural quantifiers without noting them. Such inferences are natural and intuitive, and draw parallels to inferences invoking the logic of singular quantifiers.

something that is so-and-so, there are some things of which something is one if and only if it is so-and-so.)[28]

These are logical truths pertaining to logical predicates of $PL^1$. They are axioms of $S_{\mathrm{PL}}$ (see Axioms 8–9 and 11–12 in Appendix). It would be useful to note some theorems of the system:

**T1**. $\forall x \forall y(x\mathbf{H}y \leftrightarrow x = y)$. (Something is one of something if and only if the former is identical with the latter.)

**T2**. $\forall x \forall y(x \approx y \leftrightarrow x = y)$. (Something is the same thing as something if and only if the former is identical with the latter.)

**T3**. $\mathbf{\Pi} xs \mathbf{\Pi} ys \mathbf{\Sigma} zs\, zs \approx [xs@ys]$.

**T4**. $\mathbf{\Pi} xs \mathbf{\Pi} ys \forall z[\mathbf{H}(z, [xs@ys]) \leftrightarrow \mathbf{H}(z, xs) \vee \mathbf{H}(x, ys)]$. (Anything is one of some things and some things (taken together) if and only if it is one of the former things or one of the latter things.)

P1 and P2 imply T1, for P1 implies the converse of P2. P2 and T1 imply T2.[29] Using the definition of the complex term '$[ys@zs]$' (Definition 3.2), we can derive T3 and T4 from P1 and P4.

## 3.3. *Higher-order plural languages and logic*

As we can build standard (i.e., singular) higher-order languages as extensions of elementary languages, we can build higher-order plural languages as extensions of first-order plural languages. Like standard higher-order languages, higher-order plural languages include higher-order predicates, function symbols, variables, and quantifiers.[30] But second-order (plural) predicates, function symbols, variables and quantifiers of plural languages differ from their cousins in standard higher-order languages in replacing (first-order) *plural* predicates, and relate to *plural* attributes. For example, a second-order variable (e.g., '$X$') that can replace first-order plural

[28] P4 is comparable to the comprehension principles for classes and properties, but differs from them in being restricted by the antecedent '$\exists y \Phi$.' We cannot strengthen it by removing this restriction, for the consequent implies the antecedent (see P1).

[29] '$xs \approx xs$' is a logical truth based on the definition of plural identity (Definition 3.1).

[30] Higher-order predicates or function symbols have argument places that can admit predicates, higher order-variables can replace predicates, and higher-order quantifiers bind higher-order variables.

predicates (e.g., the symbolic counterpart of 'be two') ranges over first-order plural attributes (i.e., plural attributes pertaining to objects).[31]

In plural languages with higher-order variables, we can formulate second-order statements that comprehend instances of schematic theses of first-order plural languages, such as P3 and P4. Using '$X$' as a 1-place second-order variable, we can formulate the principles of plural comprehension and substitutivity of plural identity as follows:

**P3***. *Substitutivity of Plural Identity*:
$\forall^2 X[xs \approx ys \wedge X(xs) \rightarrow X(ys)]$. (Any things that are the same things as some things that have a plural property must also have the plural property.)

**P4***. *Plural Comprehension*:
$\forall^2 X[\exists y X(y) \rightarrow \boldsymbol{\Sigma} xs\, \forall y\, (y\mathbf{H}xs \leftrightarrow X(y))]$. (If something has a property, there are some things of which something is one if and only if it has the property.)

## 4. Plural arithmetic

We can state Peano axioms of arithmetic informally as follows:

**Peano Axioms**

**A1**. 1 is a number.

**A2**. The successor of any number is also a number.

**A3**. 1 is not the successor of a number.

**A4**. No two numbers have the same successor.

**A5**. *Mathematical Induction*:
If 1 has a property and the successor of any number that has the property also has the property, then every number has the property.

Peano ( [ 28 ], 94) formulates these axioms in a first-order language (with expressions for classes).[32] This suggests the view that numbers are objects, which fall under the range of first-order variables. But proponents of the property view

[31] Plural second-order variables can replace singular predicates as well. Higher-order plural languages might, in addition, have the same higher-order expressions that standard higher-order languages have, such as singular second-order variables, which can replace only elementary predicates.

[32] See also Dedekind [ 29 ] and Russell ( [ 4 ], 5f). Peano regards 1 as the smallest natural number, but Russell includes 0 among natural numbers.

of numbers can take the informal statements as higher-order statements about numerical properties, such as being one, which is identified as the number 1.

Let *HPL* be a higher-order plural language that has the first-order plural predicate '**1**' for being one, the second-order predicate '*Num*' for natural number properties, the function symbol '*S*' for the successor function for plural properties, second-order variables that can replace plural predicates (e.g., '*X*'), third-order variables (e.g., '$\varphi$'), and second- and third-order quantifiers (e.g., the universal '$\forall^2$' and '$\forall^3$'). Then we can reformulate Peano axioms in HPL as follows:

**Axioms of Plural Arithmetic**

**PA1**. $Num(\mathbf{1})$.

**PA2**. $Num(X) \rightarrow Num(S(X))$.

**PA3**. $Num(X) \rightarrow S(X) \neq \mathbf{1}$.

**PA4**. $Num(X) \wedge Num(Y) \wedge S(X) = S(Y) \rightarrow X = Y$.

**PA5**. *(Higher-order) Mathematical Induction*:

$$\varphi(\mathbf{1}) \ \wedge \ \forall^2 X[Num(X) \wedge \varphi(X) \rightarrow \varphi(S(X))] \rightarrow \forall^2 X[Num(X) \rightarrow \varphi(X)].$$

Call the system with these axioms *Plural Arithmetic*. Proponents of the plural property view might take it to capture ordinary arithmetic.

Now, we can use logical expressions of plural languages to define arithmetical expressions. We can define the predicate '**1**' and function symbol '*S*' as follows:

**Definition 4.1.**

(a) $\mathbf{1}(xs) \equiv_{\text{df}} \exists x \forall y(y\mathbf{H}xs \leftrightarrow y = x)$. (There is something that is one of them such that nothing else is one of them.)

(b) $\mathbf{M}(xs, ys) \equiv_{\text{df}} \exists z({\sim}z\mathbf{H}ys \wedge xs \approx [ys@z])$. (*xs* are the same things as *ys* and something else, taken together.)

(c) $S(X)(xs) \equiv_{\text{df}} \mathbf{\Sigma} ys\ (\ \mathbf{M}(xs, ys) \wedge X(ys)\ )$. (*xs* are the same things as some things that are *N* and something else, taken together.)

These definitions yield definitions of all number predicates in terms of logical expressions of plural languages. For example, we can define '**2**', '**3**', and '**4**' as

'$S(\mathbf{1})$', '$S(\mathbf{2})$', and '$S(\mathbf{3})$', respectively. Moreover, we can define the predicate '$Num$' for natural number properties as follows:

**Definition 4.2.**

$$Num(X) \equiv_{\text{df}} \forall^3\varphi\,(\,\varphi(\mathbf{1}) \wedge \forall^2 X[\,\varphi(X) \rightarrow \varphi(S(X))\,] \rightarrow \varphi(X)\,).$$

Using these definitions, it is straightforward to show that all the axioms of Plural Arithmetic except PA4 are logical truths of higher-order plural languages.[33]

And such logical truths include PL1–PL3:

**PL1**. $\forall x\, \mathbf{1}(x)$. (Everything is one.)

**PL2**. $\mathbf{1}(xs) \rightarrow \sim S(N)(xs)$. (Any things that are one cannot be the same things as some things that are $N$ and something else, taken together.)[34]

**PL3**. $\mathbf{2}([x@y]) \leftrightarrow x \neq y$. ($x$ and $y$ are two if and only if $x$ is not identical with $y$.)

PL1 states that everything is one, as Frege thinks. And this is a logical truth resting on the definition of '**1** (**be one**)'.[35] Note that this does not mean that things that are two, for example, must also be one. PL1 does not imply the plural universal generalization '$\mathbf{\Pi} xs\ \mathbf{1}(xs)$' ('Any things are one') and is compatible with '$\mathbf{\Sigma} xs \sim\mathbf{1}(xs)$' ('There are some things that are not one'). Moreover, we can prove that any things that are two, for example, are *not* one. PL2 states that nothing instantiating being one can also instantiate another number property (e.g., being two). In particular, Russell and Whitehead, who are two, cannot be one. And PL3 shows that 'Russell and Whitehead are two' is logically equivalent to 'Russell is not identical with Whitehead.'[36] Similarly, we can show that 'Russell and Whitehead are two logicians' (=(1d)) is logically equivalent to 'Russell is a logician, Whitehead is a logician, and Russell is not identical with Whitehead' (=(2)).

Moreover, we can show that there cannot be any things that instantiate different number properties. Using '$M$' and '$N$' as restricted variables for numbers, we can formulate the thesis as follows:

**PL4**. $M(xs) \wedge N(xs) \rightarrow M = N$.

[33]PA3 is a logical truth because '$\exists x\ x = x$' is a logical truth (in classical logic). The status of PA4 is discussed in §5.

[34]Note that this is not a vacuous truth. '$\mathbf{\Sigma} xs\ \mathbf{1}(xs)$' ('There are some things that are one') is a logical truth, for the plural existential '$\mathbf{\Sigma}$' amounts to 'some one or more things' (so '$\exists x\ \mathbf{1}(x)$' implies '$\mathbf{\Sigma} xs\ \mathbf{1}(xs)$').

[35]PL1 follows from T2 (§3).

[36]Both sides of PL3 are logically equivalent to '$\mathbf{M}([x@y], y) \wedge \mathbf{1}(y)$', which is a logical truth.

We can prove this using the (higher-order) mathematical induction principle (viz., PA5). This shows that PL4 is a logical truth (of higher-order plural languages), for PA5 is a logical truth based on the definition of '*Num*' (i.e., Definition 4.2), as noted above.[37]

We can also see that (7) is a logical truth based on Definition 4.1. To formulate the statement in plural languages, it is useful to use a function symbol, '$\oplus$', defined as follows:

**Definition 4.3.** $(M \oplus N)(xs) \equiv_{\mathrm{df}} \Sigma ys \Sigma zs [xs \approx [ys @ zs] \land M(ys) \land N(zs) \land \sim \exists w (w\mathbf{H}ys \land w\mathbf{H}zs)]$.

Using this symbol, we can formulate (7) as follows:

(14) $\mathbf{\Pi} xs\ ((\mathbf{2} \oplus \mathbf{2})(xs) \rightarrow \mathbf{4}(xs))$. (If some things are two and some other things are two, then the former and the latter are four.)

Using definitions of numerals, we can directly show that this is a logical truth. But it would be useful to show this by deriving it from an arithmetical truth:

(6) $\mathbf{2} + \mathbf{2} = \mathbf{4}$. (Two plus two is four.)

We can show that (6) is a logical truth by giving a logical definition of addition. Invoking the successor function, we can give a recursive definition of addition:

**Definition 4.4.**

(a) $M + \mathbf{1} = S(M)$.

(b) $M + S(N) = S(M + N)$.

Thus we can give the usual proof of, e.g., (6):

$$\mathbf{2} + \mathbf{2} = \mathbf{2} + S(\mathbf{1}) = S(\mathbf{2} + \mathbf{1}) = S(S(\mathbf{2})) = \mathbf{4}.$$

And we can state a generalization of PL2 that invokes addition:

**PL5**. $M(xs) \rightarrow \sim (M + N)(xs)$.

This is also a logical truth, which we can use PA5 to show.

[37] To apply PL4 to show, e.g., that there cannot be any things that are both two and three, however, it is necessary to assume that 2 is not 3. Although PA4 implies this, PA4 is not a logical truth. But another logical truth that generalizes PL2 implies, e.g., that no things are both two and three (see PL5). See also the discussion of (15) below.

We can now show that (14) is a logical truth by proving that (6) implies (14). Consider PL6:

**PL6**. $\mathbf{\Pi} xs\ [\ (M \oplus N)(xs) \leftrightarrow (M + N)(xs)\ ]$.

We can use PA5 to show that this is a logical truth.[38] And (6) and PL6 imply (14).[39]

The arithmetical equality '$\mathbf{2} + \mathbf{2} = \mathbf{4}$' , we have seen, logically implies the ordinary numerical truth (14). Similarly, the inequality '$\mathbf{2} + \mathbf{2} \neq \mathbf{3}$', for example, logically implies that if some things are two and some other things are two, then the former and the latter are *not* three. PL4 and PL6 imply PL7:

**PL7**. $L + M \neq N \rightarrow\ \mathbf{\Pi} xs\ [\ (L \oplus M)(xs) \rightarrow {\sim} N(xs)\ ]$.

And this together with '$\mathbf{2} + \mathbf{2} \neq \mathbf{3}$' implies (15):

(15) $\mathbf{\Pi} xs\ (\ (\mathbf{2} \oplus \mathbf{2})(xs) \rightarrow {\sim}\mathbf{3}(xs)\ )$. (If some things are two and some other things are two, then the former and the latter are not three.)

It is notable that this does not show that (15) is a *logical* truth unless '$\mathbf{2} + \mathbf{2} \neq \mathbf{3}$' is a logical truth. And we cannot derive the inequality without using PA4, which is not a logical truth. Nevertheless, we can directly show that (15) is a logical truth. PL5 implies '$\mathbf{\Pi} xs\ (\mathbf{4}(xs) \rightarrow {\sim}\mathbf{3}(xs))$', and this implies (15).[40]

## 5. Logic and arithmetic

Much of arithmetic, we have seen, rests only on logic. We can give logical definitions of arithmetical notions in higher-order plural languages, and all the axioms of Plural Arithmetic except PA4 are logical truths based on the definitions. And those axioms, we have seen, imply a variety of truths about natural numbers: '$\mathbf{2} + \mathbf{2} = \mathbf{4}$', PL1–PL7, etc. But this falls short of proving logicism. PA4 is not a logical truth even in higher-order plural languages. And it is necessary to use PA4 to prove some arithmetical truths. One cannot prove '$\mathbf{2} + \mathbf{2} \neq \mathbf{3}$', for example, without using it.

---

[38] To see this, note that definitions of '$S$' and '+' yield the following:

$$S(N)(xs) \leftrightarrow \mathbf{\Sigma} ys\ \exists z\ (\ xs \approx [ys@z] \wedge N(ys) \wedge {\sim} z\mathbf{H}ys\ ).$$

$$[\ N(xs) \wedge \mathbf{1}(ys) \wedge {\sim}\exists z(z\mathbf{H}xs \wedge z\mathbf{H}ys)\ ]\ \leftrightarrow\ [\ (N + \mathbf{1})([xs@ys]) \wedge \mathbf{1}(ys)\ ].$$

[39] They imply the converse of (14) as well.

[40] For PL6 implies '$\mathbf{\Pi} xs\ (\ (\mathbf{2} \oplus \mathbf{2})(xs) \leftrightarrow \mathbf{4}(xs)\ )$'. See also note 37.

We can see that the main role of PA4 in Plural Arithmetic is to help to yield arithmetical inequalities: '**2** ≠ **3**', '**3** ≠ **4**', '**4** ≠ **5**', etc.[41] If so, can one defend logicism by showing that the inequalities are also logical truths? I think not. I do not think that '**3** ≠ **4**'and '**2** + **2** ≠ **3**', for example, can be considered logical truths.

Some might analyze '$M = N$' as '$\mathbf{\Pi} xs\,[M(xs) \leftrightarrow N(xs)]$' and thus '$M \neq N$' as '$\sim \mathbf{\Pi} xs\,[M(xs) \leftrightarrow N(xs)]$.' But this analysis has a serious problem. On the analysis, '**3** ≠ **4**', for example, logically implies that there are at least three objects. Thus '**3** = **4**' is true, on the analysis, if there are at most two objects.[42] Now, some might propose a modification of the analysis. Invoking modality, they might analyze '$M = N$'as follows:

(16) $\Box \mathbf{\Pi} xs[M(xs) \leftrightarrow N(xs)]$. (It is necessary that some things are *M* if and only if they are *N*.)

Proponents of this analysis might hold a version of logicism: arithmetic is reducible to *modal plural logic*, the logic of languages that extend plural languages by adding modal expressions (e.g., the necessity symbol '□'). But the modal analysis has essentially the same problem as its plain cousin. Consider an instance of (16):

(17) $\Box \mathbf{\Pi} xs[\mathbf{3}(xs) \leftrightarrow \mathbf{4}(xs)]$.

On this analysis, the arithmetical truth '**3** ≠ **4**' implies (18):

(18) $\Diamond \mathbf{\Sigma} xs\mathbf{3}(xs)$. (It is possible that there are some objects that are three.)

Although this is weaker than its plain cousin (viz., '$\mathbf{\Sigma} xs\mathbf{3}(xs)$'), it is not a logical truth of modal plural languages just as the plain cousin is not a logical truth of (non-modal) plural languages.

Now, can one defend the modal analysis by holding that some arithmetical truths reach beyond logic? Although I think it is right to reject logicism, I do not think the analysis captures arithmetical equalities (and inequalities). The analysis rests on the modal analysis of property identity that yields (19):

(19) $P = Q \leftrightarrow \Box \mathbf{\Pi} xs[P(xs) \leftrightarrow Q(xs)]$. (Properties are identical if and only if they are necessarily co-extensive.)

---

[41] It also implies '**2** ≠ **4**', '**2** ≠ **5**', etc. (Note that it is not necessary to use PA4 to show '**1** ≠ **2**'. This follows from PA3 and is a logical truth.) These simple inequalities yield more complex inequalities, such as '**2** + **2** ≠ **3**', which follows from '**2** + **2** = **4**' and '**3** ≠ **4**.'

[42] This analysis lies under the analysis of addition I proposed in Yi ([ 1 ], 87):

$$L + M = N \equiv_{\text{df}} \mathbf{\Pi} xs[(L \oplus M)(xs) \leftrightarrow N(xs)].$$

(See also Hardy ([ 30 ], 106).) This analysis is equivalent to the analysis of equality, for '$\mathbf{\Pi} xs[(M \oplus N)(xs) \leftrightarrow N(xs)]$' is logically equivalent to '$\mathbf{\Pi} xs[(L + M)(xs) \leftrightarrow N(xs)]$' (see PL6). And it has the same problem: '$\mathbf{\Pi} xs[(\mathbf{2} \oplus \mathbf{2})(xs) \leftrightarrow \mathbf{3}(xs)]$' and thus '**2** + **2** = **3**' are true on the analysis, if there are at most two objects.

But this is not correct. Different properties might be necessarily co-extensive. Any possible triangular object is trilateral and *vice versa*, but being triangular and being trilateral are not the same property. It is the same with some plural properties. Being three sides of a triangular object and being three sides of a trilateral object are not the same property, but any possible objects that have the former property must also have the latter property and *vice versa*. The same holds for being such that nothing is one of them (i.e., $\lambda xs \sim \exists y\, yHxs$) and not being the same things as themselves (i.e., $\lambda xs \sim xs \approx xs$).[43]

So I think the modal analysis (16), like its plain cousin, fails to capture arithmetical equality. But I think it is useful to consider modality to clarify the contents of arithmetic.

We can derive arithmetic equalities (e.g., '$\mathbf{2} + \mathbf{2} = \mathbf{4}$'), as we have seen (§4), using only axioms of Plural Arithmetic based on definitions of '$\mathbf{1}$', '$S$', and 'Num'. To derive most inequalities (e.g., '$\mathbf{2} \neq \mathbf{3}$', '$\mathbf{2} + \mathbf{2} \neq \mathbf{3}$'), however, it is necessary to invoke PA4, which goes beyond logic. With the plural language definitions of numerals, I think we can intuitively see the truth of PA4 as we can intuitively see that being triangular and being trilateral, for example, are not the same property: being the same things as *some one thing* and something else taken together (i.e., being two) and being the same things as *some two things* and something else taken together (i.e., being three), for example, are not the same property unless being one and being two are. Still, it would help to clarify the basis of the axiom if we can prove it. And we can do so if we assume that there are infinitely many objects, for this implies that all the number properties are instantiated. So in their analysis of arithmetic, Whitehead and Russell [ 17 ] propose the axiom of infinity, which states that there are infinitely many objects. And one can use the standard versions of set theory to prove PA4, for they imply the existence of infinitely many sets. But it is not necessary to assume the existence of *sets* to clarify the basis of arithmetic. Nor is it necessary to assume that there actually are infinitely many objects. It is sufficient to assume that any natural number *might* have been instantiated.

It is straightforward to see that (18) implies '$\mathbf{3} \neq \mathbf{4}$.'[44] Similarly, '$\Diamond \boldsymbol{\Sigma} xs \mathbf{4}(xs)$' implies '$\mathbf{4} \neq \mathbf{5}$', and so on. So we can prove PA4 by assuming the generalization of (18) for all number properties:

---

[43]Note that the modal analysis of equality is equivalent to the modal analysis of addition, which analyzes '$L + M = N$' as '$\Box \boldsymbol{\Pi} xs[(L \oplus M)(xs) \leftrightarrow N(xs)]$.' (Hodes ([ 7 ], 149) proposes a similar analysis of addition. See also Putnam [31 ], Hellman [32 ], and Chihara [ 33 ] for modal analyses of mathematics.) Thus this analysis of addition has the same problems as the modal analysis of equality.

[44]'$\mathbf{3} = \mathbf{4}$' is false because it implies '$\Box \boldsymbol{\Pi} xs[\mathbf{3}(xs) \leftrightarrow \mathbf{4}(xs)]$', which contradicts (18) (by PL5). Note that this proof does not assume the modal analysis of '$\mathbf{3} = \mathbf{4}$'. This logically implies '$\Box \boldsymbol{\Pi} xs[\mathbf{3}(xs) \leftrightarrow \mathbf{4}(xs)]$' (substitutivity of identity), while the analysis accepts the converse as well.

(20) $\forall^2 N \Diamond \Sigma xs N(xs)$. (Any natural number might have been instantiated.)

Note that this does not imply (21):

(21) $\Diamond \forall^2 N \Sigma xs N(xs)$. (It is possible that all natural numbers are instantiated.)

This implies the possibility of co-existence of infinitely many objects, but (20) does not. (20) is true if there is a possible world with one object, another with two objects, yet another with three objects, and so on. In the Aristotelian terminology, then, (20) implies the *potential* but not the *actual* infinity. Using this terminology, we can say that the potential infinity of objects is *sufficient* to ground arithmetic.

This does not mean that arithmetic *implies* that there is a potential infinity of objects. But this seems in a sense assumed in arithmetic. In making the assumption, we take a substantial step beyond logic. This departure from logic might be considered an entry into the domain of mathematics proper. I leave it for another occasion to consider the basis for this move.


## Acknowledgments

The work for this paper was supported in part by a SSHRC Insight Grant [Grant No. 435-2014-0592], which is hereby gratefully acknowledged. I would like to thank two anonymous referees for Proceedings of the 14th and 15th Asian Logic Conferences for comments on an earlier version, and C. Lee and Y. El Gebali for their help in LaTeX typesetting. I am solely responsible for the remaining errors and infelicities.


## Appendix: Partial axiomatization of plural logic

Here I formulate the axiomatic system $S_{\text{PL}}$ for first-order plural languages that yields a partial yet substantial characterization of first-order plural logic. Let $\mathscr{L}$ be a first-order plural language. The system $S_{\text{PL}}$ (for $\mathscr{L}$) has one rule of inference: modus ponens. To formulate its axioms, say that a sentence $\Phi$ of $\mathscr{L}$ is a *closure* of a sentence $\Psi$ of $\mathscr{L}$, if $\Phi$ is either $\Psi$ itself or $Q_1 \nu_1\, Q_2 \nu_2 \ldots Q_n \nu_n \Psi$, where $Q_1$, $Q_2, \ldots, Q_n$ are universal quantifiers and $\nu_1, \nu_2, \ldots, \nu_n$ are variables suitable for $Q_1, Q_2, \ldots, Q_n$, respectively.[45] Then the axioms of $S_{\text{PL}}$ are the closures of instances of the following:

[45]Greek letters are used as metavariables for $\mathscr{L}$: '$\Phi$' and '$\Psi$' for sentences (open or closed); '$\pi$' for predicates and '$\pi^n$' for $n$-place predicates; '$\tau$' and '$\mu$' for terms of any kind; '$\varsigma$' and '$\sigma$' for singular terms; and '$\upsilon$' for singular variables, '$\omega$' for plural variables, and '$\nu$' for variables of any kind. (Although plural variables (e.g., '*xs*') result from attaching the italicized '*s*' to singular variables, the metavariables for the former (e.g., '$\omega$') do not contain the italicized '*s*'.) '$Q$' is used as a metavariable for quantifiers (singular or plural). And the results of adding numeral subscripts to '$Q$' or the Greek letters are also

**Axioms of $S_{\text{PL}}$**

**Axiom 1**. Truth-functional tautologies.

**Axiom 2**. $\Phi(\upsilon/\varsigma) \rightarrow \exists\upsilon\Phi$.

**Axiom 3**. $\forall\upsilon[\Phi \rightarrow \Psi] \rightarrow [\forall\upsilon\Phi \rightarrow \forall\upsilon\Psi]$.

**Axiom 4**. $\Phi \rightarrow \forall\upsilon\Phi$, where $\upsilon$ does not occur free in $\Phi$.

**Axiom 5**. $\Phi(\omega/\tau) \rightarrow \mathbf{\Sigma}\omega\Phi$.[46]

**Axiom 6**. $\mathbf{\Pi}\omega[\Phi \rightarrow \Psi] \rightarrow [\mathbf{\Pi}\omega\Phi \rightarrow \mathbf{\Pi}\omega\Psi]$.

**Axiom 7**. $\Phi \rightarrow \mathbf{\Pi}\omega\Phi$, where $\omega$ does not occur free in $\Phi$.

**Axiom 8**. $\exists\upsilon\, \upsilon\mathbf{H}\tau$.

**Axiom 9**. $\varsigma\mathbf{H}\sigma \rightarrow \varsigma = \sigma$.

**Axiom 10**. $\varsigma = \sigma \rightarrow \varsigma \approx \sigma$.

**Axiom 11**. *Substitutivity of Plural Identity*:
$\mu \approx \mu_1 \wedge \pi^{\text{n}}(\tau_1, \ldots, \tau_{\text{i-1}}, \mu, \tau_{\text{i+1}}, \ldots, \tau^{\text{n}}) \rightarrow \pi^{\text{n}}(\tau_1, \ldots, \tau_{\text{i-1}}, \mu_1, \tau_{\text{i+1}}, \ldots, \tau^{\text{n}})$.

**Axiom 12**. *Plural Comprehension*:
$\exists\upsilon\Phi \rightarrow \mathbf{\Sigma}\omega\forall\upsilon[\upsilon\mathbf{H}\omega \leftrightarrow \Phi]$, where $\omega$ does not occur free in $\Phi$.

This system assumes the usual definitions of universal quantifiers in terms of existential quantifiers (or *vice versa*) and the definition of plural identity (Definition 3.1). Singular and plural definite descriptions can be introduced via contextual definitions (see Yi ([ 22 ], 244ff)). See 3.2 for the definition of conjunctive plural terms (e.g., '[$x$@$ys$]') formed by the term-connective '@'. (The definition yields T3 and T4 in §3.2.)

Note that the above formulation of $S_{\text{PL}}$ differs from my earlier formulations in having Axiom 10 instead of the principle of substitutivity of singular identity (the

used as metavariables of the same kind. A variable $v$ is said to be *suitable for* a quantifier $Q$, if $Q$ and $v$ are both singular or both plural.

[46] $\Phi(\mu/\tau)$ is the result of properly substituting $\tau$ for $\mu$ in $\Phi$.

singular cousin of Axiom 11). Given other axioms, however, these are equivalent.[47] See Yi ([ 6 ], 179ff; [ 1 ], 57f; [ 22 ], §7) for presentations of essentially the same system.

## References


1. Yi, B.-U., *Understanding the Many* (London, England & New York, NY: Routledge, 2002).
2. Frege, G., *Die Grundlagen der Arithmetik* (Breslau: Köbner, 1884). Translated by J. L. Austin as *The Foundations of Arithmetic*, 2nd rev. ed. (Oxford: Basil Blackwell, 1980).
3. Russell, B., *Principles of Mathematics* (Cambridge: Cambridge University Press, 1903). 2nd ed. (London, UK & New York, NY: Norton, 1937).
4. Russell, B., *Introduction to Mathematical Philosophy* (London: George Allen & Unwin, 1919). 2nd ed. (London: George Allen & Unwin, 1920)
5. Frege, G., *Posthumous Writings*, edited by H. Hermes, F. Kambartel, and F. Kaulbach (Oxford: Basil Blackwell, 1979).
6. Yi, B.-U., Is two a property?, *Journal of Philosophy* **96** (1999): 163–190.
7. Hodes, H. T., Logicism and the ontological commitments of arithmetic, *Journal of Philosophy* 81 (1984): 123–149.
8. Yi, B.-U., Abstract nouns and resemblance nominalism, *Analysis* **74** (2014): 622–629.
9. Yi, B.-U., Nominalism and comparative similarity, *Erkenntnis* **83** (2018): 793–803.
10. Yi, B.-U., Review of Friederike Moltmann, *Abstract Objects and the Semantics of Natural Language*, *Mind* **124** (2015): 958–964.
11. Maddy, P., Perception and mathematical intuition, *Philosophical Review* **89** (1980): 163–196.
12. Maddy, P., *Realism in Mathematics* (Oxford: Oxford University Press 1990).
13. Kim, J., The role of perception in a priori knowledge: some remarks, *Philosophical Studies* **40** (1981): 339–354.
14. Yi, B.-U., Numerical cognition and mathematical knowledge: The plural property view, in S. Bangu (ed.), *Naturalizing Logico-Mathematical Knowledge: Approaches from Psychology and Cognitive Science* (London, England & New York, NY: Routledge, 2018), pp. 52–88.
15. Mill, J. S., *A System of Logic: Ratiocinative and Inductive* (Honolulu, HI: University Press of the Pacific, 2002). Reprinted from the 1891 edition.
16. Maddy, P., Psychology and the a priori sciences, in S. Bangu (ed.), *Naturalizing Logico-Mathematical Knowledge: Approaches from Psychology and Cognitive Science* (London, England & New York, NY: Routledge 2018), pp. 15–29.
17. Whitehead, A. N. and Russell, B., *Principia Mathematica*, 3 Vols. (Cambridge: Cambridge University Press, 1910–13). 2nd ed. (Cambridge: Cambridge University Press, 1925–27).
18. Yi, B.-U., The logic of classes of the no-class theory, in N. Griffin and B. Linsky (eds.),


[47] For the convenience in axiomatization, we can define the singular identity predicate '=' as a form of the plural identity predicate '≈' (see T2 in §3.2) and remove Axiom 10. But I think singular identity is more basic than plural identity.

*The Palgrave Centenary Companion to Principia Mathematica* (New York, NY & Basingstoke, UK: Palgrave Macmillan, 2013), pp. 96–129.

19. Yi, B.-U., Is there a plural object?, in A. J. Cotnoir and D. Baxter (eds.), *Composition as Identity* (Oxford: Oxford University Press, 2014), pp. 169–191.
20. Yi, B.-U., Is composition identity?, Unpublished manuscript.
21. Yi, B.-U., The logic and meaning of plurals, Part I, *Journal of Philosophical Logic* **34** (2005): 459–506.
22. Yi, B.-U., The logic and meaning of plurals, Part II, *Journal of Philosophical Logic* **35** (2006): 239–288.
23. Yi, B.-U., Numbers and relations, *Erkenntnis* **49** (1998): 93–113.
24. Linnebo, Ø., Plural Quantification, in E. N. Zalta (ed.), *The Stanford Encyclopedia of Philosophy*, Summer 2017 ed., URL = <https://plato.stanford.edu/archives/sum2017/entries/plural-quant/>, 2017.
25. McKay, T., *Plural Predication* (Oxford: Oxford University Press, 2006).
26. Oliver, A. and Smiley, T., *Plural Logic*, 2nd ed. (Oxford: Oxford University Press, 2016).
27. Rayo, A., Word and objects, *Noûs* 36 (2002): 436–464.
28. Peano, G., *Arithmetices principia, nova methodo exposita* (Turin: Bocca 1889). Translated as "The principles of arithmetic, presented by a new method" (trans. J. van Heijenoort) in J. van Heijenoort (ed.), *From Frege to Gödel: A Source Book in Mathematical Logic, 1879–1931* (Cambridge, MA: Harvard University Press, 1967), pp. 83–97.
29. Dedekind, R., *Was sind und was sollen die Zahlen?* (Brunswick: Vierwg, 1888). Translated as 'The nature and meaning of numbers' in W. W. Beman (ed), *Essays on the theory of numbers* (New York, NY: Dover, 1963), 31–115.
30. Hardy, G. H., *A Mathematician's Apoogy* (Cambridge: Cambridge University Press, 1940). Canto edition (1992).
31. Putnam, H., What is mathematical truth?, *Journal of Philosophy* **64** (1967): 5–22.
32. Hellman, G., *Mathematics Without Numbers* (Oxford: Oxford University Press, 1989).
33. Chihara, C., *Constructibility and Mathematical Existence* (Oxford, UK: Oxford University Press, 1990).

# Author Index

Printed in the United States
By Bookmasters